RATES OF EVOLUTION

How fast is evolution, and why does it matter? Rates of evolution and whether evolution is gradual or punctuated are actively debated topics among biologists and paleontologists. This book compiles and compares examples of evolution from laboratory, field, and fossil studies, analyzing them to extract their underlying rates. It concludes that while change is slow when averaged over many generations, change is rapid on the generation-to-generation time scale of the evolutionary process. Chapters cover the history of evolutionary studies, from Lamarck and Darwin in the nineteenth century to the present day. An overview of the statistics of variation, dynamics of random walks, processes of natural selection and random drift, and effects of scale and time averaging are also provided, along with methods for the analysis of evolutionary time series. Containing case studies and worked examples, this book is ideal for advanced students and researchers in paleontology, biology, and anthropology.

PHILIP D. GINGERICH is Professor Emeritus of Earth and Environmental Sciences, Ecology and Evolutionary Biology, and Anthropology at the University of Michigan, as well as Curator Emeritus of Vertebrate Paleontology. His honors include the Henry Russel Award, Distinguished Faculty Achievement Award, and Collegiate Professorship at the University of Michigan; the Charles Schuchert Award from the Paleontological Society; the André Dumont Medal from the Belgian Geological Society; and the Romer-Simpson Medal from the Society of Vertebrate Paleontology. He is a fellow of the American Academy of Arts and Sciences and the American Philosophical Society, and was previously president of the Paleontological Society.

Philip Gingerich, renowned among paleontologists for his research on the evolution of mammals, has been a leading authority on rates of evolution for more than three decades. His analyses of evolution on different time scales have been critical to understanding this important, sometimes controversial, subject. *Rates of Evolution: A Quantitative Synthesis* will provide insights and statistical approaches that will interest a broad range of researchers and students working in evolutionary biology and paleontology.

– Professor Douglas Futuyma, Stony Brook University

This book is a deeply thought-out, scholarly and lucid account of how to connect measurements of contemporary evolution with evolution as revealed in the fossil record. Rigorous and quantitative throughout, it will be a stimulating primer for professional evolutionary biologists. There is no other book like it.

– Professor Peter Grant, Princeton University

Using evidence from many fields of biology, paleontology, and beyond, Gingerich's *Rates of Evolution* is a comprehensive synthesis of a pillar of the evolutionary paradigm. This book is a sophisticated analysis of quantitative empirical data integrated with evolutionary theory. It is destined to be an authoritative reference and much-cited classic in evolutionary biology.

– Professor Bruce MacFadden,
Florida Museum of Natural History, University of Florida

RATES OF EVOLUTION

A Quantitative Synthesis

PHILIP D. GINGERICH

University of Michigan

CAMBRIDGE
UNIVERSITY PRESS

University Printing House, Cambridge CB2 8BS, United Kingdom

One Liberty Plaza, 20th Floor, New York, NY 10006, USA

477 Williamstown Road, Port Melbourne, VIC 3207, Australia

314–321, 3rd Floor, Plot 3, Splendor Forum, Jasola District Centre, New Delhi – 110025, India

79 Anson Road, #06–04/06, Singapore 079906

Cambridge University Press is part of the University of Cambridge.

It furthers the University's mission by disseminating knowledge in the pursuit of
education, learning, and research at the highest international levels of excellence.

www.cambridge.org
Information on this title: www.cambridge.org/9781107167247
DOI: 10.1017/9781316711644

First published 2019

Printed in the United Kingdom by TJ International Ltd. Padstow Cornwall

A catalogue record for this publication is available from the British Library.

Library of Congress Cataloging-in-Publication Data
Names: Gingerich, Philip D., author.
Title: Rates of evolution : a quantitative synthesis / Philip D. Gingerich.
Description: New York, NY : Cambridge University Press, 2019. | Includes bibliographical
references and index.
Identifiers: LCCN 2018051985 | ISBN 9781107167247 (hardback)
Subjects: LCSH: Evolution (Biology)
Classification: LCC QH366.2 .G524 2019 | DDC 576.8–dc23
LC record available at https://lccn.loc.gov/2018051985

ISBN 978-1-107-16724-7 Hardback

Additional resources for this publication at www.cambridge.org/evolution

In memory of my father, Orie Jacob Gingerich (1919–1982),
who encouraged my interest in numbers and
introduced me to statistics

Contents

Preface

I was educated in the light of evolution's Modern Synthesis. My professor when I first became interested in paleontology and evolution was Glenn L. Jepsen, who hosted the 1947 Princeton Conference on Genetics, Paleontology, and Evolution. Jepsen, Ernst Mayr, and George Gaylord Simpson co-edited the resulting volume on *Genetics, Paleontology, and Evolution* (Jepsen et al., 1949). I was fortunate to meet architects of the Modern Synthesis early in my education. Starting field work as an undergraduate helped me see the evidence and logic of Darwin's "slow and gradual modification" of species "through descent and natural selection" (Darwin, 1859, p. 312).

My doctoral dissertation was a study of population variation in the common European and North American Paleocene mammal *Plesiadapis*, with application of the classic stratigraphic principles of superposition, faunal succession, and correlation to understand its evolution through time. Large samples of *Plesiadapis* were available in the Princeton University collection, resulting from 40 years of field work that Jepsen started when he himself was a student.

I was well into my dissertation research in 1972 when fellow Yale graduate student Peter Dodson and I visited the American Museum of Natural History. At the end of the day we met Niles Eldredge. Peter knew Niles and they talked about Niles' forthcoming theory of "punctuated equilibria" developed with Harvard professor Stephen Jay Gould. On the trip back to New Haven I questioned Peter about punctuated equilibria, and was surprised at what I learned. Peter explained that according to punctuated equilibria, new species do not evolve where their ancestors are found, and they do not arise from any perceptible transformation of those ancestors. The history of life is rather one of "homeostatic equilibria" perturbed, "punctuated," rarely by random events of speciation. When published, punctuated equilibria was promoted as an alternative to phyletic gradualism. I was a graduate student with a substantial investment in an empirical dissertation documenting lineages changing through time in the fossil record.

With new expectations inspired by the theory of punctuated equilibria, I set aside my dissertation research and went to work in the Peabody Museum basement measuring *Hyopsodus* molar teeth. *Hyopsodus* is the most common fossil mammal in the North American early Eocene, and it is known from thousands specimens in the Yale collection, recovered from many levels in a thick sequence of strata exposed in the Bighorn Basin of Wyoming. *Hyopsodus* molars do not grow: They erupt at a definitive size related to the body size of the full-grown animal. If punctuated equilibria were true, I expected to find that the size of *Hyopsodus* molars in a species would remain relatively stationary from stratum to stratum through time. The only change should come when new species appeared, and these too should then remain stationary for the remainder of their existence. What I found, however, was a pattern of molar size changing and branching through time as the species themselves changed, branched in a gradual pattern reminiscent of the diagram in Charles Darwin's *Origin of Species*. The *Hyopsodus* study was published as a research article in *Nature* in 1974.

Gould and Eldredge responded in various ways. They first speculated that the overall increase in tooth size and body size in *Hyopsodus*, *Haplomylus*, and *Pelycodus* in the study interval might mean that early Eocene climate cooled through time, causing larger-bodied, northern populations to move south into Wyoming (Gould and Eldredge, 1977, pp. 130–131). Then they proposed that guidelines be redrawn to feature segments of stasis, averaging change when necessary to achieve this. Parallel change was regarded as stasis (p. 131). Gould and Eldredge's most interesting observation was documented in a table of rates (in percent change per million years), where they wrote, echoing J. B. S. Haldane (1949):

We must consider the characteristic rates of supposed gradualistic events. When this is done, one cardinal fact emerges: they are too slow to account for most important evolutionary phenomena, particularly for adaptive radiations and the origin of new morphological designs... How can we view a steady progression yielding a 10% increase in a million years as anything but a meaningless abstraction?

(Gould and Eldredge 1977, p. 133)

How indeed? Knowing of Haldane's study of rates, and realizing that punctuated equilibria had no explicit scale, my interest in evolution turned to rates and scales. This book, long in gestation, is the result.

This brings me to a short, provocative, important study on the west coast of Britain by Lewis Fry Richardson (1961) and Benoit Mandelbrot (1967). In this, Richardson and Mandelbrot showed, counterintuitively, that the coast has no fixed length. Its length is variable and depends on the length of the ruler, the scale, used to measure it. Richardson used log-log plots of total length versus measurement scale to quantify the dependence, and found a slope of -0.25. Mandelbrot interpreted this as a "fractional dimension" of 1.25. The coastline of Britain is thus more complicated, "rougher," than a one-dimensional Euclidean line, but it is less complicated, smoother, than the line one might draw to fill a two-dimensional area on a Euclidean plane (something we all did as children with crayons in a coloring book).

I fear that no reader will understand this book on rates of evolution without first understanding Richardson's and Mandelbrot's analysis of length, scale, and the coast of Britain. Maybe I am wrong. Maybe for some it will prove easier to understand what is presented here, and this will then help in understanding Richardson and Mandelbrot. To be provocative I will say in advance that there are very few smooth one-dimensional lines anywhere in nature, and *no* one-dimensional evolutionary lineages. Even stasis is not smooth.

Illustration and modeling go hand in hand. Every model has an underlying geometry that can be illustrated, and the best way to affirm that a model works is to illustrate it. This is especially important for models that involve dynamic simulation. The best way to confirm that a simulation works is to make it draw itself. Underlying geometry and model affirmation are two reasons for many of the illustrations presented here.

Finally, I will end with an impression. When I was an undergraduate I majored in geology and paleontology but took many courses in biology. There was an understanding at the time that paleontologists such as George Gaylord Simpson, Glenn Jepsen, and many others, were authorities on the history of life. There was an understanding too that biologists such as Theodosius Dobzhansky, Ernst Mayr, and many others, were authorities on the process of evolution. In the heyday of the Modern Synthesis there were conferences like the Princeton Conference on Genetics, Paleontology, and Evolution where paleontologists and biologists met to learn from each other.

Now science is a bigger enterprise, there is more specialization, and there is seemingly less communication between paleontologists and biologists. As a consequence, paleontologists understand less about the process of evolution from a biological point of view, and biologists understand less about the history of life from a paleontological point of view. Paleontologists seem happy to invent processes of evolution to fit the histories of life that they study, and biologists seem happy to invent histories of life to fit the processes they study. Neither benefits from the constraint of the other. It would be better to communicate more widely. Here, returning to my youth, I attempt to bridge paleontology and biology. Hopefully readers benefit from the reciprocal illumination.

Acknowledgments

First and foremost, my anthropologist–statistician wife of many years, B. Holly Smith, aided and encouraged completion of this study. Long-time colleagues in paleontology and evolutionary biology at the University of Michigan, especially Daniel C. Fisher and Philip Myers, listened as my ideas developed and raised important questions along the way. PhD student William C. Clyde, now at the University of New Hampshire, understood what I was doing and worked to improve it. Former postdoctoral scholar P. David Polly, now at Indiana University, helped and encouraged completion. I thank all five. I owe a debt too to the successive cohorts of students who forced me to think critically about rates in the courses and seminars I taught on fossil mammals and evolution.

Sources of rate information are cited in the text, but some colleagues contributed unpublished information as well. Jennifer Babin-Fenske, Coordinator of Earth-Care, City of Greater Sudbury, provided information on whirligig beetles. Michael A. Bell, Stony Brook University, provided measurements and other information on three-spine sticklebacks. Mary Bomberger Brown at the University of Nebraska provided information on cliff swallows. Sandra Carlson at the University of California, Davis, provided information about generation times in brachiopods. Nicholas M. Caruso, Virginia Tech University, provided measurements and information on generation times in plethodontid salamanders. Alan Cheetham, United States National Museum, provided ages and Mahalanobis D values for *Metrarabdotos* species samples. The late Douglas Falconer, University of Edinburgh, provided measurements of the mice in his selection experiments. Theodore Garland helped with branch lengths and other information needed in Chapter 13. Ann E. McKellar, Environment Canada, provided a copy of her data compilation on body weight variability in mammals. Philip Myers, University of Michigan, helped with understanding Galápagos rats. Kathleen M. Scott, Rutgers University, provided a copy of her large set of ungulate postcranial measurements. Trisha Spanbauer, University of Nebraska, provided measurements of *Cyclostephanos andinus*. Yoel E. Stuart, University of Texas, provided measurements and information on *Anolis* species. Jane M. Taylor, University of Western Australia, provided measurements of Australian rabbits. Peter Taylor, University of Venda, provided measurements and information on South African vlei rats published with Aluwani Nengovhela. Finally, Edward Theriot, University of Texas, provided measurements and information about generation times in the diatom *Stephanodiscus yellowstonensis*.

Scott Steppan, Florida State University, reviewed a manuscript version of Chapter 13, and Gene Hunt, United States National Museum, reviewed a manuscript version of Chapter 14. Both provided helpful perspectives.

This project was initiated with fellowship support from the John Simon Guggenheim Foundation. Grants from the United States National Science Foundation supported field work yielding the evolutionary time series of Paleocene and early Eocene mammals analyzed in Chapter 9.

1

Introduction

The history of science is important for understanding the development of ideas. This is certainly true in evolutionary studies. Four nineteenth-century naturalists framed a thesis and an antithesis that still concern us now, two centuries later.

1.1 Thesis and Antithesis

Jean-Baptiste Lamarck in France proposed a radical thesis involving "transformation" of species to explain the diversity of life on Earth and changes seen in the fossil record. Charles Lyell in England, citing Carl Linnaeus of Sweden, countered Lamarck with a reactionary antithesis consistent with his own uniformitarian view of earth history: creation imbues species with characteristics that never vary. Charles Darwin in England, influenced by the other three, had the last word. All made lasting contributions to our understanding of evolution. The full history is complicated, but, simplifying, we shall focus on Lamarck, Lyell, and Darwin.

1.1.1 Jean-Baptiste Lamarck

À mesure que les individus d'une de nos espèces changent de situation, de climat, de manière d'être ou d'habitude, ils en reçoivent des influences qui changent peu à peu la consistance et les proportions de leurs parties, leur forme, leurs facultés, leur organization même ... à la suite de beaucoup de générations qui se sont succédées les unes aux autres ... se trouvent à la fin transformés en une espèce nouvelle, distincte de l'autre. [As the individuals of one of our species change their situation, climate, manner of being or habit, they receive influences that gradually change the consistency and proportions of their parts, their form, their faculties, and their organization itself ... after

many generations, these individuals are at length transformed into a
new species, distinct from the first.]

Jean-Baptiste Lamarck, 1809, Philosophie Zoologique,
Tome Premier, *pp. 62–63*

Jean-Baptiste Lamarck was born in 1744, in Picardy in northern France. He was
born with the title *chevalier* or knight. Lamarck's career included military service,
which ended with an injury at the age of 22. Medical studies followed, and then a
ten-year career as a botanist. Lamarck published a three-volume *Flore Française* in
1778. When the *Museum National d'Histoire Naturelle* in Paris was founded in
1793, Lamarck was appointed *enseigner*, not of botany, his field of specialization,
but of invertebrate paleontology. Lamarck embraced the new position and published *Système des Animaux sans Vertèbres* in 1801, and then his two-volume
Philosophie Zoologique in 1809. Lamarck died in Paris in 1829 at the age of 85.

Lamarck is famous for two ideas. The first is our focus here, the idea of
transformation little by little, "*peu à peu,*" of one species into another in the course
of "*beaucoup de générations.*" Thus Lamarck (1809, pp. 77–78) inquired, rhetoric-
ally, is it not possible for species known only as fossils to have changed and
become the species we see alive today? Transformation is expressed too in his
assertion that "it is not the form, either of the body or its parts, that gives rise to the
habits and manner of living of animals; but, on the contrary, the habits, manner of
living, and all other influential circumstances that have, in time, given rise to the
form of the body and its parts in animals. With new forms, new facilities have been
acquired, and little by little nature has succeeded in forming animals as we now see
them." (Lamarck, 1809, p. 268).

Lamarck's second idea is thoroughly discredited and often used to denigrate all
of his writings. This is the idea, expressed in two "laws," of the inheritance of
acquired characteristics. The first law attributed the physiological development
of an organ in an individual to its use. The second law was a conjecture that
developments resulting from use and disuse were heritable, somehow, and passed
from individuals to their descendants (1809, p. 235). In one of his examples,
Lamarck proposed that continual striving to browse in treetops, sustained for a
long time by all giraffes, was sufficient to explain their long legs and necks (1809,
pp. 256–257). Remember that little was known of heredity in 1809 beyond the
close resemblance of parents and offspring.

1.1.2 Charles Lyell

We must suppose, that when the Author of Nature creates an animal or
plant, all the possible circumstances in which its descendants are destined to live are foreseen, and that an organization is conferred upon it

which will enable the species to perpetuate itself and survive under all
the varying circumstances to which it must be inevitably exposed.
Charles Lyell 1832, Principles of Geology, Volume II, *pp. 23–24*

Charles Lyell was born in 1797 in Forfarshire, in the east central lowland of
Scotland. Lyell was educated in classics and started his career as a lawyer at the
age of 23, but within a few years he gave up this profession in favor of travel and
geological studies. Lyell is best known for *Principles of Geology*, published in
three volumes in 1830, 1832, and 1833. The full title of all three volumes is
*Principles of Geology, Being an Attempt to Explain the Former Changes of the
Earth's Surface by Reference to Causes Now in Operation*. Lyell was an empiricist
who traveled widely and spent his life promoting the "uniformitarian" view
expressed in the subtitle of *Principles*. Lyell revised *Principles* regularly, and the
twelfth edition was published posthumously. Lyell died in London in 1875 at
the age of 77.

Here we are concerned with the first edition of Lyell's *Principles*, specifically
the first edition of the second volume, which was published in January of 1832.
Volume I was a review of changes in the physical or inanimate world. Volume II
provided a parallel review of progress in what Lyell called "animate creation." On
the title page of volume II, Lyell quoted Playfair (1802, §412, pp. 469–470):
"A change in the animal kingdom seems to be part of the order of nature, and is
visible in instances to which human power cannot have extended." Did Lyell really
believe change in the animal kingdom is part of the order of nature?

Lyell answered this question on the very first page of the volume II text, where
he proposed to inquire "whether species have a real and permanent existence in
nature; or whether they are capable, as some naturalists pretend, of being indefinitely modified in the course of a long series of generations?" Lyell never repeated
or explained the Playfair quotation but avidly pursued the French "pretenders"
Jean-Baptiste Lamarck and his younger colleague Étienne Geoffroy Saint-Hilaire.
In Lyell's favor, Gavin de Beer (1960) and the *Oxford English Dictionary* credit
him with the first use of the word "evolution" in the English language in the sense
in which it is now widely employed in biology. Lyell's use of evolution is found in
a passage in chapter 1 of *Principles of Geology, Volume II* (1832, p. 11) where he
questioned Lamarck's fanciful belief in "gradual evolution" from sea to land.

Étienne Geoffroy Saint-Hilaire published a *Mémoire* in 1828 exploring the
possibility that "antediluvian" beings gave rise to animals living in modern times
(Geoffroy Saint-Hilaire 1828). In volume II of *Principles*, Lyell translated Geoffroy Saint-Hilaire as saying that "there has been an uninterrupted succession in
the animal kingdom effected by means of generation, from the earliest ages of the
world up to the present day" and "ancient animals whose remains have been

preserved in the strata, however different, may nevertheless have been the ancestors of those now in being" (Lyell, 1832, p. 2). Lyell dismissed Geoffroy Saint-Hilaire's ideas by noting that they were not generally accepted, and then focused his refutation on Jean-Baptiste Lamarck and on Lamarck's *Philosophie Zoologique* (Lamarck, 1809).

Lyell criticized Lamarck's concept of species, writing that "the majority of naturalists agree with Linnaeus in supposing that all the individuals propagated from one stock have certain distinguishing characters in common which will never vary, and which have remained the same since the creation of each species" (Lyell 1832, p. 3). Lamarck had argued that individuals change, little by little, and after many successive generations are transformed into a new and distinct species (paraphrasing Lyell, 1832, p. 5, and the full quotation cited above). "*Peu à peu*," little by little, is a phrase that appears often in *Philosophie Zoologique*, but Lyell's 1832 uniformitarianism seemingly excluded such changes to life on Earth.

Lamarck's novel thesis of species changing and transforming little by little in response to the environment was contradicted by Lyell's reactionary antithesis, citing Linnaeus, of species created by an "Author of Nature" with their organization conferred upon them. If we relax Lyell's "creation" of species and "organization conferred upon them" to represent only distinct origins and characteristic essences, we see that the conflicting views of Lamarck and Lyell, developed in the eighteenth and nineteenth centuries, survive and divide us still.

1.1.3 *Charles Darwin and the* Origin of Species

> I do believe that natural selection will always act very slowly, often only at long intervals of time, and generally on only a very few of the inhabitants of the same region at the same time. I further believe, that this very slow, intermittent action of natural selection accords perfectly well with what geology tells us of the rate and manner at which the inhabitants of this world have changed.
>
> *Charles Darwin 1859,* Origin of Species*, pp. 108–109*

Charles Darwin was born in 1809 in Shrewsbury, Shropshire, in the West Midlands of England. He started medical school at the age of 16 but neglected medical studies in favor of natural history. Darwin's professional career began in 1831 when, at age 22, he embarked as "naturalist" on the survey ship *H. M. S. Beagle* for a nearly five-year round-the-world voyage. The ship left Plymouth Sound in England on December 27, 1831, and returned to Falmouth Harbour on October 2, 1836. The *Journal of Researches* resulting from the voyage was published in 1839, and soon republished as the now-classic *Voyage of the Beagle*. Darwin is best known for the *Origin of Species* or, to give the book its full title: *On the Origin of Species by Means of Natural Selection or the Preservation of Favoured Races in the Struggle for Life*. This was published in 1859 when Darwin was 50 years old. Darwin died at Down House, Kent, in 1882 at the age of 73.

Darwin took the then newly published first volume of Charles Lyell's *Principles of Geology* with him on the *Beagle*, and we know that he read much of it before the ship's first stop in the Cape Verde Islands off the west coast of Africa. Darwin received the second volume of Lyell's *Principles* some ten months later when the ship was docked in Montevideo in South America (Darwin's own copy survives, inscribed "Montevideo, November 1832"). This is important because Lyell's second volume deals extensively with the transmutation of species and what we today call 'evolution.' As outlined above, the second volume of *Principles* opens with a critique of evolutionary ideas expressed by the French zoologists Jean-Baptiste Lamarck and Étienne Geoffroy Saint-Hilaire.

A naturalist at sea has little to do but read and think. This explains how Darwin was able to read much of the first volume of Lyell's *Principles* before reaching the Cape Verde Islands. It also means he had ample time to read and reread the second volume of *Principles* before reaching the Galápagos Islands in the Pacific some three years later. It is the second volume in which Lyell presented Lamarck's thesis of species changing and transforming little by little in response to the environment, and his own antithetical objections to such transformation. Lyell's writing is so clear and forceful that both alternatives are impressed on any reader, and both were surely clear to Darwin.

The *Beagle* reached the Galápagos Islands on September 15, 1835 and departed on October 20, 1835. In the Galápagos, Darwin visited the islands of Chatham (now San Cristobal), Charles (Santa María), Albemarle (Isabela), and James (Santiago). He was initially more interested in mockingbirds than finches, and was able to collect mockingbirds from all four islands. Later, while organizing his collections, Darwin recognized that the islands of Charles and James had similar mockingbirds that differed from those on both Chatham and Albemarle, where each had its own characteristic form. Darwin combined these observations with reported differences in the tortoises of different Galápagos Islands, and reported differences in the foxes of the eastern and western Falkland Islands that he had visited in 1834. In the "Ornithological Notes" that Darwin wrote in 1836 while still at sea, he reasoned that "archipelagos will be well worth examining; for such facts [differences found on neighboring islands] would undermine the stability of species."

Darwin started a "Red Notebook" in 1836 while he was still at sea. This was initially a catalogue of places visited by the *Beagle* during the last year of the voyage. Then in 1837 he added various geological ideas to the Red Notebook, and in March of that year some notes on "transmutation" or speciation. On page 130 of the Red Notebook he wrote: "The same kind of relation that common ostrich bears to the Petisse [common rhea to lesser rhea]; extinct Guanaco to recent [extinct llama to recent llama]: in former case [geographic] position, in latter time (or changes consequent on lapse) being the relation.— As in first cases distinct species inosculate, so must we believe ancient ones: ∴ not gradual change or degeneration. From circumstances: if one species does change into another it must be per saltum — or species may perish … Inosculation alone shows not gradation." The transcription here is from Sandra Herbert in Barrett et al. (1987). Darwin's notes seem conflicted because "inosculate" would imply gradation. However, Darwin's "if one species does change into another it must be per saltum" is perfectly clear. He had not yet made the metaphorical examination of archipelagos that would undermine the perceived stability of species.

Darwin started the first of the notebooks he devoted to transmutation of species, his "Notebook B" later that year, during the summer of 1837. The early pages were seemingly inspired, to some extent, by his grandfather Erasmus Darwin's *Zoonomia* (Darwin, 1794–1796). Darwin wrote on page 3 of Notebook B: "Seeds of plants sown in rich soil, many kinds, are produced, though new individuals produced by buds are constant, hence we see generation [sexual reproduction] here seems a means to vary, or adaptation." Then on page 7: "Animals, on separate islands, ought to become different if kept long enough apart, with slightly different circumstances. — Now Galápagos Tortoises, Mocking birds; Falkland Fox …" Darwin's branching diagrams first appear on pages 26 and 36. On the latter he wrote: "Case must be that one generation then should be as many living as now. To do this and to have many species in same genus (as is) requires extinction." Then on page 37 he wrote: "With respect to extinction we can easily see that variety of ostrich, Petise [lesser rhea] may not be well adapted, and thus perish out." The transcription here is from David Kohn in Barrett et al. (1987). The notes are telegraphic but give us, already in the summer of 1837, an outline of a theory for the origin and evolution of species. This involved: (1) the production of variation, (2) geographic separation or isolation, and (3) differential survival based on adaptation.

With this outline, why did it take a prolific author like Darwin 22 years to flesh out his theory of natural selection in the *Origin of Species*? First, much of the evidence was subtle and beyond the experience of potential readers, so it required careful presentation; and second, the evidence supported and extended the thesis of Lamarck and contradicted the reactionary antithesis of Lyell, who was by then Darwin's friend and patron. Interestingly, and for whatever reason, Darwin did not use the word "evolution" in any form in the *Origin* until the closing sentence, where he wrote that "endless forms most beautiful … have been, and are being, evolved."

1.2 Evolutionary Synthesis of the 1930s and 1940s

Much happened in the years following publication of the *Origin of Species*. New discoveries in paleontology and systematic biology continued, but the greatest progress was made in our understanding of inheritance.

1.2.1 Progress in Understanding Inheritance

One of the most interesting and transformative discoveries in genetics came in the 1860s, shortly after publication of the *Origin of Species*. The Moravian biologist and friar Gregor Mendel determined that each phenotypic trait in his fertilized experimental peas was represented by two and only two underlying genetic factors (or gene alleles). These factors segregate, assort independently, and dominate or subordinate, leading to characteristic combinations of phenotypes in subsequent generations. Mendel's study (Mendel, 1866) was published in a Moravian journal that must not have been widely seen because the study had no impact until the year 1900.

In another development, August Weismann (1883) recognized that inheritance in multicellular organisms depends on germ cells or gametes that are separated and effectively insulated from the somatic cells comprising the rest of the body. Thus Weismann eliminated the possibility of any direct Lamarckian inheritance of characteristics acquired through use or disuse.

Hugo de Vries was the first of later geneticists to discover Mendel's 1866 publication, which he reported in a footnote to a similar study of his own. A botanist, de Vries went on to develop a "theory of mutations" to explain discontinuous phenotypic variation. He considered genetic mutations leading to new phenotypes sufficient to explain the origin of species (de Vries, 1901).

The focus on Mendelian inheritance soon led to cytological identification of chromosomes as carriers of genetic material and determiners of sex. Genes and gene alleles are more cryptic and were necessarily recognized by what they do functionally. Modern genetics emerged as breeding experiments began to be carried out using a variety of organisms: *Oenothera lamarckiana*, *Zea mays*, *Drosophila melanogaster*, *Mus musculus*, *Cavia porcellus*, and others.

1.2.2 Microevolution and Macroevolution

Theodosius Dobzhansky (1937) initiated the Modern Synthesis of evolution when he published *Genetics and the Origin of Species*. This made new developments in genetics, expressed in the technical writings of R. A. Fisher, J. B. S. Haldane, T. H. Morgan, Sewall Wright, and others, accessible to a broad range of evolutionary biologists.

Dobzhansky regarded evolution as any change in the genetic composition of populations, recognizing that this might have small or large consequences depending in part on the time involved. This led him to distinguish "microevolution" and "macroevolution," borrowing the terms from Philiptschenko (1927). Dobzhansky wrote (1937, p. 12): "There is no way toward an understanding of the mechanisms of macro-evolutionary changes, which require time on a geological scale, other than through a full comprehension of the micro-evolutionary processes observable within the span of a human lifetime."

Dobzhansky's single sentence highlights three important contrasts in evolution: (1) minor change versus major change, (2) short timescale versus long timescale, and (3) process versus pattern. The problem in each instance is to understand how one side of each contrast is related to the other, that is: how minor changes are related to major changes; how short timescales are related to long timescales; and how processes are related to patterns.

Three years after Dobzhansky's book appeared, Richard Goldschmidt published *The Material Basis of Evolution* (Goldschmidt, 1940). Here, he contradicted Dobzhansky and argued that "the facts of microevolution do not suffice for an understanding of macroevolution" (Goldschmidt 1940, p. 8). Goldschmidt proposed the word "macromutation" (p. 182) for systemic mutations leading to what he had previously labeled "hopeful monsters." He went on to claim that a "monstrosity [hopeful monster] appearing in a single genetic step might permit the occupation of a new environmental niche and thus produce a new type [new species] in one step" (Goldschmidt, 1940, p. 390). Finally, he concluded: "species and the higher categories originate in single macroevolutionary steps as completely new genetic systems" (Goldschmidt, 1940, p. 396).

1.2.3 *"Modern Synthesis" of the Twentieth Century*

Advances in the study of inheritance answered some questions about evolution and evolutionary change, but others remained. In the 1930s and 1940s scholars interested in evolution made a conscious effort to broaden their communication and comprehension. We have already mentioned Theodosius Dobzhansky and his *Genetics and the Origin of Species*. The Modern Synthesis took its name from a book by Julian Huxley (1942) titled *Evolution: The Modern Synthesis*. Another book on *Systematics and the Origin of Species* by Ernst Mayr (1942) brought systematics, phenotypic variation, and biogeography into the emerging synthesis.

The fourth book commonly included in the Modern Synthesis is *Tempo and Mode in Evolution* by paleontologist George Gaylord Simpson (1944). While exemplary in many ways, Simpson's book shows that the "Modern Synthesis" was a partial synthesis at best. Simpson proposed the term "quantum evolution" for "the relatively rapid shift of a biotic population in disequilibrium to an equilibrium distinctly unlike [its] ancestral condition . . . [Quantum evolution] may be involved in either speciation or phyletic evolution . . . It is . . . believed to be the dominant and most essential process in the origin of taxonomic units . . . such as families, orders, and classes. It is believed to . . . explain the mystery that hovers over the origins of such major groups" (Simpson, 1944, p. 206).

Thus, at the end of the Modern Synthesis we are still left to wonder at the relationship of microevolution and macroevolution, to wonder at the relationship of micromutation and macromutation, and to wonder at an origin of species and higher taxa that is said to involve a mysterious "quantum" origin followed by a new equilibrium.

1.3 Quantification of Rates

Charles Darwin recognized that rates are important but he seemingly made no attempt to quantify what he meant by fast and slow. Dobzhansky's (1937) only explicit consideration of rate was his characterization of some groups of organisms as having an unlimited store of variation and evolving rapidly, while he characterized others, such as the "living fossil" brachiopod *Lingula*, as being conservative and showing no change through epochs of geological time.

Huxley (1942, p. 56) discussed evolutionary rates briefly, observing that "no rate of hereditary change hitherto observed in nature would have any evolutionary effect in the teeth of even the slightest degree of adverse selection." Mayr (1942, p. 297) wrote of different rates of evolution in different groups or in different periods within the same group, noting that "an animal group that is searching for a new 'adaptive peak' may undergo rapid evolution, but as soon as this peak has been reached evolution may begin to stagnate."

George Gaylord Simpson was the only Modern Synthesis author to attempt a real quantification of rates, approaching the problem from a paleontological point of view. J. B. S. Haldane followed up on this, taking both a paleontological perspective and a more general biological view.

1.3.1 George Gaylord Simpson (1944)

Simpson's (1944) *Tempo and Mode* was explicitly about rates of evolution. How fast, he asked, do animals evolve in nature? "It is the first question that the geneticist asks the paleontologist" (Simpson, 1944, p. 3). Geneticists envisioned evolution as change in the genetic composition of populations, and Simpson considered that it might be desirable to define evolutionary rates in terms of genetic change per year, per century, or through some other unit of absolute time.

However, he recognized that genetic change was unknowable in extinct populations. Simpson assumed, as a compromise, that morphological change should be proportional to genetic change, and rates based on morphology should be similar to rates of any underlying genetic modification.

Change can be studied in individual morphological characteristics, yielding what Simpson called "unit-character" rates. Change can be studied in a number of related morphological characteristics, yielding what he called "character-complex" rates. Finally, change can be studied in morphological characteristics representing whole animals, yielding what he called "organism" rates. Unit characters can be studied in relation to each other, yielding what Simpson called "relative" rates, or they can be studied in relation to time, yielding what Simpson called "absolute" rates.

In *Tempo and Mode* Simpson started by comparing rates of change in tooth shape in the evolution of horses. He measured "paracone height" or the crown height of a tooth, measured on an upper molar, and he measured "ectoloph length" or the anteroposterior crown length measured on the same molar. Simpson then defined "hypsodonty" as the ratio of paracone height to ectoloph length. In other words, hypsodont teeth are relatively high crowned. When horses first appeared in the fossil record at the beginning of the Eocene epoch of Cenozoic time they had low-crowned molars: The crowns were only about one half as high as they were long. Later, in the Miocene epoch, acquisition of progressively more high-crowned hypsodont molars enabled horses to chew and digest more abrasive vegetation. Horses today have molars with crowns about three times higher than their anteroposterior length. This is a sixfold increase in terms of proportion.

The first table of data in *Tempo and Mode* listed measurements of paracone height, ectoloph length, and hypsodonty for teeth of five genera and species of Cenozoic-era horses. Simpson combined these in a figure (Simpson's figure 2), assuming first that ectoloph length represented overall body size and then, provocatively, that logarithms of both might increase at a constant rate (that is, both, on a proportional scale, might increase steadily in relation to time). If this "orthogenetic" assumption were true, then the spacing of species should be proportional to the geological time separating species. The geological timescale was poorly constrained in 1944, but Simpson's result was sufficiently different from expectation that he rejected the idea of ectoloph length and body size changing at a constant rate. Comparison of paracone height to ectoloph length showed that if one increased at a constant rate then the other did not. Plotting paracone height and ectoloph length against geological time, as best it was known, suggested that neither increased at a constant rate.

Simpson was seemingly unaware that the Russian paleontologist Alexei Petrovich Pavlov calculated an evolutionary rate for horses some dozen years earlier

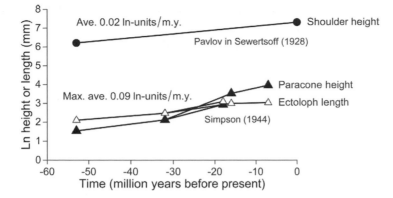

Figure 1.1 Rates of evolution in horses calculated by Alexei Petrovich Pavlov (in Sewertzoff 1931) and by George Gaylord Simpson (1944). *Y*-axis is a proportional or logarithmic scale, in natural log (ln) units. Average rates are given by the slopes of lines connecting successive samples. Note that most lines parallel the rate of 0.02 ln units per million years found by Pavlov, but Simpson's highest rate of 0.09 ln units per million years, during the evolution of high-crowned or hypsodont teeth in horses, is about four times as fast. Rates here are extremely slow, generally only a few millimicrons per year, or millimeters per million years, on timescales ranging from 7 to 53 million years. Geological ages of samples on the *x*-axis are updated from MacFadden (1992)

(reported in Sewertsoff, 1931). Pavlov assumed that the earliest horses stood about a half-meter high at the shoulder whereas recent horses stand about 1.5 m high, a difference of a full meter. He further assumed that the earliest horses lived a minimum of 2.5 million years before present, meaning that a 1.0 m difference in size evolved in 2.5 m.y. From this Pavlov calculated that the average rate of increase was about 40 cm per million years, or about 0.4 micrometers per year, a rate Sewertzoff considered unimaginably slow. Now a realistic estimate for the time involved would be closer to 53 m.y., rather than 2.5 m.y., making the increase in stature even slower at 0.02 micrometers per year. Sewertzoff concluded (1931, p. 303): "Even if we assume that there were times ('*Stillstandsperioden*') when evolution did not progress, and other times of more rapid change, the course of the evolutionary process remains very slow."

Simpson's (1944) principal innovation for evolutionary rates was recognition that measurements of unit characters should be transformed to logarithms for comparison (what is important is not their difference in measurement units but their difference in proportion). Simpson was content to compare slopes graphically, as shown here in Figure 1.1. This figure, modified from Simpson's figure 4, with the addition of Pavlov's shoulder heights), shows that ectoloph length and shoulder height increased in parallel, indicating similar rates. It also shows that the slopes for paracone height are a little steeper than those for both ectoloph length and shoulder height, indicating more rapid change.

The comparisons Simpson made led to four conclusions concerning character evolution (Simpson, 1944, p. 12), (1) the rate of one character may be a function of another even when the rates are different, (2) rates for one character or a combination of characters may change markedly even when the direction of evolution remains the same, (3) rates for two or more characters within a lineage may change independently, and (4) two closely related lineages may become differentiated when rates for their characters differ, with no change in the direction of their evolution.

Following this, it seems fair to say that Simpson (1944, pp. 16–17) foundered in uncertainty about what to measure, how to compare measurements, and the weighting to use to obtain a representative average rate. In the end Simpson fell back on what he called "organism" rates, reasoning that "subjective judgment of the total difference between organisms is ... more reliable than any objective measurement yet devised ... For such purposes genera are the most useful units." He then reasoned that if genera are comparable, "organism rates of evolution would be proportional to the reciprocals of the durations of the genera in question" and "for a sequence of successive genera, a more reliable value would be obtained by dividing number of genera by total duration." For horse evolution Simpson calculated an average rate of 0.18 genera per million years, or reciprocally 5.6 million years per genus. Simpson's "organism" rates are now generally referred to as taxonomic rates.

1.3.2 J. B. S. Haldane (1949)

J. B. S. Haldane was a geneticist and evolutionary biologist who achieved fame in many fields. Haldane's (1949) reaction to Simpson's (1944) study of evolutionary rates was first to question the objectivity of genera and their utility for quantifying rates. Haldane then shifted his focus to measurable change in the morphological characteristics of biological populations. Like Simpson, Haldane represented change in terms of proportion, and time in millions of years. Haldane (1949) went a step farther and coined the term *darwin* for change in proportion by a factor of *e*, base of the natural logarithms, per year, writing:

$$\frac{1}{x}\frac{dx}{dt} = \frac{d}{dt}(\log_e x) = \frac{\log_e x_2 - \log_e x_1}{t} = \frac{\log_e(x_2/x_1)}{t} \tag{1.1}$$

Simpson's *Hyracotherium* had a paracone height of $x_1 = 4.67$ mm and *Mesohippus* had a paracone height of $x_2 = 8.36$ mm. Assuming the two were separated by $t = 16$ million years, Haldane then calculated the rate of change to be (with ln now representing $\log_e$):

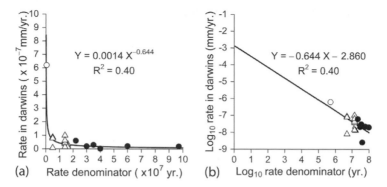

Figure 1.2 Rates of evolution from the fossil record of humans, horses, and dinosaurs calculated by Haldane in his *darwin* rate unit, representing factors of e per year on million-year scales of time (Haldane, 1949). (a) Scatter of rates and rate intervals (rate denominators) plotted on arithmetic axes. (b) Scatter of the same rates and intervals plotted on logarithmic axes. Open circle is one rate for human evolution in the Pleistocene epoch, open triangles are rates for horse evolution in the Cenozoic era, and filled circles are rates for dinosaur evolution in the Mesozoic era of geological time.

$$\frac{\ln x_2 - \ln x_1}{t} = \frac{\ln (8.36) - \ln (4.67)}{16000000} = \frac{2.123 - 1.541}{16000000}$$
$$= 0.000000036 = 3.60 \times 10^{-8} \, \text{darwins} \qquad (1.2)$$

Haldane made the calculation for all 12 of Simpson's comparisons of successive species. He called the darwin a "factor of e per million years" (Haldane, 1949, p. 55), but for the example given it is really a factor of e per year on a 16 million-year timescale.

Haldane's darwin unit has been widely used to represent evolutionary rates, reflecting the high regard people have for Haldane and for Darwin. However, there is nothing intuitive about factors of e, nor about change by a factor of e per year. And there is nothing convenient about a darwin rate with seven leading zeros. Nor did Haldane's invention of the term *millidarwin* as a substitute for 10^{-9} darwins really help.

In addition to the rates Haldane (1949) calculated for horses, he calculated six rates of size change for dinosaur evolution and one rate for change in the shape of the skull in human evolution. Haldane's 12 rates for horses had a median of 4.05×10^{-8}. His six rates for dinosaurs had a median of 2.15×10^{-8}. The one rate for humans was 6.2×10^{-7}. All 19 rates are shown graphically in Figure 1.2. Haldane realized that the rate he calculated for humans was more than an order of magnitude higher than the medians for horses and dinosaurs. If he had made the comparison of rates in Figure 1.2, he might have recognized that there is a problem with such rates.

Haldane wrote (1949, p. 51): "As biologists we would like to be able to measure time in generations." He solved this by multiplying a rate in darwins by generation time in years to yield factors of e per generation. As a worked example, the rate of change in paracone height from *Hyracotherium* to *Mesohippus* is 3.60×10^{-8} darwins (Eq. 1.2). Multiplied by an estimated three-year generation time yielded an evolutionary rate of 1.08×10^{-7} factors of e per generation.

Haldane also wrote (p. 52): "The use of the standard deviation as a yardstick has a certain interest because, on any version of the Darwinian theory, the variation within a population at any time constitutes, so to say, the raw material available for evolution." Here he brought the sample coefficient of variation $V = s/\bar{x}$ to bear. For *Hyracotherium* to *Mesohippus* the mean value of V is 0.055. The initial rate of 3.60×10^{-8} darwins divided by V becomes a rate of 6.55×10^{-7} standard deviations per year (or the inverse, 1.53×10^{6} years to evolve through one standard deviation). The generation-time rate of 1.08×10^{-7} factors of e per generation divided by V becomes a rate of 1.96×10^{-6} standard deviations per generation (or the inverse, 5.09×10^{5} generations to evolve through one standard deviation).

The important point is that when Haldane (1949) settled on a rate that interested him, a rate he could use to relate morphology to genetics, it was not a rate in darwins but a rate in standard deviations per generation. Thus some years ago Gingerich (1993) turned Haldane's tribute to Darwin into a tribute to Haldane himself by proposing an evolutionary rate unit to replace the darwin. The new unit was a *haldane*, which is simply Haldane's favored increase or decrease in units of standard deviation per generation.

Haldane seemingly dove down a rabbit hole when he realized that a half-million generations might be required for natural selection to move a phenotypic mean through a phenotypic standard deviation. He wrote (1949, p. 56): "Such calculations are extremely rough, but they suggest the remarkably small order of magnitude of the selective 'forces' which are at work if natural selection is largely responsible for evolution ... The slowness of the rate of change also makes it clear that agencies other than natural selection cannot be neglected because [such rates] are extremely slow by laboratory standards or even undetectable during a human lifetime." Haldane then raised the possibility that mutation rather than selection might determine the course of evolution.

1.4 Punctuated Evolution from Lyell to Modern Times

When Niles Eldredge and Stephen Jay Gould published their theory of punctuated equilibria in 1972, they introduced a new word, "punctuated," to the lexicon of evolution, but the concept was not new. Eldredge and Gould presented punctuated equilibria as "an alternative to phyletic gradualism," and wrote that "the history of life ... is not one of stately unfolding, but a story of homeostatic equilibria, disturbed ... by rapid and episodic events of speciation" (Eldredge and Gould 1972, p. 84). It was, in other words, another alternative to the Lamarck–Darwin thesis outlined above: an alternative to *peu-à-peu*, little-by-little, step-by-step evolutionary change. We can tabulate adverse reactions over the years to the Lamarck–Darwin thesis, and representative examples of these are listed in Table 1.1.

The first edition of Carl Linnaeus' *Systema Naturae* (1735) was, in some sense, the accepted wisdom of naturalists early in the eighteenth century: Species were simply part of an early creation, placed on Earth to reproduce and perpetuate themselves. Linnaeus softened this belief in later editions, responding to Georges Louis Buffon and other advocates who emphasized reproduction as a criterion for species cohesion and recognition. Linnaeus eventually admitted that perhaps only orders were created, with genera and species resulting from intermixtures of these.

Enter Lamarck in the nineteenth century with a thesis later taken up by Darwin: Individuals change, little by little, and after many successive generations male and female individuals are transformed into new and distinct species (Lamarck, 1809). Reaction to Lamarck was swift. It came first from a colleague in the Paris *Muséum National d'Histoire Naturelle*, Georges Cuvier, the father of vertebrate paleontology. Cuvier's experience with fossils in the Paris Basin taught him that species are *formes perpétuées dans ces limites,* while the Earth suffered repeated *révolutions et catastrophes* (Cuvier, 1812, 1817, 1825). Reaction came too in the antithesis of Lyell (1832) outlined above.

Darwin (1859) followed Lamarck in accepting the slow action of existing causes leading to the origin of new and distinct species. However, Darwin made the important substitution of populations for individuals as the functional units of evolution, and he made the crucial substitution of natural selection for the inheritance of acquired characteristics as the mechanism leading to change. Darwin was not dogmatic and admitted from the beginning that some species might remain unchanged for long periods of time.

Table 1.1 includes 69 studies published after the *Origin of Species* (Darwin, 1859) that argue against Darwinian step-by-step slow gradual change. Each introduced new terminology or brought some new wrinkle to the argument. The first of the 69 studies was published in 1863 and the last was published in 2014. Thirty-nine of the 69 studies (57%) were by paleontologists, and 30 (43%) were by biologists including botanists and zoologists (28%), geneticists (9%), and developmental biologists (7%).

Table 1.1 *Samples of eighteenth- through twenty-first-century literature that contradict Lamarck's and Darwin's advocacy of gradual evolutionary change.*

Characterization of evolutionary patterns	Author
"Si opera Dei intueamur, omnibus satis superque patet, viventia singula ex ovo propagari, omneque ovum producere sobolem parenti simillimam. Hinc nullae species novae hodienum producuntur." [If you look at God's work, everything is clear enough, living things propagate from an egg, and all the offspring resemble the parent. For this reason, no new species are produced.]	Carolus Linnaeus (1735, p. 2)
"Renouvellement des faunes" [renewal of faunas]	Georges Cuvier (1812, p. 81)
Species as "formes perpétuées sans excéder ces limites" [forms perpetuated without exceeding their limits] and "formes fixes qui ne se produisent ni ne se changent" [forms fixed that neither extend themselves nor change themselves]	Georges Cuvier (1817, pp. 19–20)
"Révolutions du globe" [global revolutions], "révolutions et catastrophes" [revolutions and catastrophes], and "révolutions subites" [sudden revolutions]	Georges Cuvier (1825, title and pp. 7–8)
In the formation of races "the quantity of divergence diminishes . . . in a very rapid ratio"	Charles Lyell (1832, pp. 37)
"Fixité des espèces" [fixity of species]	Contradicted by Isidore Geoffroy Saint-Hilaire (1836, p. 606)
"Renouvellement des faunes" [renewal of faunas] and "vingt-sept créations distinctes" [twenty-seven distinct creations]	Alcide d'Orbigny (1851, p. 251)
"Transmutation without transition" and "passing from species to species 'Natura fecit saltum'" [Nature makes leaps]	Thomas Henry Huxley (25 June 1859 letter to Charles Lyell, sent five months before publication of *Darwin's Origin of Species*; letter published in L. Huxley, 1900, pp. 185, 186)
"Species are produced and exterminated by slowly acting and still existing causes, and not by miraculous acts of creation and catastrophes" while "a number of species . . . keeping in a body might remain for a long period unchanged"	Charles Darwin (1859, pp. 487–488)
"Persistence and constancy . . . of the distinctive characters . . . most concerned in the existence and habits of species . . . we nowhere meet with intermediate forms."	Hugh Falconer (1863, p. 79)
"Aufblüzeit, das erste Stadium der Phylogenese"[blooming-time or youth, the first	Ernst Haeckel (1866, p. 321–322)

Table 1.1 (*cont.*)

Characterization of evolutionary patterns	Author
stage of phylogenesis], and "Blüthezeit, das zweite und mittlere Stadium der Phylogenese" [flowering-time or maturity, the second and typical stage of phylogenesis]	
"Change . . . sudden and considerable" and "departures from parental type, probably sudden and seemingly monstrous"	Richard Owen (1868, pp. 795–797)
"Stability of types" and "changes in jerks"	Francis Galton (1869, pp. 368 and 369)
"Mutation . . . höchst constant, stets sicher wieder zu erkennen" [species differing from earlier and later forms . . . most constant, always recognized with certainty]	Wilhelm Waagen (1869, p. 186)
"Modifications appearing at once," "sudden successive manifestations," and "stability of species"	St. George Jackson Mivart (1871, pp. 109 and 127)
"Constanz, mit völlig scharfe Ausprägung der neuen Art" [constancy, with sudden expression of new species], "Constanzperiode" [stasis-time] and "Variabilitätsperiode" or "Variationsperiode" [change-time]	August Weismann (1872, pp. 43 and 52)
"New and widely distinct species may be suddenly evolved from preexisting forms without the intervention of connecting links."	Thomas Meehan (1875, p. 12)
"Constanz der Species als unbestrittene Grundlage der Systematik" [constancy of species as an unchallenged assumption of systematics]	Melchior Neumayr and Carl Maria Paul (1875, p. 94)
"Grundformen früzeitig in viele Arten sich entwickeln" [major groups that evolve many species early] and "Zwischenformen nur höchst selten sich finden, also wohl nur kurze Zeit hindurch bestanden haben" [intermediate forms that are rarely found, and probably only existed for a short time]	Karl Ernst von Baer (1876, p. 431)
"Saltatory evolution" involving "more or less stable equilibrium," resisting change as long as effectual, followed by "sudden change . . . upon which the tendency to equilibrium may reassert itself"	William Healey Dall (1877, pp. 136)
"Expression points" or "periods of metamorphosis"	Edward Drinker Cope (1887, p. 112)
"Definite positions of organic stability," with a "leap from one position of organic stability to another, or as we may phrase it, through 'transilient' variation"	Francis Galton (1894, pp. 364 and 368)

Table 1.1 (*cont.*)

Characterization of evolutionary patterns	Author
"Metakinese bedeutet Umschüttelung, und . . . mit einem Umbildungsprozess . . . tiefgreifende Umgestaltung einer Form" [metakinesis means shaking, and with a transformation process, profound transformation of a form]	Otto Jaeckel (1902, pp. 34–35)
"Sideways" speciation, "mutability," and "ordinary constancy"	Hugo de Vries (1905, pp. 393, 396).
"Law of saltation . . . that saltation is a constant phenomenon in nature, a vera causa of evolution, one can no longer deny"	Henry Fairfield Osborn (1905, p. 230)
"Saltative palingenesis" [phylogeny skipping stages in ontogeny]	Sydney Buckman (1909, p. vii)
"Law of the persistence of fluctuating variation" [phyletic stability of a variable characteristic]	Henry Shaler Williams (1910, p. 306)
"Anpassungsperioden an neue Lebensbedingungen sind daher wohl in der Regel sehr rasch erfolgt und diese Zeiten sind es, in denen eine ruckweise Entwicklung einsetzte" [Periods of adaptation to new living conditions are usually very fast and these are times of spasmodic development]	Othenio Abel (1918, pp. 119–120)
"Disturbance provoking a saltation, and so giving a new centre to fluctuation"	Francis A. Bather (1920, pp. 71–72)
"Periodische oder diskontinuierliche Entwicklung" with "Virenzperioden von beschränkten Dauer wie die der Invirenz" [periodic or discontinuous evolution, with greening- or verdancy-periods of more limited duration than those of inverdancy]	Rudolf Wedekind (1920, p. 20)
"Per saltum species mutations" and "cataclysms"	G. Udny Yule (1924, pp. 22–23)
"The birth of . . . species . . . is a catastrophic process . . . saltus . . . well exemplified in paleontology"	Lev Berg (1926, p. 388)
"Sprunghaftigkeit der Stammesentwicklung" [abruptness of phyletic-evolution]	Roland Brinkmann (1929, p. 54)
"Conditions in nature are often such as to bring about the state of poise among opposing tendencies on which an indefinitely continuing evolutionary process depends"	Sewall Wright (1931, p. 158)
"Explosiver Entwicklungsphase" [explosive evolution phase] followed by a "gerichteter Weiterbildungsphase" [directed training phase] or "allmählich Stabilisierungsphase" [gradual stabilization phase]	Karl Beurlen (1932, p. 75)
"Hopeful monster"	Richard Goldschmidt (1933, p. 547)
"Das Wesen der saltierenden Entwicklung beruht darauf, daß gewisse Merkmale sich schnell umwandeln, um dann für längere Zeit konstant	Rudolf Kaufmann (1933, p. 25)

Table 1.1 (*cont.*)

Characterization of evolutionary patterns	Author
zu bleiben." [The essence of saltational speciation is that certain characteristics transform quickly, and then remain constant for a long time.]	
"Plötzlichen grossen Mutationsschritten" [sudden large mutation-steps]; "Typostrophe, Typogenese, Typostase" [form-turning, form-origin, form-stasis]	Otto Schindewolf (1936, p. 85; 1950, p. 238)
"Some evolutionary changes can be by sudden and discontinuous jumps"	Conrad Waddington (1939, p. 241)
"Canalization" or "canalized developmental reactions," and "constancy of the wild type"	Conrad Waddington (1942, pp. 563, 564)
"Explosive Formenaufspaltung am Beginn von Stammesreihen" [explosive diversification of form at the beginning of phylogenetic lineages] and "Phase der explosiven Formbildung" [phase of explosive morphogenesis]	Bernard Rensch (1943, pp. 24 and 50)
"Progressive evolution" compared to "still-stand evolution" or "stagnation"	Ivan Schmalhausen (1943, p. 283)
"Bradytely" [unusually slow or low-rate change], "tachytely" [unusually fast or high-rate change], and "quantum evolution" [rapid "disequilibrium" shift of a population or species from one state of equilibrium to another]	George Gaylord Simpson (1944, pp. 131–132, 193, and 206)
"Every species passes through an episode of rapid evolution but may become stabilized thereafter and persist unaltered for a long time"	Frederick Zeuner (1946, p. 379).
"Proliferation of genera is multiple and accelerated . . . a burst pattern"	G. Arthur Cooper and Alwyn Williams (1952, p. 326)
"Sprungtheorie . . . Phasen der Sprünge . . . eingeschaltet zwischen Phasen der relativ langsamen Normalentwicklung" [Saltation theory with phases of jumps intercalated between phases of relatively slow normal evolution]	Walter Zimmermann (1953, pp. 496–497)
"Genetic revolution" to reach a "new state of equilibrium"; "phylogenetic saltation"; "effect of selection will normally be to stabilize [the] phenotype" and "resist change . . . in the face of new selection."	Ernst Mayr (1954, pp. 170, 176; 1963, p. 296)
"Genetic homeostasis" or the "tendency of a Mendelian population as a whole to retain its genetic composition"	I. Michael Lerner (1954, p. 81)
"Evolving series . . . may remain quiescent, almost dormant, for long periods and then enjoy brief bursts of evolution"	T. Neville George (1958, p. 410)

Table 1.1 (*cont.*)

Characterization of evolutionary patterns	Author
"In man . . . selection is . . . largely ineffective, preserving an existing equilibrium"	J. B. S. Haldane (1958, p. 18)
"Process of anagenesis is one of rapid morphological change . . . preceded and followed by stasigenesis"	Peter Sylvester-Bradley (1959, p. 196)
Adaptation "may involve a quantum shift followed by a period of slower adjustment and stabilization"	Verne Grant (1963, p. 565)
"Bursts in evolution"	Michael House (1963, p. 499)
"Trilobites . . . are well represented in practically all the zoogeographical provinces of the world; yet species transitions are not forthcoming, and there is little intraspecific evolution to be observed. In any given area, species appear as immmgrants, have a shorter or longer run, and then vanish. The situation suggests that new species arose comparatively rapidly, but once established, tended to continue without any change."	Björn Kurtén (1965, pp. 344–345)
"Bursts of cladogenetic activity" or "cladogenetic phases" followed by "stasigenetic interludes"	Alan Cheetham (1968, p. 56; 1971, p. 3)
"Great majority of species do not show . . . evolutionary change at all. These species appear . . . without obvious ancestors . . ., are stable once established, and disappear . . . without leaving any obvious descendants."	Hendrik J. Mac Gillavry (1968, p. 70)
"Establishment of a new species was . . . so rapid by comparison with the period of its relatively stable existence that it is . . . reasonable to regard it as a 'jump'"	V. N. Ovcharenko (1969, p. 62)
"History of life is . . . a picture of 'punctuated equilibria' . . . a story of homeostic equilibria, disturbed . . . by rapid and episodic events of speciation."	Niles Eldredge and Stephen Jay Gould (1972, p. 84)
"Faunistic revolutions or caesuras"	Jost Wiedmann (1973, p. 160)
"Epochal events . . . lead to evolutionary leaps that may occur only once in the history of a species"	Herbert Kubitschek (1974, p. 107)
"Large mutations" corresponding to "archetypical modifications"	Søren Løvtrup (1974, pp. 415–416)
"Speciational events may be set in motion and important genetic saltations toward species formation accomplished by a series of catastrophic, stochastic genetic events."	Hampton Carson (1975, pp. 87–88)
"Rectangular patterns of phylogeny"	Steven Stanley (1975, p. 646)
"The fossil record . . . long periods of quiescence separated by short periods of explosive evolution"	Derek Ager (1976, p. 131)

Table 1.1 (*cont.*)

Characterization of evolutionary patterns	Author
"Stable evolutionary set" [of genes]	Richard Dawkins (1976, p. 93)
"Reticulate speciation with eruptive and stabilized phases"	Peter Sylvester-Bradley (1977, p. 46)
"Long-term equilibria in [biotic] communities . . . interrupted by destabilizing revolutions"	Geerat Vermeij (1977, p. 255)
"Punctuated gradualism"	Björn Malmgren et al. (1983, p. 387)
"Évolution polyphasée"	Jean Chaline (1984, p. 794)
"Episodic molecular clock"	John Gillespie (1984, p. 8009)
"Structural complexity entered in a grand burst at the Cambrian explosion"	Stephen Jay Gould (1985, p. 2)
"Turnover-pulse hypothesis"	Elisabeth Vrba (1985, p. 232)
"Punctuational phyletic change"	John Turner (1986, p. 204)
"Organotaxism"	Bernard Michaux (1988, p. 406)
"Stepwise (or staircase) pattern in which well-established species . . . undergo rapid change in between long intervals of approximate stasis or slow phyletic change."	Steven Stanley (1989, p. 159)
"Coordinated stasis"	Carlton Brett and Gordon Baird (1995, p. 307)
"Macroevolutionary bursts"	Josef Uyeda et al. (2011, p. 15908)
"Early burst niche-filling pattern"	Roger Benson et al. (2014, p. e1001853)

Fifty-three of the 69 post-*Origin* studies (77%) advocated some form of saltation or punctuated equilibria in the evolution of species. These species-level studies overlap considerably, but 24 of the 69 (35%) mentioned saltation, 17 (25%) advocated a combination of change and stasis, four (6%) mentioned both fast and slow change, and eight (12%) were focused on stasis without addressing how change takes place. Sixteen of the 69 post-Darwinian studies (23%) were concerned with change at a broader faunal level and mentioned bursts of diversification.

Table 1.1 shows that there has been strong post-*Origin* support for a punctuated equilibria model of evolution among paleontologists since Falconer (1863), and this continues to the present day. Support among biologists goes back to Huxley (1859) and Galton (1869). Studies advocating saltation at the species level out-number studies on a faunal scale by about 3 to 1, but faunal-scale studies too continue to the present day. The punctuated equilibria hypothesis of Eldredge and Gould is a current example of saltation studies, and the destabilizing revolution, turnover pulse, and coordinated stasis hypotheses of Vermeij (1977, Vrba (1985), and Brett and Baird (1995) are current representatives of faunal-scale studies.

1.5 Thesis and Antithesis: Quantitative Comparison

J. B. S. Haldane's calculation of very slow rates of evolution, described above, is a reminder that slow change was built into the Lamarck–Darwin thesis of gradual evolution. Haldane's reaction to the slow rates, raising the possibility that mutation determines the course of evolution, is a reminder of the antithesis ascribed to Lyell and Linnaeus. The thesis and the antithesis have long histories, and both thrive in some form today.

Neither Darwin in 1859 nor Eldredge and Gould in 1972 provided numbers to support their respective claims for the predominance of gradual change within and between species, or numbers to support equilibria within species and punctuated change between species. But fortunately each set of authors illustrated the visions in their minds in what have become iconic diagrams—shown in Figure 1.3.

Darwin provided a temporal scale for the vertical axis of his illustration (Figure 1.3a). He wrote that intervals between his horizontal lines, lines labeled with Roman numerals, "represent each a thousand generations," and added "it would have been better if each had represented ten thousand generations" (Darwin, 1859, p. 117). The letters, A, B, etc. on Darwin's horizontal axis were said to represent "species of a genus large in its own country" (Darwin, 1859, p. 116). He then added "these species are supposed to resemble each other in unequal degrees, as is so generally the case in nature ... represented in the diagram by the letters standing at unequal distances". Darwin's horizontal axis was not scaled explicitly, but there is a scale implicit in the morphological distance that separates closely related species.

Eldredge and Gould (Figure 1.3c) labeled their vertical axis "time," and labeled their horizontal axes (here combined on a single axis) "morphology." Time is unscaled in their illustration, but horizontal lines have been added with Roman numerals to represent the passage of time, mimicking the lines in Darwin's diagram. The spacing of these lines could represent thousands of generations or millions of years, or some other interval. The morphological scale in the Eldredge and Gould diagram is, again, implicitly the distance that separates closely related species.

It is not possible to recover quantitative rates of change from diagrams so ambiguously scaled as those in Figures 1.3a and 1.3c, but it is possible to recover distributions of relative rates that are consistent within each diagram. Figure 1.3b is a histogram of relative rates for Figure 1.3a, where rates reflect the angular deviation of each of Darwin's dotted lines from vertical for each of his successive intervals of time. Most of Darwin's lines are vertical, and the statistical mode for relative rates is the mode of stasis. Further, most of the remaining rates are relatively low rates that fall near the mode of stasis.

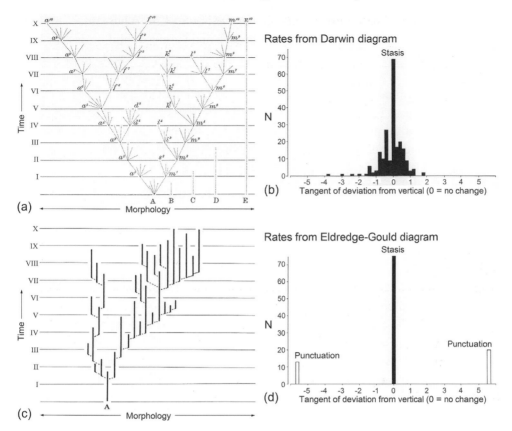

Figure 1.3 Graphical representations of evolution by Darwin (1859) and by Eldredge and Gould (1972), showing the contrasting rate distributions implicit in each. (a) Portion of the sole illustration in Darwin's *Origin of Species* (Darwin, 1859) showing the morphologies of species A, B, etc., and the morphologies of their descendants as these changed or stayed the same through time. (b) Histogram of rates for Darwin's illustration, where each rate is the trigonometric tangent of the deviation from vertical for each lineage segment in each time interval. (c) Eldredge and Gould's illustration showing the morphology of one species, here labeled A, and its descendants through time (adapted from fig. 10 in Eldredge and Gould 1972). Vertical lines are species in "equilibrium" (stasis) through time, and dashed lines show "punctuated" speciation events. (d) Histogram of rates for Eldredge and Gould's illustration, with rates calculated as before. Note: (1) all descendants of a species are connected vertically through time; (2) stasis – little or no change, near or at zero rate – predominates in Darwin's illustration as it does in Eldredge and Gould's, and (3) rates in Darwin's illustration are approximately normally distributed about the central stasis mode, while rates in Eldredge and Gould's diagram are distinctly trichotomous, with positive and negative punctuation modes distinctly different from the central stasis mode. Figure is modified from Gingerich (1984), reproduced by permission of Oxford University Press

Figure 1.3d is a histogram of relative rates for Figure 1.3c, where rates similarly reflect the angular deviation of each of Eldredge and Gould's solid or dashed lines from vertical. Most of Eldredge and Gould's lines are vertical, and again a statistical mode is the mode of stasis. However, here the histogram is trimodal, with additional modes for positive and negative punctuated change distinct from the mode of stasis. The punctuation-mode histogram bins are unshaded because the angles the relative rates reflect are arbitrary and the positions of these bins are unknown. Punctuation-mode rates are relatively high rates that fall relatively far from the mode of stasis.

Gould and Eldredge (1977, p. 116) promoted the idea that "stasis is data" in a review following publication of punctuated equilibria — "each case of stasis has as much meaning for evolutionary theory as each example of change." This became a motto, and then indeed a mantra. Species in stasis are data, yes, but stasis does not tell us how change takes place within a species, or when one species becomes another. Are rates of change in evolution clustered around the mode of stasis as it appears in Figure 1.3b, or are rates of change well separated from this mode as it appears in Figure 1.3d? This is a question that seems testable empirically.

1.6 Outline of the Following Chapters

How can we reconcile the Lamarck–Darwin thesis of little-by-little, step-by-step evolutionary change through time with the Lyell–Linnaeus antithesis supported by a century and a half of no-change "equilibrium" patterns reported by paleontologists studying the fossil record? Lyell is named first in labeling the antithesis because Lyell was the author reacting to Lamarck and he was the one inspiring Darwin. Linnaeus is included because Lyell cited him for support.

In Chapters 2 through 6 we consider the importance and scale of variation in nature, the time available for evolution, random walks as evolutionary models, temporal scaling in theory, and natural selection as a process. Then in Chapters 7– 9 we quantify and compile empirical evolutionary rates from laboratory experiments, field studies, and the fossil record. These are compared and combined in Chapter 10 to reconcile the evolutionary thesis of Lamarck and Darwin with the antithesis so clearly stated by Lyell and endorsed by later authors. This synthesis has consequences for quantitative models that are widely employed in paleontology and evolutionary biology, and some consequences are considered in Chapters 11 through 14. Finally, Chapter 15 is a summary with conclusions emphasized.

1.7 Summary

1. Jean-Baptiste Lamarck published *Philosophie Zoologique* in 1809. In this he advanced the thesis that species can be transformed into new species, *peu à peu*, little by little, over the course of many generations.
2. Charles Lyell published the second volume of *Principles of Geology* in 1832. He countered Lamarck's thesis with a reactionary antithesis, arguing that Linnaeaus and most naturalists believe distinguishing characteristics of species have remained the same from the time each species was created.
3. During the voyage of the *Beagle* young Charles Darwin had ample time to read and think about Lyell's representation of Lamarckian evolution and about Lyell's reaction to it. In *the Origin of Species*, published in 1859, Darwin sided with Lamarck on the transformation of species over the course of many generations.
4. Microevolution and macroevolution were named to distinguish minor from major change, short time scales from long time scales, and processes from patterns. In some sense the act of naming microevolution and macroevolution institutionalized their dichotomization.
5. The "Modern Synthesis" of evolution in the mid-nineteenth century was at best a partial synthesis, with "quantum evolution" proposed to describe and explain rapid shifts of biotic populations from one equilibrium state to another.
6. J. B. S. Haldane calculated very slow rates of evolutionary change from the fossil record and raised the possibility of micromutation as an alternative to natural selection to explain long-term evolution.
7. Punctuated equilibria is the current incarnation of quantum evolution and a host of other characterizations of evolution involving phases of static equilibrium interrupted by shorter intervals of rapid change.
8. Stasis is data, in some sense, but stasis is stasis and does not tell us how change takes place
9. The challenge before us is reconciliation of the Lamarck–Darwin thesis of slow gradual change with the Lyell–Linnaeus antithesis involving a mysterious origin of species followed by stasis.

2

Variation in Nature

No one supposes that all the individuals of the same species are cast in the very same mould. These individual differences are highly important for us, as they afford material for natural selection to accumulate, in the same manner as man can accumulate in any given direction individual differences in his domesticated productions.

Charles Darwin, Origin of Species*, 1859, p. 45*

Charles Darwin started the *Origin of Species* with two chapters on variation, the first on variation under domestication, and the second on variation in nature (Darwin, 1859). He followed this with publication of a full two-volume treatise on the *Variation of Animals and Plants under Domestication* (Darwin, 1868). The first volume of *Variation* is a descriptive species-by-species survey of domesticated animals and plants, but the second volume is more thematic in dealing with inheritance, cross-breeding, selection in domestication, variation, and finally Darwin's theory of pangenesis as a mechanism of inheritance.

Darwin (1868; 1875) summarized the subject of variation in terms of laws, but given what he knew concluded:

When we reflect on the several foregoing laws, imperfectly as we understand them, and when we bear in mind how much remains to be discovered, we need not be surprised at the intricate and to us unintelligible manner in which our domestic productions have varied, and still go on varying.

Darwin, 1875, v. 2, p. 348

Darwin was concerned with the origin of variants and varieties as a step toward the origin of species. He emphasized that individual differences are essential for natural selection but never made much progress finding patterns in the variation that he studied so carefully.

2.1 Quantification of Human Variation

Some of Darwin's contemporaries were more gifted quantitatively. These included the Belgian polymath Adolphe Quetelet, the American statistician Ezekiel Elliott, and Darwin's cousin Francis Galton. All three focused their studies on humans, a large species living in large populations that are both accessible and easy to measure.

2.1.1 Adolphe Quetelet

Adolphe Quetelet (1846) combined measurements of chest circumference for a number of local militias in Scotland, taken from the *Edinburgh Medical and Surgical Journal* for April 1817. The journal reported observations for 5,758 men, measured to the nearest inch (imperial inch, 25.4 mm), and Quetelet grouped all of the measurements by inch. He then constructed a table showing the number (frequency) of men in each group, and calculated the weighted mean of the measurements. The mean value for chest circumference based on Quetelet's table is 39.84 inches (1.012 m), with a probable error or median deviation of 1.381 inches (35.08 mm). The mean falls within the most frequent or modal group, and observations differing from the mean decrease in number as their difference increases (Figure 2.1a). In Quetelet's time, "probable error" was the value used to characterize the dispersion of observations about the mean value. The probable error is equal to 0.6745 times the standard deviation (see below).

Quetelet also studied measurements of stature for 100,000 military conscripts from France compiled by Antoine-Audet Hargenvilliers in 1817 and published by Louis Villermé (1829). Here the original measurements were in French inches (*pouce*, 27.07 mm), but this time the groups were centered on midpoints of the successive one-inch ranges. No increments were reported for statures less than 58 inches nor for statures greater than 65 inches, reducing the sample size from 100,000 to 68,890 and making calculation of relevant statistics problematical. Villermé reported a mean of 59.67 inches (1.615 m), to which Quetelet added a probable error of 1/33 times the mean. Graphing frequencies for measurements reported to the nearest inch yields the distribution shown in Figure 2.1b, where it appears that a mean of 60.1 inches (1.626 m) fits the distribution better than that reported by Villermé. Quetelet's probable error should then be 1.821 inches (49.29 mm). Quetelet did not graph either of the distributions shown in Figure 2.1 but showed equivalents in tabular form.

Quetelet (1846) found variation in the empirical measurements of chest circumference and stature satisfying because their patterns mimicked the distributions of error recorded in astronomy and physics. He believed in natural laws, among them

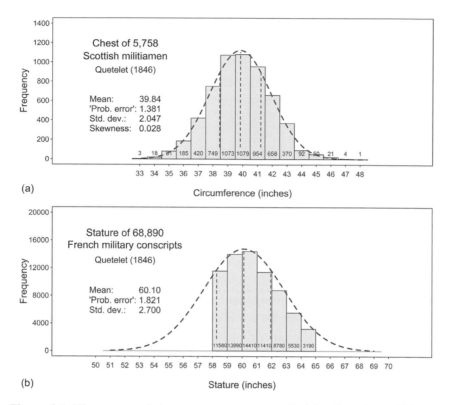

Figure 2.1 Histograms of the measurements compiled by Quetelet (1846). (a) Chest circumference of Scottish militiamen, based on measurements in imperial inches published anonymously in Edinburgh in 1817. (b) Standing height or stature of French military conscripts in French inches published by Villermé (1829; censored by failure to report shorter and taller statures). Mean stature for the French recruits would be 64.1 inches on an imperial scale. Numbers of men in each category are shown near the base of the corresponding column. Note the bell-shaped distributions of variation in both sets of measurements: variation considered to mimic distributions of error in astronomical measurements. Probable error was an early measure of dispersion, superseded by the standard deviation.

conservation of a human "type," and he went so far as to call such laws divine (Quetelet, 1870, p. 21). Quetelet is famous for his interest in social science or "social physics" (*physique sociale*) and for his concept of the "average man" (*l'homme moyen*; Quetelet, 1835). He accepted distributions of "error" in human measurements as mathematical confirmation of the applicability of physics-like laws to both the human type and the average man. However, Quetelet's idea of an "average man" or human "type" was misguided and the antithesis of Darwin's subsequent emphasis that individual differences are important for natural selection.

Before leaving Quetelet (1846), it is interesting to note his preoccupation with human giants and dwarfs. He went so far as to calculate, from the sample of French conscripts, that a population with a mean stature of 1.62 meters might be expected to include Frenchmen ranging in stature from 1.21 to 2.03 meters (deviations of ±0.41 meters from the mean). The smaller men Quetelet considered to be dwarfs, and larger men to be giants. Herschel (1850, p. 27) challenged this, citing examples, and argued that "the 'probable' deviation of nature's workmanship from her universal human type cannot possibly be less than double that resulting from the French measurements." If we follow Herschel, then Quetelet's range of human stature should reach 0.80 meters for dwarfs and 2.44 meters for giants. We need not worry about Quetelet's or Herschel's numbers here but will return to the subject of dwarfs and giants.

2.1.2 Ezekiel Elliott

Ezekiel Elliott was an American statistician and a government delegate to the International Statistical Congress that met in Berlin in 1863. There he presented a study of American military statistics (Elliott, 1863; 1865). Elliott included measurements of the stature of a large sample of Civil War volunteers. Here again measurements were in inches, and the measurements were grouped by the inch (imperial inch, 25.4 mm). Following Quetelet (1846), Elliott constructed a table showing the number of men in each group. Observations differing from the mean decreased in number as the difference increased (Figure 2.2a), as Quetelet had shown for chest circumference and, less well, for stature. Elliott did not graph the distribution of variation for his large sample, but he did construct a graph like that in Figure 2.2a for a smaller and more manageable set of measurements.

Elliott was impressed by Quetelet, with his emphasis on the human type, and by the regularity of variation about the type:

Statistical researches, conducted by M. Quetelet of Belgium, have established the fact, previously contested, of the existence of a human type, and that the casual variations from it are subject to the same symmetrical law in their distribution as that, which the doctrine of probabilities assigns to the distribution of errors of observation. In the accompanying tables, showing the distribution of heights and of measurements of the circumference of chests of American soldiers, the conclusions of this eminent statist and mathematician are strikingly confirmed.

Elliott, 1863, p. 14

The distributions are the same for variation within a species and for errors of astronomical observation, but it is a mistake to conflate variation and error. We shall return to this below.

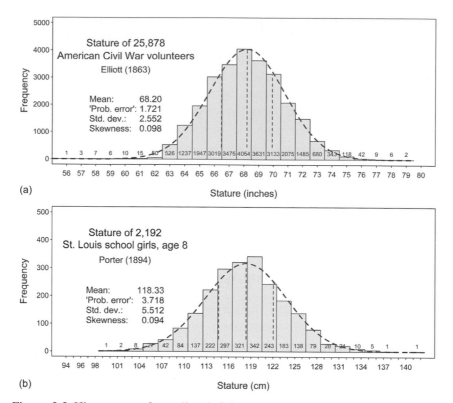

Figure 2.2 Histograms of standing heights in (a) American Civil War recruits (Elliott, 1863); and (b) eight-year-old St. Louis school girls (Porter, 1894). Measurements were reported to the nearest inch or centimeter, respectively, and numbers in each category are shown near the base of the corresponding column. Porter's school girl measurements were analyzed by Karl Pearson (1895), who was interested in their skewness. Probable error was an early measure of dispersion, superseded by the standard deviation.

Elliott analyzed both the physical measurements of Civil War volunteers and their ages. He was more perceptive than his contemporaries in distinguishing a law of error for ages based on *differences* from a law of error for human forms ("types") based on *proportions*:

According to the law, already stated, which appears to obtain with the distribution of the ages of the volunteers, the differences of the numbers at consecutive equidistant ages are very nearly in equi-rational or [arithmetical] progression. — In the distribution of the representatives of a type, where the law assigned by the theory of probability strictly holds, the quotients (not the differences) of the proportionate numbers at consecutive equidistant points of measurement are in equi-rational progression.

Elliott, 1863, p. 15

This represents, in cryptic form, the kernel of a key insight of Francis Galton to be developed below.

2.2 Probability and the Law of Error

We need some background to understand Quetelet's fascination with what he called the "binomial law", "law of probability," or "mathematical law of size." Quetelet used all three names for what is essentially the same concept; others have called this the "law of error."

Start with a coin. It has two sides, two "faces," and a negligible edge separating these. Call one side a tail (T), and the other side a head (H). If you flip a coin in the air, it will land and fall flat, with a tail facing upward or a head facing upward. On a fair coin the probabilities of these alternative events are equal. We can note the possible outcomes of one coin flip as a "T" for tail or an "H" for head.

What happens if we flip two coins? There is one way we can get two tails: TT. There are two ways to get a tail and a head: TH if the tail comes up first and the head comes up second, or HT if the head comes up first and the tail second. And there is one way to get two heads: HH. We can summarize this by calling TT, TH, HT, and HH the four possible *permutations* resulting from flipping two coins. Two tails, a tail and a head, and two heads, are the corresponding three *combinations*. We may not know what permutations went into the combinations we see, but we know that for two coins we can expect the permutations to be related to the combinations in the proportions 1:2:1.

If we flip three coins, then we may find any of eight permutations, TTT, TTH, THT, HTT, THH, HTH, HHT, and HHH, in one of four combinations of three heads, two heads and a tail, one head and two tails, or three tails. Without knowing the history, again we don't know what the permutations were, but we can expect the permutations to be related to the combinations in the proportions 1:3:3:1.

Extending this logic, we can expect flipping ten coins to yield combinations of heads and tails in the proportions 1:10:45:120:210:252:210:120:45:10:1, for a total of 1,024 permutations (Figure 2.3). Each combination has a probability or relative likelihood or "density" equal to the number of permutations yielding the combination, divided by the total number of permutations. In Figure 2.3, the combination with the greatest probability is five tails and five heads, for which the probability is $252/1024 = 0.246$. This indicates too that the combination of five tails and five heads has the greatest likelihood in relation to other combinations.

The permutations and combinations of a coin-flipping exercise are interesting for many reasons, but the point to be made here is illustrated in Figure 2.4. Flipping a single coin yields two permutations and two combinations, one in each case a tail and one a head. Graphing yields a simple symmetrical histogram of one permutation for each combination. Graphs for the permutations and combinations of two and three coins are slightly more complicated, but the first inkling of the relationship of interest emerges in a graph of the permutations and combinations for four

Ten coins   Permutations   Probabilities

Combinations		Permutations	Probabilities
T T T T T T T T T T		1	0.001
T T T T T T T T T H		10	0.010
T T T T T T T T H H		45	0.044
T T T T T T T H H H		120	0.117
T T T T T T H H H H		210	0.205
T T T T T H H H H H		252	0.246
T T T T H H H H H H		210	0.205
T T T H H H H H H H		120	0.117
T T H H H H H H H H		45	0.044
T H H H H H H H H H		10	0.010
H H H H H H H H H H		1	0.001
		1024	1.000

Figure 2.3 Permutations and combinations of tails (T) and heads (H) for ten coins. There is one permutation yielding the combination "all tails"; ten permutations of "nine tails and one head" (the head can be in any of 10 positions in the row); etc. The numbers of permutations that can yield the observed proportion of tails and heads in each combination are listed to the right of the coins, along with the associated probability for each combination. The latter is calculated as the number of permutations in a particular combination (e.g., 1, 10, 45, etc.), divided by the total number of permutations in the experiment (1,024).

coins. The graph for four coins has vertical bars differing by small, then large, then small amounts flanking the modal value. This becomes increasingly clear for permutations representing larger numbers of coins, and the permutations for 9 or 10 coins follow the bell-shape of a "normal" distribution. In a binary coin-flipping experiment like this the permutations for each combination are known as the "binomial coefficients" for a given number of trials. Historically, the mathematical shape of what we today call a "normal distribution" was derived by finding the limiting shape of the continuous curve of permutations for an infinite number of coins.

The same distribution and curve can be derived by rolling dice (Figure 2.5). One die has six sides or faces, with negligible edges separating these. The sides are generally numbered from 1 to 6. If you roll a die on a table, it will land and fall flat, with one of the numbers facing upward. On a fair die the probabilities of the alternative events are equal. We can note the possible outcomes for rolling one die by recording the resulting number: There is one way to score a "1," one way to score a "2," etc. Each has the same probability, and graphing yields a simple symmetrical histogram of one permutation for each combination (Figure 2.5a).

A graph for the permutations and combinations of two dice is slightly more complicated because there is one way to achieve a score of two, by rolling a one on each die ("1 + 1"). A combination score of three can be achieved in two ways, by rolling "1 + 2," or "2 + 1." The modal combination score is seven, which can be achieved six ways, as "1 + 6," "2 + 5," "3 + 4," "4 + 3," "5 + 2," or "6 + 1." The

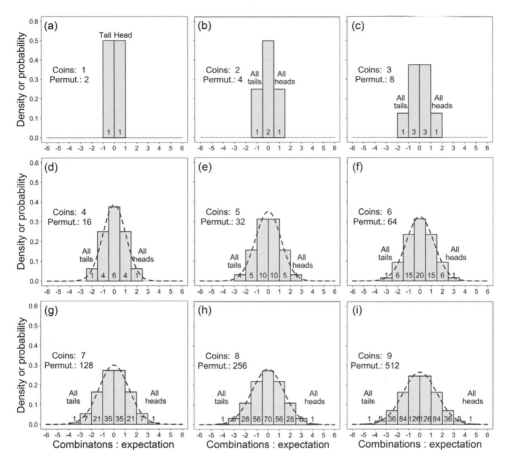

Figure 2.4 Histograms of permutations for increasing numbers of coins and combinations in a coin-tossing experiment. For n coins, the number of combinations of tails and heads increases as $n + 1$, but the number of permutations increases as 2^n. Note that the histogram of permutations converges to a normal curve (dashed line) as the number of coins increases.

sequence of successive permutations for each combination forms a symmetrical pyramid rising on one side to the mode and descending on the other (Figure 2.5b).

The relationship of interest emerges clearly in a graph of the permutations and combinations for three dice. Now the permutations for each combination form the series 1:3:6:10:15:21:25:27:27:25:21:15:10:6:3:1, for a total of 216 permutations. These are no longer pyramidal on a graph but now approximate a normal distribution (Figure 2.5c), with the heights of the vertical bars approximating a normal curve (dashed line). The fit to a normal curve for permutations of three dice is even better than the fit for permutations of 9 or 10 tosses of a coin. Here again, the mathematical shape of the normal curve can be derived by finding the limiting shape of the continuous curve for an infinite number of dice.

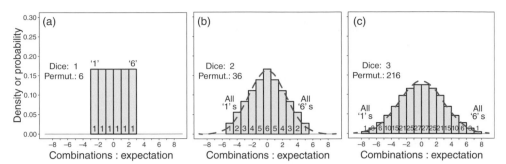

Figure 2.5 Histograms of permutations for increasing numbers of dice and combinations in a dice-rolling experiment. For n dice, the number of combinations of scores "1" through "6" increases as $5n + 1$, but the number of permutations increases as 6^n. Note that the histogram of permutations converges rapidly to a normal curve (dashed line) as the number of dice increases.

The importance of this exercise is that it shows how normal distributions of variation can be derived from the relative frequencies of small differences accumulating in the simple combinations expected by chance. The distributions of variation we see in biological populations reflect the contributions of many small genetic differences in constituent individuals.

2.3 The Normal Curve

"Normal" is an adjective, derived from Latin, that is commonly used in European languages to represent a "norm" or expected occurrence. As we have seen, this bland label is assigned to curves and distributions like those in the preceding figures. Common usage means it should come as no surprise that "normal" has no fixed point of origin in the history of statistics. Gustav Fechner (1860, p. 125) wrote of a "*normaler Fehlervertheilung*" (normal distribution of error). Quetelet (1869, p. 36) wrote of "*déviations d'une grandeur normale*" (deviations from a normal size; translated from Herschel's 1850, "deviations from a standard"). Charles Peirce (1873, p. 206), who read Fechner, wrote of comparing an observed curve of errors to "the normal least-squares curve." Wilhelm Lexis (1877, p. 34*ff.*), who read both Fechner and Quetelet, repeatedly compared "*normaler Dispersion*" (normal dispersion) to a dispersion greater or less than normal. Normal is an adjective that grew slowly into a name. By the time Francis Galton wrote *Natural Inheritance*, normal variability was the title of a chapter, the name of a curve, and the name of a distribution (Galton, 1889). In this he was followed by Karl Pearson (1894) and many others.

A normal distribution is sometimes called a Gaussian distribution, named for Carl Friedrich Gauss. Gauss (1809) is the one credited with inferring the form of

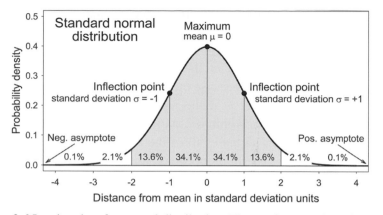

Figure 2.6 Landmarks of a normal distribution. The maximum value of the curve on the vertical axis gives the "location" of the distribution, which is the mean value of the probability density under the curve (here 0 on the horizontal axis). Left and right inflection points of the curve give the "dispersion" of the distribution. Each inflection point is one standard deviation from the mean (±1 on the horizontal axis). The curve here is standardized to a unit integral or unit area under the curve (making the total probability density = 1). Note that 68.3% of the area under the curve lies within ±1 standard deviation of the mean, 95.4% lies within ±2 standard deviations, 99.7% lies within ±3 standard deviations, and virtually all lies within ±4 standard deviations. Reducing the standard deviation will narrow the dispersion, and increasing the standard deviation will broaden it, but the probability density within a given standard deviation range does not change.

the error curve, which he wrote as $\varphi(\Delta) = (h/\sqrt{\pi}) \cdot e^{-h \cdot h \cdot \Delta \cdot \Delta}$ (parentheses and multiplication symbols added), where *phi* denotes a function of delta, *h* is a measure of precision (or dispersion), *pi* is the familiar ratio of circumference to diameter of a circle, and *e* is the base of natural logarithms. Abraham de Moivre published something similar in 1733 as an approximation to the binomial distribution, as did Laplace in 1810 in the form of the central limit theorem (Stigler, 1986).

2.3.1 Landmarks of a Normal Curve

If we look at any continuous normal curve (Figure 2.6), we see that it is symmetrical, with three landmarks familiar from calculus. The most prominent landmark is the highest point on the distribution. This point, the maximum value of the curve, is also, from symmetry, the *mean* value of the distribution on the horizontal axis, a parameter represented by μ. The mean specifies the location of the curve. Secondary landmarks are the two inflection points equidistant from the mean. These differ from the mean on the horizontal axis by minus one and plus one *standard deviation*, a parameter represented by σ, and the square root of the *variance* σ^2.

For a distribution of constant size (standard area under the curve), the standard deviation specifies the dispersion of the distribution. There is no negative or positive limit to the distribution because the extreme values are asymptotic to the horizontal axis. Note that 68.2% of the area under the curve lies within ± 1 standard deviation, 95.5% lies within ± 2 standard deviations, 99.7% lies within ± 3 standard deviations, and virtually all of the area under the curve lies within ± 4 standard deviations. These proportions hold for any normal curve or distribution.

The general form of the probability density function comprising the normal curve and normal distribution is:

$$f(x) = \frac{1}{\sigma\sqrt{2\pi}} e^{-\frac{(x-\mu)^2}{2\sigma^2}} \tag{2.1}$$

The curve is fit to an empirical sample by substituting sample values $\bar{x}$ and s for the parametric mean μ and standard deviation σ, respectively. For $\mu = 0$ and $\sigma = 1$, Equation 2.1 simplifies to:

$$f(x) = \frac{1}{\sqrt{2\pi}} e^{-\frac{x^2}{2}} \tag{2.2}$$

Both equations yield normal distributions standardized to a unit integral or a unit area under the curve, making the total probability density equal to one as well.

2.4 Logarithms and Coefficients of Variation

Logarithms and exponentials of the same base are inverse or mirror transformations. That is, $\log_{10} x$, or simply "log x" here, is the inverse of 10^x. $\log_e x$, or "ln x" here, is the inverse of e^x. $\log_2 x$ is the inverse of 2^x. The choice of bases is not completely arbitrary but depends on the range of numbers being compared. $\log_{10}$ is most useful when the range spans several orders of magnitude. $\log_2$ is most useful when the range spans several doublings but lies within an order of magnitude.

Napierian or natural logarithms, $\log_e$ or ln, are "natural" because they have special properties. First, the natural logarithm for a number $a > 0$ can be defined as the area under the curve $y = 1/x$ as x increases (or decreases) from 1 to a (with the area being negative for $a < 1$). Second, the slope of the curve $y = \ln x$ is $1 / x$, and the slope is 1 for $x = 1$ when and only when the base is e (y itself is 0 at this point, as it is for logarithms of all bases). Natural logarithms are "natural" too with respect to the population or sample variation that interests us here.

For normal distributions of different sizes and shapes, the standard deviation s increases in proportion to the mean $\bar{x}$ as well as the dispersion. The ratio of the two, $V = s/\bar{x}$, is commonly called the coefficient of variation (and sometimes multiplied by 100 to represent s as a percentage of $\bar{x}$). The ratio is considered a measure of dispersion independent of the mean. It is a special property of natural logarithms that the variance of ln transformed measurements approximates the squared coefficient of variation V^2 (Lewontin, 1966), from which it follows that the standard deviation of ln-transformed measurements is a close approximation to the coefficient of variation V. The opposite is also true: The coefficient of variation V is a close approximation to the standard deviation of ln-transformed measurements. For this reason, natural logarithms of base e are the preferred transformation for measurements in biology rather than $\log_2$ or $\log_{10}$.

Empirically, linear measurements of organisms commonly have $V \approx 0.05$, meaning that for linear measurements a standard deviation will be about 0.05 units on a natural log scale. Measurements of area commonly have $V \approx 0.10$, and a standard deviation will be about 0.10 units on a natural log scale. And finally, volumetric measurements of organisms such as weight commonly have $V \approx 0.15$, and a standard deviation will be about 0.15 units on a natural log scale. Yablokov (1974) provides extensive documentation. Note that for organisms that vary in size but have similar shapes, the coefficients of variation 0.05, 0.10, and 0.15 are proportional to the dimensions of the measurement: 1, 2, and 3 (Schmalhausen, 1935; Yablokov, 1974; Lande, 1977).

2.5 Arithmetic Normality versus Geometric Normality

Arithmos is the Greek word for number, and *arithmetic* is the science of counting and calculation that involves addition and subtraction (and multiplication and division), moving forward and backward on what is commonly called a "number line." An arithmetic progression is one that involves successive numbers differing by equal amounts, such as 1, 2, 3, 4; or -10, -20, -30, -40.

Ge or *gaia* is the Greek word for Earth, and *geometry* is the science of measurement, shape, and proportion. It started as measurement of land on the Earth's surface but progressed to measurement and comparison of sizes and shapes of all kinds. The emphasis on proportion is evident in geometric progressions, such as 1, 2, 4, 8; or 1, 10, 100, 1000; or 0.050, 0.135, 0.368, 1.000.

A number line and arithmetic operations on it are relatively easy to relate to our everyday experience of addition and subtraction, and slightly more complicated versions of addition and subtraction in the form of multiplication and division. Most of our measuring devices, such as rulers and weighing balances, are calibrated arithmetically and we read them by *counting* in units of convenience.

This works well enough for comparisons that are inherently linear, yielding numbers in arithmetic progression, but it does not work well for comparisons that are inherently areal or volumetric, yielding numbers in geometric progression. Geometry is called geometry because it is different from arithmetic. It involves arithmetic, but it involves more and is an extension of arithmetic.

2.5.1 Francis Galton's Giants and Dwarfs

The extraordinary limits [to the height of man], beyond which are found monstrosities, seem to me difficult to fix … When we suppose the number of observations infinite, we may carry the differences to equally infinite distances from the mean, and find the corrresponding probabilities. *This mathematical conception evidently cannot agree with that which is in nature.*
                    Adolphe Quetelet 1846 [1849], p. 102; italics added

The ordinary law of Frequency of Error, based on the arithmetic mean, … asserts that the existence of giants, whose height is more than double the mean … implies … *the existence of dwarfs, whose stature is less than nothing at all.*
                    Francis Galton 1879, p. 367; italics added

Quetelet took 5 feet 4 inches (1.62 m) as an average human height, and he accepted 1 foot 5 inches (0.43 m), exaggerated or not, as the height of the smallest dwarf. This is a difference of 3 feet 11 inches. Quetelet then added 3 feet 11 inches to the average height and predicted the limit to the size of giants to be 9 feet 3 inches (2.82 m). Quetelet recognized, intuitively at least, that there was some disagreement between the symmetry of his mathematical expectation and the asymmetry of limits actually observed in nature.

Galton exaggerated slightly in claiming giants to exist that are twice as tall as the average person, but he was clever in making an important point. Quetelet's "average man" was 1.62 meters or 162 cm tall. If a giant could be more than double this mean, say 162 + 164 = 326 cm, then symmetry of the normal curve would imply that a dwarf could be 162 − 164 = −2 cm tall: a Galtonian dwarf "whose stature is less than nothing at all." The conundrum is illustrated graphically in Figure 2.7a. However, experience now interferes — because we do not see people of negative stature, nor do we see people that we might consider to be approaching zero or negative stature. Galton's exercise demonstrates that the "normal" curve of human stature cannot be symmetrical. In arithmetic terms, there are fewer standard deviations to the lower limit of human stature than there are to the upper limit.

The geometric equivalent of Figure 2.7a is illustrated in Figure 2.7b. The two graphs have the same arithmetic axes, but vertical-axis values in Figure 2.7b are

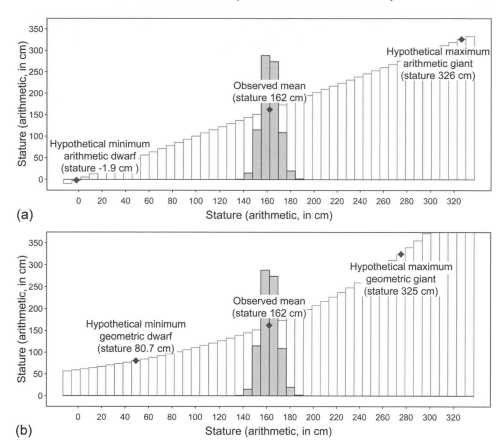

Figure 2.7 Comparisons of human stature in hypothetical human giants and dwarfs. On an arithmetic scale (a), where stature is added and subtracted in equal amounts, the postulated existence of giants more than double the mean implies, as Francis Galton argued, the existence of dwarfs whose stature is negative or "less than nothing at all." In contrast, on a geometric scale (b), when stature is added and subtracted in equal proportions, the existence of giants more than double the mean implies, more plausibly, the existence of dwarfs whose stature is less than half the mean. Vertical bar widths and step heights are standard deviations, with heights in (b) calculated geometrically. The observed mean and standard deviation are those of Quetelet's French military conscripts, 162 cm and 7.28 cm, and extreme statures assuming arithmetic normality are -1.8 and 326 cm. Equivalents assuming geometric normality, expressed on a proportional (natural-log) scale, are 5.09 and 0.045, with extreme statures of 4.39 and 5.78. Note that the existence of a geometric giant is more likely than the existence of an arithmetic giant because the doubling defining a geometric giant begins fewer standard deviations from the mean (15.5 versus 22.5; neither is really likely).

scaled geometrically. The vertical steps increase and decrease not by equal amounts but by equal proportions. There are fewer standard-deviation steps to the upper limit of 325 or 326 cm, and hence fewer steps to the lower limit when both are scaled proportionally. Fewer standard-deviation steps to these limits, 15.5 in Figure 2.9b versus 22.5 in Figure 2.7a, means that the geometric limits are more likely. Importantly, in the geometric case a lower limit of one halving matches an upper limit of one doubling.

Galton (1879) and the Cambridge mathematician Donald McAlister (1879) recognized that distributions of error and variation in terms of proportion are just as plausible as distributions of error and variation in terms of amount. Both men recognized that the appropriate measure of central tendency in such a case is the geometric mean (exponentiated mean of the logarithms of measurements) rather than the arithmetic mean of the raw observations. McAlister (1879) then showed that the ordinary law of error applies to the logarithms of measurements in the geometric case just as it does to raw measurements in the arithmetic case.

Neither Galton nor McAlister used the term "lognormal," but their work implicitly introduced the concept. Both surely recognized that geometric normality leads to an expectation of asymmetry, not symmetry, on an arithmetic scale. However Galton, oddly, in his subsequent work followed a facile path and ignored lognormality, writing, for example, "it was found that the distribution of stature was sufficiently normal to justify our ignoring any shortcomings in that respect" (Galton, 1889, p. 117). And "had I possessed better data, I should have tried the geometric mean throughout" (Galton, 1889, p. 119).

Empirical distributions of biological variation like those in Figures 2.1 and 2.2 here, and similar distributions published by Weldon (1893; 1895) and others, motivated Karl Pearson to develop an application of moments, borrowed from physics and mechanics, to investigate normality and departures from normality. He hoped, optimistically, to factor complex curves into normal components. The first moment of a normal curve is the mean, and the second moment is the variance. Pearson focused on the standard deviation, the positive square root of the second moment, as the most appropriate measure of dispersion (Pearson, 1894), and then on a standardized third moment as a measure of asymmetry or skewness (Pearson, 1895). Right or positive skewness was common if not ubiquitous in the empirical distributions studied by Quetelet, Galton, Weldon, Pearson, and others.

It may seem surprising that little attempt was made to analyze the distributions as lognormal rather than normal, but this requires substantial samples structured appropriately; it requires goodness-of-fit tests that were not available at the end of the nineteenth century; and it also requires computational power that was not yet available. Pearson's final word on lognormality, published in his biography of Galton, reads as an epitaph:

I am unaware of any comprehensive investigation being ever undertaken to test the "goodness of fit" of [Galton and McAlister's] geometric mean curve to actual observations. McAlister gives no numerical illustration, and I do not think Galton ever returned to the topic. It would still form the subject of an interesting research, but *I fear the Galton-McAlister curve would be found wanting*.

Karl Pearson 1924, pp. 227–228; italics added

2.5.2 Empirical Support for Lognormality

One comparison of the normality and lognormality of biological variation started as an attempt to use a large set of measurements to reject one or the other hypothesis. Gingerich (2000) analyzed an extraordinary set of human measurements published in a series of monographs resulting from the mid-twentieth-century All India Anthropometric Survey. This professional government survey involved measurement of numerous homogeneous sets of 50 adult human males of the same caste and village, from villages broadly distributed across political states and geographic regions of India. These were compiled and finally published, state by state, in volumes issued by the Anthropometric Survey of India. Gingerich (2000) chose to analyze two of the larger samples from the states of Maharashtra (Basu et al., 1989) and Uttar Pradesh (Banerjee and Basu, 1991). The former provides 14 measurements and 14 indices for 6,869 individuals and the latter provides the same measurements and indices for 7,766 individuals.

The usual approach to the problem of normality is to consider arithmetic normality as a null hypothesis, H_0, test this against a set of measurements, fail to reject the null hypothesis, and then proceed as if variation is arithmetically normal because the hypothesis was not rejected. However, failure to reject normality as a null hypothesis rarely means anything because the samples employed are usually too small to provide the statistical power required for rejection (Gingerich, 1995).

Empirical distribution function (or EDF) tests are among the most powerful non-parametric goodness-of-fit tests for normality (D'Agostino, 1986). Each test is based on the fit of a stepped cumulative empirical curve to a model cumulative normal curve with parameters estimated from the empirical sample. Lilliefors' version of the Kolmogorov–Smirnov goodness-of-fit test involves a supremum statistic representing the maximum deviation from expectation for all steps of an EDF. Cramer–von Mises and Anderson–Darling tests employ quadratic statistics representing sums of differently weighted squared differences. A full set of original measurements or indices is required for computation of these statistics, and the goodness-of-fit is calculated first for the original measurements or indices, and then for the logarithmically transformed measurements or indices. Details are given in Gingerich (2000).

Goodness-of-fit tests of normality that treat arithmetic normality or geometric normality (lognormality) as a null hypothesis often fail because: (1) as mentioned above, small sample sizes mean neither hypothesis can be rejected or (2) when sample sizes are large, both hypotheses can be rejected. The appeal of a model is its generality, and empirical distributions rarely fit a general model exactly. Thus, large samples often provide the power to reject all models. In Gingerich (2000) measurements of low variability behaved differently from measurements of higher variability. Stature is a low-variability measure with relatively small coefficients of variation on an arithmetic scale and small standard deviations on a geometric scale. The shapes of the distributions on the two scales are similar (Figure 2.8), and the large Maharashtra and Uttar Pradesh samples taken as a whole generally failed to reject either of the alternatives as a null hypothesis. For measurements of higher variability such as body weight, involving larger coefficients of variation and larger standard deviations (Figure 2.9), the large Maharashtra and Uttar Pradesh samples generally forced rejection of both hypotheses. Hence the large-sample tests failed in different ways, depending at least in part on the variability of the measurement or index being examined. Whatever the reason, most large-sample tests failed to distinguish normality from lognormality.

An alternative approach to the normality versus lognormality problem is to compare the two as alternative hypotheses and ask which is better supported by the empirical information at hand. This is a classic likelihood solution (Edwards, 1972; 1992) to a problem where ordinary hypothesis testing fails. Alternative hypotheses (H_1, H_2, etc.) are tested not by comparison to some statistical-model critical value, but by comparison of the hypotheses to each other to see which has greater relative likelihood or support. Support is the difference in the natural logarithms of the probabilities associated with each hypothesis (arithmetic minus geometric) or, equivalently, the natural logarithm of the likelihood ratio of probabilities favoring one hypothesis over the other (arithmetic over geometric). Support scores are additive. A positive support score favors arithmetic normality, and a negative support score favors geometric normality.

Large-sample support scores for the Maharashtra measurements total -11.63 (for 12 of 14 measurement scores) and for the Uttar Pradesh measurements total -25.56 (12 of 14 scores). Large-sample support scores for the Maharashtra indices total -51.60 (6 of 14 index scores) and for the Uttar Pradesh indices total -55.24 (6 of 14 scores). Some measurement and index scores are missing because the probabilities required for their calculation are too small to be computed.

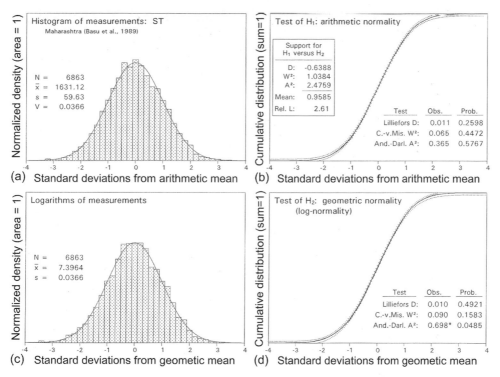

Figure 2.8 Likelihood comparison of arithmetic and geometric normality of human stature, based on a sample of 6,863 adult men from Maharashtra (Basu et al., 1989; Gingerich, 2000). (a and b) Normalized density histogram and cumulative empirical distribution function (EDF; stepped line) for raw measurements. (c and d) Normalized density histogram and cumulative EDF (stepped line) for ln-transformed measurements. Goodness-of-fit statistics are Lilliefors D, Cramer-von Mises W^2, and Anderson–Darling A^2 (D'Agostino, 1986). Support for hypothesis H_1 (arithmetic normality) relative to H_2 (geometric normality or lognormality) is given by the log–likelihood ratio, the natural logarithm of the ratio of probabilities for the corresponding test statistic (e.g., $-0.6388 = \ln[0.2598/0.4921]$). Mean support of 0.9585 for all three tests suggests that H_1 (arithmetic normality) is about 2.61 times more likely than H_2 (geometric normality) for these measurements. The only goodness-of-fit statistic with a probability less than the critical value for significance ($\alpha < 0.05$, asterisk) is Anderson–Darling A^2. Subsamples are more homogeneous and their relative likelihoods are more tractable computationally (see text). Figure is modified from Gingerich (2000), reproduced by permission of Elsevier Publishing

Fortunately, both the Maharashtra and the Uttar Pradesh surveys were carried out caste by caste and village by village, and then published as a collection of smaller and more homogeneous subsamples preserving this information. Most subsamples include measurements and indices for 50 individuals. The best way to take advantage of all of the information is to calculate support scores for each

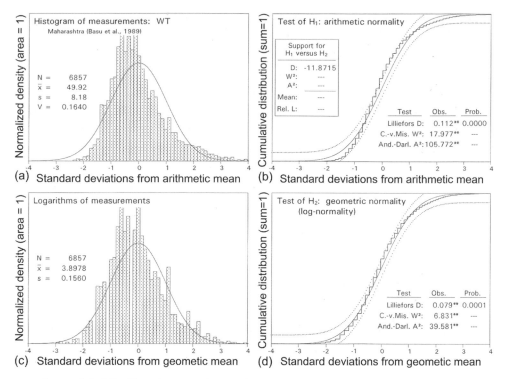

Figure 2.9 Likelihood comparison of arithmetic and geometric normality of human body weight, based on a large sample of 6,857 adult men from Maharashtra (Basu et al., 1989; Gingerich, 2000). (a and b) Normalized density histogram and cumulative empirical distribution function (EDF; stepped line) for raw measurements. (c and d) Normalized density histogram and cumulative EDF (stepped line) for ln-transformed measurements. Goodness-of-fit statistics are Lilliefors D, Cramer--von Mises W^2, and Anderson–Darling A^2 (D'Agostino 1986), as in Figure 2.8. All test statistics for H_1 and H_2 have probabilities much less than the critical values for significance ($\alpha < 0.05$ and $\alpha < 0.01$, double asterisk), indicating that the empirical distributions do not fit either model. Probabilities are too small to compute in four of the six tests, compromising calculation of mean support. Subsamples are more homogeneous and their relative likelihoods are more tractable computationally (see text). Figure is modified from Gingerich (2000), reproduced by permission of Elsevier Publishing

measurement or index for each subsample and then add these together. Pooled support scores for 143 Maharashtra samples of measurements total -288.90 (14 of 14 scores) and for 153 Uttar Pradesh samples of measurements total -136.40 (14 of 14 scores). Pooled support scores for 143 Maharashtra samples of indices total -539.08 (14 of 14 scores) and for 153 Uttar Pradesh samples of indices total -338.50 (14 of 14 scores).

For Maharashtra 5 of 14 subsample support sums for measurements are positive and 9 are negative. Two subsample support sums for indices are positive and

12 are negative. For Uttar Pradesh 4 of 14 subsample support sums for measurements are positive and 10 are negative. One subsample support sum for indices is positive and 13 are negative.

There is some variation in likelihood support scores, but for large samples studied to date, whether analyzed as a whole or divided into homogeneous subsamples, virtually all support scores are negative. Thus, empirically, geometric normality is favored over arithmetic normality. This supports the Elliott (1863), Galton (1879), and McAlister (1879) claims that distributions of variation should be studied in terms of proportion, and supports the Galton–McAlister application of the "law of error" to the logarithms of measurements. Deficiencies of the arithmetic methods of measurement we use to study geometric phenomena are easily compensated by transforming measurements to logarithms.

2.6 Applications of Normality and Lognormality

The normality of biological variation has a number of important consequences, following first from normality itself, and then from the geometric nature of normality.

2.6.1 Phenotypes and Genotypes

The classical Greek words *phaino* and *phaneros*, meaning manifest and evident, are the roots of common English words such as phenomenon and phenomenal. Phenomena are *observable*, perceived through the senses rather than inference or intuition. *Phaenotypus* was introduced in biology by the Danish botanist Wilhelm Johannsen (1909, p. 123), and "phenotype" is the name commonly given to the observable form and behavior of a single organism or a statistical population of organisms. The word was introduced to distinguish what is seen (the phenotype) from what is unseen (Johannsen's *Genotypus* or genotype) in an organism or population, reflecting how genetic inheritance and development were understood at the time.

Johannsen (1909, p. 130) was most impressed by the divergent phenotypes of sexually dimorphic organisms, and he reasoned that the way the phenotypes are manifest says nothing ("*absolut nichts*") about the underlying genotype. Johannsen argued that phenotypic differences can be seen where no genotypic differences exist, and genotypic differences exist where no phenotypic differences can be seen. More is known today, and one is reluctant to criticize someone writing a century ago, but Johannsen was wrong to think phenotypes tell us nothing about underlying genotypes. Normal distributions of variation like those in Figures 2.1–2.2 and 2.8–2.9 are constructed from the addition of many genetic differences of small effect. Additive genetic variance underlies virtually all of the polygenic, normally distributed traits of interest in quantitative evolutionary studies.

2.6.2 Quetelet's "Average Man"

Measurement error and biological variation are both normally distributed, and hence "error" and "variation" are commonly conflated. However, they are not the same. One reflects the error inherent in repeated observation of a system (possibly combined with variation in behavior of the system itself). The other reflects true variation in a system (possibly combined with error in observation of the system). This is important in considering the meaning of constructs like Quetelet's (1835) "average man." There is an average value for any characteristic we can measure, but what is its meaning?

The mean value for a normal distribution of error is taken to represent the true value being measured, and there is only one expected or "normal" value. On the other hand, the mean for a normal distribution of natural variation is just one of many expected values, and the whole of the observed distribution is what is "normal." If the chest circumference in Figure 2.1a were an error distribution, the expected value would be the mean value of 39.84 inches. However, it is not an error distribution but a distribution of natural variation, and the expected value is the whole bell-shaped exponential curve centered on the mean.

2.6.3 Species Comparisons

It is sometimes challenging to compare the variability of traits in one biological species or population with the variability of traits in another because the variability of the traits so often depends on the size of the organisms involved: standard deviations depend on their associated means. In the example of Figure 2.10a the white-tailed deer *Odocoileus virginianus* on the right side of the chart has a much broader range of variation in cranial length than does the deer mouse *Peromyscus maniculatus* on the left side of the chart. The two distributions of variation are very different in shape, with standard deviations of 1.53 and 20.45, respectively, and no consistent expectation. The problem is more difficult of course when attempting to understand how questionably identified organisms might group into species. One commonly accepted solution is to consider variability in relation to the mean by calculating a coefficient of variation: the standard deviation divided by the mean. When we do this for the deer mouse and the deer we see that the coefficients of variation for cranial length, 0.078 and 0.073, are closely comparable, and we can safely expect other species to have similar coefficients of variation.

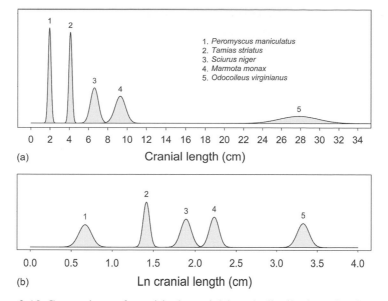

Figure 2.10 Comparison of empirical cranial length distributions for five mammalian species commonly found in Michigan. (a) Cranial length on the arithmetic scale of measurement, where small species have a relatively narrow range of dispersion and large species have a much greater range. (b) Cranial length on a geometric scale following transformation of measurements to base-*e* natural logarithms. Note that ln-transformed measurements have distributions that are similar across species, facilitating interpretation of species differences and species boundaries. The ranges of ln-transformed linear measurements like those shown here average about 0.3–0.4 units on an ln scale (± 3 = 6 standard deviation units). The same standardization can be achieved with base-2 and base-10 logarithms, but base-*e* is preferred because one standard deviation closely approximates the ordinary coefficient of variation. (Lewontin, 1966)

An equivalent and more powerful approach to standardization is to compare the species on a logarithmic geometric scale rather than the arithmetic scale of measurement. This comparison is shown in Figure 2.10b, where now the distributions for all five species are much more similar. Natural-log transformation of measurements is preferred because the resulting standard deviations approximate the coefficients of variation just calculated (0.078 and 0.073, respectively; see Lewontin, 1966, for an analytical explanation). Base-2 and base-10 logarithms are equally effective in standardizing variation and making species comparable, but they do not have the advantage of practical equivalence to the coefficient of variation.

Why is the variability of a trait proportional to the size of the trait? This is an open question that may be related to the generation of variation, or to functional limits governing interactions within a population, or both. It is important to acknowledge, first, that variability is proportional to size and, second, that logarithms provide a simple way to standardize this proportionality.

2.6.4 Allometry

The paleontologist Henry Fairfield Osborn (1925) introduced what he called the "principle of allometry" to emphasize the importance of "change of proportion" in vertebrate evolution. Julian Huxley (1924, 1932) first called this "heterogony," following Albert Pézard, but later adopted the word "allometry" (Huxley and Teissier, 1936). Allometry is a biological equivalent of geometry in mathematics — each is given a name to distinguish it from simple additive arithmetic.

In a study of fiddler crabs, Huxley (1924) found that the variable y, representing the weight of the large chela of males, was related to x, body weight less the weight of the chela, by the relation:

$$\log y = k \log x + \log b \tag{2.3}$$

Huxley then expressed this in what he called its "simplest" form as:

$$y = bx^k \tag{2.4}$$

It is debatable whether a power function is simpler than a linear equation, although it may have seemed so in Huxley's day when logarithmic transformation required book-length tables. Equations 2.3 and 2.4 are alternative representations of the same relation, one geometric and one arithmetic. One involves logarithms and the other does not.

Logarithmic transformation converts ranges of equivalent proportion to ranges of equivalent size. Exponentiation does the opposite, transforming ranges of equal size to ranges of equal proportion. The mouse-to-elephant simulations in Figure 2.11 show the effects of logarithmic and exponential transformation. The curved progression of small-to-large species in Figure 2.11a, each of equivalent coefficient of variation, becomes straight and uniform when measurements are transformed to proportions, logarithmically, in Figure 2.11b. The straight and uniform series of species in Figure 2.11b, each of equivalent range, becomes a curved progression of small-to-large species when proportions are transformed to measurements, exponentially, in Figure 2.11a.

The ellipses representing species in the simulation of Figure 2.11b have 95% confidence ranges of about 0.20 units on both the length (x) and width (y) axes. A 95% confidence range corresponds to ± 2 standard deviations (Figure 2.6), for a full range of four standard deviation units. The standard deviations of tooth length and width in the simulation are each 0.05, and the coefficients of variation V are 0.05 as expected for a linear measurement. We would expect a standard deviation

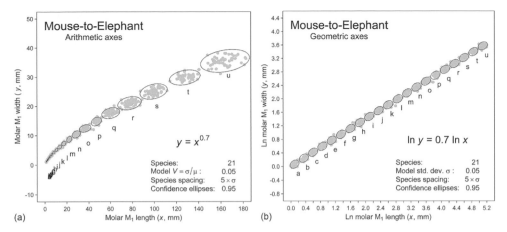

Figure 2.11 Relative sizes of 21 simulated "mouse-to-elephant" model species, here labeled $a - u$, represented by measurements of first lower molars. (a) Species on arithmetic axes representing the scales of measurement. (b) Species on geometric or allometric axes representing the scales of functional relationship. Ellipses enclose 95% confidence regions for 50 individuals drawn at random from bivariate normal populations. Means μ of successive species increase by the Hutchinsonian ratio 1.28, and standard deviations in A and B are $\sigma = 0.05 \cdot \mu$ and $\sigma = 0.05$, respectively. The within-species correlation between x and y is constant at $\rho = 0.3$. Note the consistent size ranges and uniform spacing of species when plotted on geometric axes. This consistency simplifies comparison and interpretation, and is itself an indication that the underlying functional relationships of molar length and width are geometric. Figure is modified from Gingerich (2014), reproduced by permission of John Wiley and Sons Publishing

of tooth crown area (length × width) to be about 0.10, and a standard deviation of tooth crown volume (length × width × height) to be about 0.15. It would be difficult to see such regularity and consistency in the very same tooth sizes when plotted on arithmetic axes, as they are in Figure 2.11a.

2.6.5 Limiting Similarity

In 1958, G. Evelyn Hutchinson delivered a classic presidential address to the American Society of Naturalists. His title, "Homage to Santa Rosalia," recalled a happy day spent collecting water bugs on Monte Pellegrino in Sicily (Hutchinson, 1959). Two species were present in a pond, one small and one a little larger. This started Hutchinson thinking about why there are so few, and so many, species in nature. He then asked himself about "limiting similarity." What morphological difference is required for closely related species to occupy adjacent niches at the same level in a food web? Hutchinson found, empirically, that closely related species living sympatrically differ in linear measurements by a factor averaging about 1.28, which he considered the minimum difference necessary to fill different niches. Hutchinson expressed this as 1.28/1.00, or 1:1.28, and 1.28 is now known as a "Hutchinsonian ratio."

But what if the measurement comparing species is two dimensional instead of linear? What if it is three dimensional? We can investigate this as Hutchinson (1959) investigated the limiting similarity for linear measurements. Hutchinson's first example, comparison of males of two sympatric mustelid species found in Great Britain, employed measurements published by Miller (1912). The smaller of the species, *Mustela nivalis*, is the British weasel, and the larger of the species, *M. erminea*, is the stoat or short-tailed weasel. Miller (1912) provided measurements for crania of 12 male *M. nivalis* and 12 male *M. erminea*, including condylobasal length, zygomatic breadth, and occipital depth (Miller's measurements are in tables starting on his pages 392 and 408). The male *M. nivalis* crania average 39.5 mm in length, and the male *M. erminea* crania average 50.5 mm in length, yielding the 1.28 ratio that Hutchinson reported.

If we multiply Miller's measurements of cranial length by zygomatic breadth, we have a measure of cranial area. This averages 859.1 mm^2 for crania of male *M. nivalis*, and 1474.1 mm^2 for crania of male *M. erminea*, yielding a Hutchinsonian ratio for area of 1.72. If we multiply Miller's measurements of cranial length by zygomatic breadth by occipital depth (height), we have a measure of cranial volume. This averages 9,046 mm^3 for crania of male *M. nivalis*, and 19,530 mm^3 for crania of male *M. erminea*, yielding a Hutchinsonian ratio for volume of 2.17. Results are shown graphically in Figure 2.12. Hutchinsonian ratios for lengths, areas, and volumes are clearly sensitive to the dimension of the form being represented.

The Hutchinsonian ratios of 1.28 for lengths, 1.72 for areas, and 2.17 for volumes are sensitive to dimension. If we take the natural logarithms of these, we find that a ratio of 1.28 for length measurements is equivalent to separation by 0.25 units on a natural log scale; a ratio of 1.72 for areas is equivalent to separation by 0.54 units on a natural log scale; and a ratio of 2.17 for volumes is equivalent to separation by 0.77 units on a natural log scale. These separations of 0.25:0.54:0.77 are almost exactly in the proportions 1:2:3, showing again their sensitivity to the dimensions being measured.

If a standard deviation is equivalent to 0.05 units on a natural log scale, as modeled above, then a separation of 0.25 for limiting similarity of cranial length would be equivalent to a separation of 5 standard deviations. If a standard deviation is equivalent to 0.10 units on a natural log scale, then a separation of 0.54 units for limiting similarity of cranial area would be equivalent to a separation of 5 standard deviations. And finally, if a standard deviation is equivalent to 0.15 units on a natural

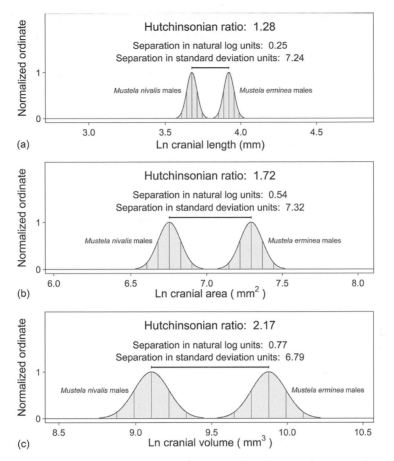

Figure 2.12 Quantification of limiting similarity for closely related species occupying adjacent niches at the same level in a food web. Example is from Hutchinson (1959), extended to include differences in (a) cranial length, (b) cranial area, and (c) cranial volume. Measurements from Miller (1912) compare male weasels, *Mustela nivalis*, and male stoats, *M. erminea*, sympatric in Great Britain. Differences between species quantified as Hutchinsonian ratios (1.28, 1.72, and 2.17) or in natural log units (0.25, 0.54, and 0.77) remain proportional to the dimensions (1, 2, and 3) of the lengths, widths, and volumes involved. Vertical lines within normal curves show standard deviations (see Figure 2.6; here averaged across the two species). Note that quantification in standard deviation units removes the effect of dimension and yields a consistent measure for limiting similarity of about seven standard deviation units in all three cases.

log scale, then a separation of 0.77 units for limiting similarity of cranial area would again be equivalent to a separation of 5 standard deviations. Empirically, the weasels and stoats studied by Miller and Hutchinson are a little less variable than modeled above, the standard deviations average 0.03, 0.07, and 0.11 rather than 0.05, 0.10, and 0.15, and the limiting similarity for weasels and stoats is close to seven standard deviations for lengths, areas, and volumes. This example is important because it illustrates how standard deviation units incorporate dimensionality and eliminate its effect, making standard deviations the preferred units for expressing limiting similarity. For this reason standard deviation units represent differences between variable populations more effectively than any ratios or differences on a proportional scale. This comes up again later when we consider how to quantify evolutionary rates.

2.7 Summary

1. Darwin emphasized variation, individual differences within species, as "highly important" for natural selection. Variation is indeed essential for the process to work.
2. Attempts to quantify biological variation in the nineteenth century focused on resemblance of this variation to distributions of measurement error in astronomy and physics.
3. Measurement error is generated independently in astronomical and other physical observations, but in biology variable populations evolve from variable populations. Quetelet's "average man" was misguided: The expectation in biology is not an average over and over, but a full distribution of variation in each successive generation.
4. Normal distributions can be generated by summing permutations across combinations of flipped coins or rolled dice. By analogy, normal distributions of variation in biological populations reflect "additive genetic variance" and the sum of many small differences in constituent individuals.
5. Arithmetic is mathematics based on counting, and geometry is mathematics based on proportion. Theoretically and empirically the normality of biological variation is geometric rather than arithmetic: biological variation is lognormal rather than normal, and individual differences are differences of proportion. Logarithms employed to transform counts to proportions can be chosen to reflect halving and doubling ($\log_2$), standardized deviations (ln or $\log_e$), or orders of magnitude ($\log_{10}$).
6. Allometry is the biological equivalent of geometry, and the allometric equation in simplest form is a linear equation relating dependent log Y to independent log X, or dependent ln Y to independent ln X.
7. Finally, comparisons of populations in standard deviation units incorporate dimension and remove its effect, making standard deviations the preferred units for expressing the similarities and differences of variable populations.

3

Evolutionary Time

The time we experience in our everyday lives is measured in minutes, hours, days, and years. All are astronomical. Units related to days are based on the Earth's rotation about its axis, and units related to years are based on the Earth's orbital cycling of the sun. Days are divided into hours, minutes, and seconds, while years are aggregated as decades, centuries, and millennia. The moon's orbit around the Earth defines months and years on some calendars. Earth's rotation does not divide the moon's orbit evenly, nor does it divide the sun's orbit evenly. Consequently, we have leap years and other corrections, and we tolerate the imprecision.

Days, months, and years are calibrations of an underlying theory of time that is metronomic in the sense that the beat is steady and the rhythm is precise. Units are added in equal measure repeatedly, over and over again. Time is directional: units are always added and never subtracted. Time is quintessentially arithmetic, based on counts, and it parallels if not defines a counting number line. The units can be any size but once specified they are always the same.

Time in biology is similarly arithmetic. Units of interest may be stages in the life history of individual organisms. The stages include gestation, birth, growth to maturity, times of reproduction, average age of reproduction or generation time, senescence, and age at death. Stages of life history are typically measured, depending on the organism, in minutes, hours, days, months, or years. If we take generation time to represent the functional length of an individual life, then longer lineages can include tens, hundreds, thousands, and even millions of generations.

3.1 Time and Scale in Biology

Theoretical ecology, and theoretical science more generally, relates [patterns] that occur on different scales of space, time, and organizational

53

complexity. Understanding patterns in terms of the processes that produce
them is the essence of science.

Simon A. Levin, The problem of pattern and scale
in ecology, *1992, p. 1944*

Sometimes we are able and choose to observe evolutionary time series generation
by generation, and sometimes we are limited by the record at hand to observe these
on longer scales of time. Scale is important. The importance is widely recognized
in comparative biology (Peters, 1983; Schmidt-Nielsen, 1984; Calder, 1984; Reiss,
1989), and it is also recognized in ecology (Wiens, 1989; Levin, 1992).

Scale has received less attention in evolutionary studies, but here too patterns
are observed on different scales of time. A challenge of evolutionary biology is to
relate the patterns and explain them in terms of process. To be explicit, the
timescale of an *observation* is the interval of time over which the observation is
made. This is not necessarily the same as the timescale of the *process* underlying
the observation.

We can use an example from Schneider (1994) to illustrate the difference
between timescales of observation and the timescale of a process. Guano is the
name given to nitrogen-rich bird droppings that accumulate on nesting sites,
especially those of fish-eating sea birds. Measurement of an 80-year deposit year
by year as it accumulates will give us a record on an annual timescale. A single
measurement of the whole deposit after 80 years will give us a record on an 80-
year timescale. Nature being what it is, the annual measurements are not likely to
be constant and the 80-year measurement may be less than the sum of the annual
measurements. An annual measurement and an 80-year measurement are numbers
on different measurement scales.

The temporal scale of a process is the time step or interval over which the
process is repeated. If guano accumulation depends on an annual cycle of nesting
in birds, then the process is annual whether we observe it annually or not. If, on the
other hand, the birds nest twice a year, then the process of guano accumulation is
semiannual, half-yearly, no matter the scale of our observation. The timescale of an
observation and the timescale of a process are not necessarily the same.

3.2 Timescale of Natural Selection

As biologists we should like to be able to measure time in generations.

J. B. S. Haldane, Suggestions as to quantitative measurement
of rates of evolution, *1949, p. 51*

A generation is called a *generation* because it is the timescale of generation of a
new set of descendent organisms and the timescale of generation of new variation.

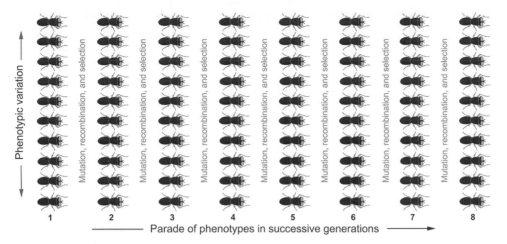

Figure 3.1 Parade of hypothetical phenotypes in successive generations. Phenotypes in each generation (1, 2, etc.) must be measured to quantify their variation, and phenotypes must be measured to detect and document any change from generation to generation.

It is the average time between production of parental and parentally derived phenotypes, or in other words, the timescale of expression of new genotypes. In terms of cell division, this is the time between the binary fission or division-and-recombination that gives rise to a parental generation and the binary fission or division-and-recombination that gives rise to the succeeding generation. Each new generation represents a new cycle of exposure to natural selection through its functional phenotype and, taken in sequence, successive generations represent a parade of new phenotypes (Figure 3.1).

Generations need not be discrete and may be overlapping, but the generation nevertheless represents the fundamental scale of time in the process of evolution. New variation in a new generation often involves genetic mutation or recombination, or both, making the variation heritable, but some variation is also induced environmentally in which case it is not inherited. The parade of phenotypes exposes each generation to selection, and generation times are consequently basic units for quantifying and modeling evolution by natural selection. Alternatively, in the absence of selection, each generation is a basic unit for quantifying and modeling random change.

3.3 Generation Times

Generation times are recognized in different ways in different organisms. Bacterial generations involve simple cell division, and hence are equated with the doubling

Table 3.1 *Definitions of generation time, with the gray squirrel as an example*

Definitions of generation time	Symbol	*Sciureus carolinensis* (years)
1. Mean age at reproduction for a female in a stationary population, where x is the age class and $k(x)$ is the expectation of female offspring at age x	$T_m = \Sigma x\, k(x)$	3.332
2. Mean age at reproduction for a cohort of females independent of population growth	$T_0 = \Sigma x\, k(x) / \Sigma k(x)$	2.794
3. Natural logarithm of net reproductive rate (ln R) divided by the intrinsic rate of increase (r) — definition preferred by Cavalli-Sforza and Bodmer	$T_1 = (\ln R)/r$	2.273
4. Mean age of mothers of new-born individuals in a population with a stable age distribution — definition preferred by Charlesworth	$T = x\, e^{-rx}\, k(x)$	2.676

Source: Cavalli-Sforza and Bodmer, 1971; Charlesworth, 1980

times of bacterial populations during phases of exponential growth. Bacterial generations are often measured in minutes.

Generation times in metazoans are more complicated. Brian Charlesworth (1980) compared four definitions (Table 3.1). The alternatives yield generation times for the gray squirrel *Sciurus carolinensis* ranging from 2.273 to 3.332 years, with the longer exceeding the shorter by nearly 50%. Charlesworth's preferred measure of generation time, the mean age of mothers of new-born individuals in a population with a stable age distribution, is intermediate in yielding a generation time of 2.676 years. Given the range of values based on alternative definitions, rounding to the nearest tenth of a year is often justified.

3.3.1 Generation Times across Taxa

Cecil Yarwood (1956) compiled generation times for a broad range of organisms of different sizes and showed that generation times increase progressively with increasing body weight across the 36 species he studied (including seven viruses). The nonviruses ranged in size from the bacterium *Escherichia coli* with a generation time of 20 minutes to the giant redwood *Sequoia gigantea* with a generation time of 9.5 years (many of Yarwood's generation times are questionable). Regression slopes and intercepts are included in Table 3.2. The coefficients of determination, r^2, are high, 0.898 and 0.930, with and without viruses. Slopes differ with and without viruses, but the intercepts do not differ significantly.

John Bonner (1965) compiled information for 46 species, ranging from *Escherichia coli* with a generation time of 12.5 minutes to *Sequoia gigantea* with a generation time of 60 years. He used this to show that body length increases with increasing generation time. If we turn the relationship around and regress generation time on body length, we again find a high r^2 of 0.907. The slope and intercept are quite different from those of Yarwood because the units of body size measurement are different (millimeters instead of grams). Bonner was careful to focus on the generality of the relationship while acknowledging exceptional cases and substantial scatter (Bonner's intercept standard error is 0.103, for an expected range of $\pm 1.96 \cdot 0.103 = \pm 0.202$ log units).

Table A.1 in the Appendix builds on the compilations of Yarwood (1956) and Bonner (1965), listing body lengths, body weights, and generation times for 186 species representing a broad range of organisms. Addition of new information identified outliers, especially in Yarwood's data, where it appears that age at first reproduction in domestic species was used as an estimate of generation time. Age at first reproduction is always a minimum estimate for organisms that reproduce more than once. Regressions based on the compilation here are similar to those of Yarwood and Bonner in the sense that generation times increase with body size (Figure 3.2), and r^2 remains high (Table 3.2). Slopes for regression of generation time on body weight and body length in the larger data set are both higher, and intercepts are both lower than those for Yarwood's and Bonner's compilations. Standard errors are smaller for both slopes and intercepts, but scatter remains high. The general equation is:

$$\log_{10} generation\ time\ (yr) = 0.314 \cdot \log_{10} weight\ (g) - 0.519 \qquad (3.1)$$

It is relatively easy to quantify the generation times for species of bacteria because organisms in each generation reproduce just once, and reproduction itself involves simple cell division. It is more difficult to quantify the generation times for species of organisms that reproduce multiple times and have complex life histories. Then, life history characteristics of the species become important. Major studies of generation time taking life history into consideration include Millar and Zammuto (1983) for land mammals; Heppell et al. (2000) for land, volant, and marine mammals; Niel and Lebreton (2005) for birds; Goodwin et al. (2006) for fishes; and Taylor et al. (2007) for whales. Regression statistics for these studies are included in Table 3.2.

Table 3.2 *Regression of generation time on body size*

Study	Regression (log$_{10}$ axes)	Included taxa	Species (N)	Slope	Intercept	r^2	Slope SE	Intercept SE
Yarwood 1956; with viruses	Gen. time (yr) on body weight (g)	Diverse	36	0.194	−1.317	0.898	0.0112	0.1040
Yarwood 1956; without viruses	Gen. time (yr) on body weight (g)	Diverse	29	0.230	−1.360	0.930	0.0121	0.0792
This study (Table A.1; Fig. 3.2)	Gen. time (yr) on body weight (g)	Diverse	142	0.314	−0.519	0.907	0.0085	0.0414
Bonner 1965	Gen. time (yr) on body length (mm)	Diverse	46	0.804	−1.924	0.907	0.0388	0.1030
This study (Table A.1)	Gen. time (yr) on body length (mm)	Diverse	100	0.832	−1.769	0.912	0.0260	0.0791
Millar and Zammuto 1983 (Fig. 3.3)	Gen. time (yr) on body weight (g)	Land mammals	29	0.268	−0.564	0.815	0.0246	0.1000
Heppell et al. 2000	Gen. time (yr) on body weight (g)	Mammals	50	0.188	−0.049	0.648	0.0200	0.0877
Ernest 2003 (Fig. 3.3)	AFR + gest. (yr) on body weight (g)	Land mammals	752	0.266	−0.773	0.661	0.0069	0.0253
Niel and Lebreton 2005	Gen. time (yr) on body weight (g)	Birds	13	0.258	0.080	0.393	0.0967	0.2790
Goodwin et al. 2006	Gen. time (yr) on body weight (g)	Fishes	14	0.111	0.421	0.511	0.0314	0.0974
Taylor et al. 2007	Gen. time (yr) on body weight (g)	Whales	39	0.046	2.648	0.566	0.0066	0.1330

Note: AFR is age at first reproduction and *gest.* is gestation time. N sample size, r^2 the coefficient of determination, and SE the standard error of the estimate. Table A.1 is in the Appendix.

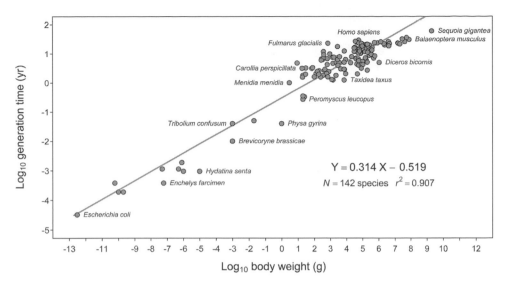

Figure 3.2 Yarwood-type plot showing the relationship of generation time to body size for a set of 142 species ranging in size from the bacterium *Escherichia coli* to the great blue whale *Balaenoptera musculus* and giant redwood *Sequoia gigantea*. Scaling coefficients are compared in Table 3.2. Measurements plotted here are listed in Table A.1.

3.3.2 Generation Times in Mammals

The Millar and Zammuto (1983) generation times for land mammals are especially interesting (Figure 3.3). These were estimated as the average age of females giving birth to all offspring, approximated from a stable life-table age distribution for the species. The Millar and Zammuto generation times can be compared to generation times calculated as the sum of age at first reproduction (AFR) plus gestation length, two quantities known for many more species (Ernest, 2003). Log AFR scales with a slope of 0.276 relative to log body weight, whereas log gestation length scales with a slope of 0.250 — the individual slopes are different. However, log AFR-plus-gestation-length as a sum plotted against log body weight scales with a slope of 0.266, which is almost exactly the scaling of Millar and Zammuto generation times. Such close agreement corroborates the use of body weight to predict generation time in land mammals based on Millar and Zammuto scaling. It also raises the possibility of estimating generation times by adding 0.220 to the intercept for AFR-plus-gestation-length scaling (dashed line in Figure 3.3). Generation time for land mammals can be predicted from body weight using the following equation:

$$\log_{10} generation\ time\ (yr) = 0.266 \cdot \log_{10} weight\ (g) - 0.773 + 0.220 \quad (3.2)$$

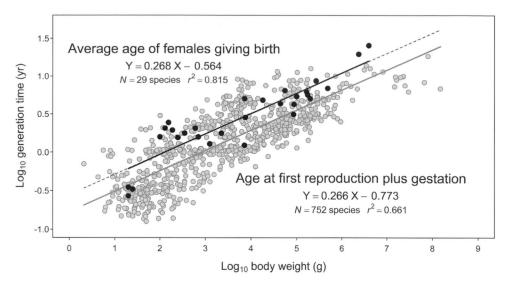

Figure 3.3 Relationship of generation time to body weight in nonvolant mammals. Generation time in a stable population is generally regarded as the average age of females giving birth, estimated from life tables (black circles and black regression line, based on information for 29 species in Millar and Zammuto, 1983). Age at first reproduction plus gestation time provides a minimal estimate of generation time (gray circles and gray regression line, based on information for 752 species in Ernest, 2003). The minimal estimates augmented by 0.220 log units (dashed line) match the life table estimates.

The intercept here is nearly the same as that in Equation 3.1, but the slope is less, indicating that land mammals have shorter generation times for their body weight than organisms in general.

3.3.3 *Generation Times in Other Vertebrates*

Regression of generation time on body weight for the Niel and Labreton (2005) data for birds yields a slope of 0.258, close to the slope for land mammals but the intercept of 0.080 is substantially higher, indicating that birds have longer generation times compared to body size. Body weight scaling slopes for the Heppell et al. (2000) generation times in mammals (weights added from Ernest, 2003 and Eisenberg, 1981), for the Goodwin et al. (2006) generation times in fishes, and for the Taylor et al. (2007) generation times in whales are all unusually low, suggesting that there is variation in the relationship of generation time to body weight in different groups of animals. In the present state it is important to base prediction of unknown generation times on closely related plants or animals.

3.4 Ecological and Evolutionary Time

Ecological steady states will be disrupted by evolutionary changes
during time intervals of the order of one-half million years. Intervals
of this order may be termed 'evolutionary time.' ... On the scale of
ecological time—that is, over intervals of the order of ten times the
length of one generation—populations may be expected to maintain an
approximately steady state.

Lawrence Slobodkin, Growth and Regulation of
Animal Populations, *1961, p. 30*

Generations are fundamental units of time for studying change by natural selection
and for studying random change as an alternative to selection, but these are not the
only scales of time involved in evolution. Just as years accumulate into centuries
and millennia, so generations accumulate into larger units of what Slobodkin
(1961) called "ecological" and "evolutionary" time. Ecological time and evolutionary time are widely used to characterize different scales of time in biology and
paleontology, and each is useful for this purpose, but there is increasing understanding that these are overlapping and nonexclusive. Evolution includes what
happens on ecological scales of time (Carroll et al., 2007), and a central argument
in this book is that differences we see in organisms over spans of ecological and
evolutionary time represent an accumulation of changes happening on a timescale
of generations.

Ecological time may be tuned to environmental change on scales of time
different from generations (El Niño cycles, for example), but evolutionary time
is time in generations. Evolution happens from one generation to the next, and
evolution happens as generations accumulate into longer intervals of time. However, as indicated above, the generation is the fundamental unit and scale of time in
the process of evolution by natural selection.

3.5 Microevolution and Macroevolution

Experience seems to show ... that there is no way toward an understanding
of the mechanisms of macro-evolutionary changes, which require time on a
geological scale, other than through a full comprehension of the microevolutionary processes observable within the span of a human lifetime.

Theodosius Dobzhansky, Genetics and the
Origin of Species, *1937, p. 12*

Microevolution is a word introduced to the lexicon of evolutionary biology by
the botanist and geneticist Reginald Gates – in a review of the English translation
of Hugo De Vries' (1901) *Die Mutationstheorie* (Gates, 1911). Gates granted
De Vries the facts of mutation, but noted that mutations only account for

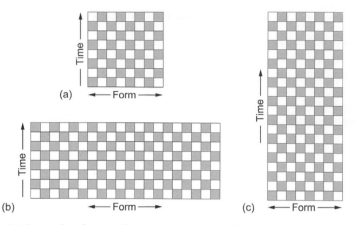

Figure 3.4 Alternative forms of an imagined evolutionary game board. (a) Time and form more or less equivalent. (b) Time more constrained than form. (c) Form more constrained than time.

"micro-evolution," while "larger tendencies" might require the intervention of other factors. Theodosius Dobzhansky (1937) regarded microevolutionary processes as those observable within the span of a human lifetime, and coined the term "macro-evolution" for the "larger tendencies" that require a geological scale of time.

Richard Goldschmidt (1940) adopted Dobzhansky's dichotomy of micro- and macroevolution, dropped the hyphens, and changed the meaning. Goldschmidt regarded microevolution as the evolution that occurred within existing species, while macroevolution included the origin of new species and the evolution of higher taxonomic categories. The distinction was not one of time but of taxonomic scale. For Goldschmidt, species and higher taxonomic categories originated in single macroevolutionary steps that involved systemic mutations leading to completely new genetic systems (Goldschmidt, 1940, p. 396).

The microevolution of Dobzhansky, occurring within the span of a human lifetime, parallels what later came to be known as "ecological time." Similarly, the macroevolution of Dobzhansky, involving time on a geological scale, parallels what later came to be known as "evolutionary time." The micro–macro dichotomy of Goldschmidt is very different. Goldschmidt's proposed micro- and macroevolutionary processes operate on essentially the same scale of time, but what is inferred regarding mutational causes and taxonomic effects is very different. Goldschmidt's idea of causal macromutations is no longer tenable, but his use of micro- and macroevolution to distinguish evolution within species from evolution between species and higher taxa persists.

3.6 Models of Evolutionary Time and Form

We generally imagine that evolution takes place as a "game of life" and investigate this on a game board or simulation space of more or less equivalent length and width (Figure 3.4a). However, imagining a board of equivalent length and width might be misleading. There could be constraints on the form that organisms can assume, or constraints on the time available for their evolution. Evolutionary game boards could be very different, and alternatives are shown, symbolically, in Figure 3.4. The idea of a game board relating form to time is introduced at this point because one axis of interest is time, but questions of constraint on form, constraint on time, or constraint on their interrelationship will be considered in detail in a later chapter.

3.7 Summary

1. Time in evolution is arithmetic, with units of equal measure added repeatedly. Time is directional in the sense that units are added but never subtracted.
2. The timescale of an observation is the interval of time over which an observation is made, but the timescale of the process generating what we observe may be different.
3. Natural selection and random change are generation-scale processes acting on the parade of phenotypes produced in each generation.
4. Generation times of organisms vary in length from minutes to decades and are closely correlated with the size of the species represented. This enables unknown generation times to be estimated from size.
5. Differences we see in organisms over spans of ecological and evolutionary time represent an accumulation of changes happening on a timescale of generations.
6. Microevolution and macroevolution parallel ecological and evolutionary time, but these terms are also used to distinguish evolution within species from evolution between species and higher taxa.
7. Evolutionary change is often imagined and modeled on a "game board" or simulation space of form and time that are equivalent in length and width. The geometry of such a simulation space is always worthy of scrutiny.

4

Random Walks and Brownian Diffusion

A random walk is a pattern of change or movement in successive steps, with the change at each step governed by chance, independent of what came before. The idea is as old as any theory of games. Random walks are a special case of repeated independent "Bernoulli trials," going back to Jacob Bernoulli's 1713 *Ars Conjectandi*. Flipping a coin yields a succession of heads and tails, which, in serial accumulation, define a random walk. Similarly, the trace of a ball that "scampers deviously" downward through the pins of Francis Galton's quincunx traces a common form of a random walk (Galton, 1889; the name *quincunx*, from Latin, refers to a pattern of four points or pins at the corners of a square, with a fifth in the center.) The serial accumulation of coin flips and the trace of a scampering ball both take place on a plane defined by the intersection of an axis of time, t, representing coin flips or the falling ball hitting pins, and an axis of displacement, d, representing the accumulation of heads versus tails or the horizontal position of the deflected ball.

Random walks are Markov chains with the Markov property that steps of the chain lack memory. Both are named for the nineteenth-century Russian mathematician Andrey Markov. A Markov process is one in which the next state of a process depends on the present state; and on the process itself – but not on any past or expected future state. Spitzer (1964), Feller (1968), Mandelbrot (1983), and Berg (1983) provide deeper and more formal treatments of random walks and their applications than can be presented here.

Random walks are widely employed as null models in most sciences. This is not because we believe that all events in science are random. Rather, we study random walks to be able to determine whether and when observed patterns differ from random. Random walks are employed as a benchmark to which observations are compared to see whether there are departures from random that require explanation.

We live at a time when random genetic drift is commonly considered a dominant force in evolution. It may be that evolution and evolutionary history are so complicated that, however deterministic the processes involved, the only patterns we see are random. This leads to two central questions: Are there patterns in evolutionary time series that require explanation when compared to a random benchmark, and how do we recognize these? Here and in the following chapter we use random-walk time series to build analytical tools for analysis of evolutionary patterns.

4.1 Rates and Axis-Labeling Conventions

Random walks are models for evolutionary time series, and this is an appropriate place to comment on the variables employed in rate calculations and how the variables are plotted graphically. Time t is the *independent* variable in random walks and in evolution, and measures such as displacement or cumulative displacement s are *dependent* variables because they increase or decrease as a function of time. This difference is straightforward, but there is potential for misunderstanding because there are two competing conventions for arranging independent and dependent variables in bivariate and multivariate plots. The competing conventions are illustrated in Figure 4.1.

One convention involves *horizontal time* and vertical dependence. Experimental scientists studying time series generally plot time on the horizontal axis and call this x, and then plot any dependent variable or variables on the vertical or y-axis (or additional axes perpendicular to the horizontal axis). We will see this later when we analyze laboratory selection experiments. Horizontal time is the arrangement

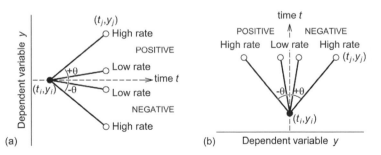

Figure 4.1 Alternative representations of independent and dependent variables. (a) Time t, the independent variable, is plotted on the horizontal axis. Dependent variable y is plotted on the vertical axis. This "horizontal time" representation is common in calculus and in experimental sciences. The tangent of angle θ measures the rate of change in y with respect to t as a slope: $(y_j - y_i) / (t_j - t_i)$ for $t_j > t_i$. (b) Time t is plotted on the vertical axis and dependent variable y is plotted on the horizontal axis. This "vertical time" representation is common in paleontology and historical sciences. Here the tangent of angle θ measures the rate of change in y with respect to t as a deviation from vertical.

that predominates in courses and textbooks on calculus. The principal advantage of the "horizontal time" arrangement of independent and dependent variables is the straightforward representation of rates of change as slopes. Positive change in the dependent variable with the passage of time yields a positive slope, and negative change with time yields a negative slope. Higher rates have steeper slopes and lower rates have shallower slopes.

In contrast, a second convention involves *vertical time* and horizontal dependence. Historical scientists often plot time on the vertical axis and call this *t*, and then plot any dependent variable or variables on the horizontal axis (or additional axes perpendicular to the vertical axis). This arrangement is conventional in paleontology, where time on the vertical axis mimics the vertical superposition of strata and hence the temporal succession of any contained fossils. Geological and paleontological experience probably explains why Charles Darwin arranged the independent and dependent axes this way in the figure he published in the *Origin of Species*. The "vertical time" convention yields rates that are deviations from vertical. Positive change in the dependent variable yields a positive deviation, and negative change yields a negative deviation. Higher rates have greater deviations and lower rates have smaller deviations.

The commonly used terms "abscissa" and "ordinate" are often interpreted in the context of horizontal time as a coordinate pair (x, y) or (abscissa, ordinate). In this case the abscissa is defined relative to the horizontal or "x" axis, and the ordinate is defined relative to the vertical or "y" axis, but the relation is really one of independence and dependence. It would make sense to use x and y, and abscissa and ordinate, when time is plotted horizontally if this were the only axis used for time. However, the use of x and y, and abscissa and ordinate, invites confusion when time is plotted vertically. Here for analysis of time series we will use t for the independent variable and s for the dependent variable, forming the coordinate pair (t, s), where t is plotted on either the horizontal or vertical axis, and s is on the remaining axis of a two-dimensional graph. Dependent variable s is a summation of cumulative displacement or cumulative difference at time t. It will be clear from context when s is used this way as a dependent summation and when s is used as a sample standard deviation.

The competing "horizontal time" and "vertical time" conventions should not cause confusion if a reader understands that there are two ways of arranging axes, and if axes are labeled properly.

4.2 Random Walks

In open country the most probable place to find a drunken man who is at all capable of keeping on his feet is somewhere near his starting point!
Karl Pearson, The Problem of the Random Walk, *1905, p. 342*

Karl Pearson (1905) introduced the random walk as a metaphor for a problem in statistics. He conceived the problem in a radial sense that is slightly different from most subsequent modeling. Pearson envisioned a random walk as involving a path of successive steps or stages from a point of origin, where the distance moved at each step is held constant and the direction of movement in each step changes randomly through a full circle of 360°. Pearson's radial and two-dimensional version of a random walk has applications beyond monitoring the movements of drunken men.

It is also common to consider a one-dimensional random walk as the trace of a body moving forward through time and varying in a single dependent dimension. Such a body moves according to the following rules (Berg, 1983):

- The body moves forward in time t by a constant time step equal to the step interval i_0. Successive steps can be named and numbered from 0 to n as $i_{0:0}$, $i_{0:1}$, ... $i_{0:n}$. The initial subscript of 0 marks the interval as a step interval. $\log_{10}(1) = 0$ for intervals involving a single step. The second subscript, optional, is the step number n.
- The body moves to the right or to the left in each time step by a step distance or displacement $\pm d_0$. In a simple random walk d_0 is constant, but this may also be a random variable. The direction of the displacement is denoted by the sign, with "+" representing a step to the right and "−" representing a step to the left. When appropriate, successive individual step displacements can be numbered from 0 to n as $d_{0:0}$, $d_{0:1}$, ... $d_{0:n}$.
- Rates are ratios. The step rate r_0, positive or negative, is $r_0 = d_0 / i_0$, which defines the relationship of d_0 and i_0. The step rate r_0 is constant when d_0 is constant. Rearranging, the rate relationship can also be expressed as $d_0 = r_0 \cdot i_0$. When appropriate, successive individual step rates can be numbered from 0 to n as $r_{0:0}$, $r_{0:1}$, ... $r_{0:n}$.
- The position of the body at any time step n is given by the successive sums s_1, ... s_n for each time step, where $t = n \cdot i_0$ (see below). The subscript is the step number n.

Initial subscripts of i, d, and r set to zero specify the timescale and interval length in $\log_{10}$ step units. For a unit time interval iterated from, say, step $i_{0:i} = 1$ to step $i_{0:i+1} = 2$, or from step $i_{0:i} = 5$ to step $i_{0:i+1} = 6$, or for any one-step interval, then $i_{0:i+1} - i_{0:i} = 1$. The initial subscript of zero reminds us that any interval, displacement, and rate labeled this way represents change on the timescale of the random walk or other generating process.

Modelers often use random walks casually without matching steps of the random walk to the timescale of repeated events in the process being modeled. Binning data into time intervals of equal duration is not enough; the intervals must reflect the timescale of the process being modeled. In evolution the timescale of the process being modeled is a timescale of reproductive generations. There may be repeated steps that influence evolution on longer timescales, but this has not been demonstrated.

Here we begin with simple random walks and then consider more complicated and realistic evolutionary random walks.

4.2.1 Simple Random Walk

A simple one-dimensional random walk is the time series that results from flipping a coin. It is a sequence of successive sums s_i drawn from repeated iteration with d_0 = 1 (or some other constant), where:

$$s_{i+1} = s_i \pm d_0 \qquad (4.1)$$

Probabilities associated with the plus and minus signs are equal at $p = q = 0.5$, making the random walk symmetric.

Note that for a given time-step interval i_0 between successive s in Equation 4.1, the corresponding step displacement d_0 determines a step rate $r_0 = d_0 / i_0$. Rate r_0 is a step rate that describes how d_0 and i_0 are related on the scale of time i_0. There is no simple random walk without a step interval i_0 and step displacement d_0, and hence there is no simple random walk without a step rate r_0. In addition, as we shall see, a random walk can have many net rates on longer scales of time. Subscripts of i, d, and r set to zero specify the timescale and interval length in $\log_{10}$ units. For a unit time step from, say, $t_i = 1$ to $t_{i+1} = 2$, or from $t_i = 5$ to $t_{i+1} = 6$, or for any one-step sequence, $t_{i+1} - t_i = 1$. Again, $\log_{10}(1) = 0$, so the subscript for a unit interval (or unit displacement, or unit rate) is 0.

The importance of step rates in random walks, and net rates on longer time-scales, is emphasized because Fred Bookstein (1987) repeatedly characterized random walks by their step variance, and then provocatively and confusingly asserted that "random walks have no rates" (Bookstein, 1987, p. 458). The step displacement d_0 is also the step standard deviation, and the square of this, d_0^2, is the step variance. Turning this around, if a random walk has a step variance, then it has a step standard deviation, a step displacement, and a step rate:

$$r_0 = d_0 / i_0 = \frac{\sqrt{d_0^2}}{i_0} \qquad (4.2)$$

Initial steps of a simple random walk are illustrated in Figure 4.2a. Imagine that the random walker is a bishop on a chess board, constrained to move diagonally as a bishop does but also constrained to move forward by a single step in time, and

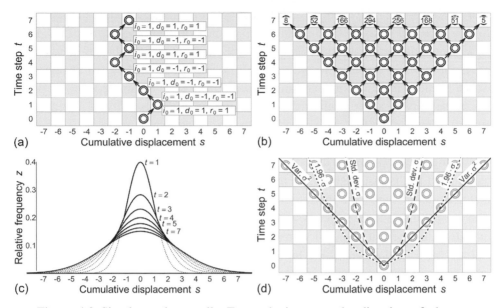

Figure 4.2 Simple random walk. For each time step the direction of change (+ or −) is random and the displacement $d_0 = 1$. (a) Path of one seven-step walk, starting at the origin ($t = 0$, $s = 0$), and ending at ($t = 7$, $s = -1$). The step interval i_0, step displacement d_0, and step rate r_0 are shown for each step. (b) Paths of 1,000 seven-step random walks, each starting at the origin. Inset numbers in the top row show the number of walks reaching each of the eight possible terminal positions. (c) Increase in dispersion of relative frequency and decrease in peak height as random walks increase in length from one to seven steps (from $t = 0$ to $t = 1$, $t = 2$, etc.). (d) Variance σ^2 of cumulative displacements s increases in proportion to t (solid lines). The standard deviation σ increases as $\sqrt{t}$ (dashed lines), and a 95% confidence interval for s increases as $1.96 \cdot \sigma$ (dotted lines). For simple random walks with a step rate $r_0 = 1$, the variance defines the positive and negative limits of dispersion at each step.

laterally to the left or right in each step. A bishop necessarily remains on squares of the same color. The random walk in Figure 4.2a begins with a step forward and to the right from time $t = 0$ to $t = 1$, with a step interval $i_0 = 1$, a step displacement d_0 of 1, and consequently a step rate $r_0 = d_0 / i_0 = 1$. The second step from $t = 1$ to $t = 2$ has the same step interval, but now the displacement d_0 and the rate r_0 are negative. The second step is part of a run of three steps forward and to the left. This is followed by a step forward and to the right, one forward and to the left, and another forward and to the right. At any step n in a simple random walk, the cumulative displacement s is given by:

$$s_n = \sum_{i=1}^{n} d_{0:i} \quad \text{where } d_{0:i} = \pm 1 \text{ (or some other constant)} \quad (4.3)$$

The successive $\pm d_0$ form a vector of displacements $\boldsymbol{d}$, and the successive s_n form a vector of sums $\boldsymbol{s}$ recording successive states of the random walk. Both have an i_0 timescale and an arbitrary length n depending on the number of iterations.

Random walks are easy to program on a computer, and they are engaging to watch because the trace of each random-walk time series is unpredictable. When the probability of a step to the right (p) is equal to the probability of a step to the left (q), then $p = q = 0.5$. As a corollary, the expected number of runs of the same sign, $(n + 1) / 2$, is just slightly greater than $n / 2$ for a random walk of n steps (including one-step runs). This is another way of saying that the mean or expected length of a run is just slightly less than two steps. We expect a random walk like that in Figure 4.2a to have about four runs, but we observe that it has five: four one-step runs and one three-step run. We expect the mean length of a run to be about two steps, but here it is only 1.4 steps.

Individual random walks are by nature idiosyncratic and unpredictable, with a wide range of possible behaviors. However, in aggregate they share some common characteristics. Figure 4.2b shows 1,000 random walks generated like that in Figure 4.2a. Each walk has a cumulative displacement s at time $t = 7$, and the numbers of walks terminating with a given cumulative displacement are inset in row seven. These numbers (8–52–166–294–256–168–51–5) differ little in terms of proportion from the corresponding row of Pascal's triangle (1–7–21–35–35–21–7–1), and closely approximate a normal curve.

Figure 4.2c confirms Pearson's inference that the single most likely place to find a random walk is where it started. However, note that the relative likelihood of such a finding decreases rapidly in the early steps of the walk. Cumulative displacements, positive and negative, expand with each step, and hence the relative frequencies associated with each displacement become more uniform. We can be more precise in saying that the standard deviation of the cumulative displacements increases in proportion to the square root of elapsed time. When $d_0 = i_0$ and $r_0 = 1$, as is true in Figure 4.2, then the standard deviation σ is not only proportional but it is also equal to the square root of time t ($\sigma = 1$ when $t = 1$; $\sigma = 2$ when $t = 4$; etc.).

The standard deviation σ and the variance σ^2 are illustrated graphically in Figure 4.2d, showing how these statistics increase with each step of the random walk. The standard deviation of cumulative displacements increases as the square root of t (or $t^{0.5}$; dashed lines in Fig. 4.2d), and a 95% confidence interval for s increases as $1.96 \cdot \sigma \cdot t^{0.5}$ (dotted lines). The variance σ^2 increases in direct proportion to t (solid lines in Fig. 4.2d). For random walks with a step rate $r_0 = 1$ the variance at each step defines the positive and negative limits of dispersion. Reaching these limits requires a sustained run of change at the step rate, $+r_0$ or $-r_0$, which is reasonably probable for a short time at the beginning of a random walk but becomes less and less probable as the walk gets longer. The probability is equal to $0.5^n = 0.5, 0.25, 0.13, 0.06, 0.03$, etc., for successive time steps n, dipping below the critical value of 0.05 in the fifth time step.

4.2.2 Evolutionary Random Walk

The difference between a simple random walk and what is here called an evolutionary random walk is an additional component of randomness. In an evolutionary random walk, successive s_i are drawn from repeated iteration of:

$$s_i + 1 = s_i \pm d_0, \text{where } d_0 = \boldsymbol{n}(\mu, \sigma) \tag{4.4}$$

The $\boldsymbol{n}(\mu, \sigma)$ are normal deviates drawn from a distribution with mean $\mu = 0$ and finite standard deviation σ. Here again successive d_0, each with its associated sign, plus or minus, form a vector of displacements $\boldsymbol{d}$, and the successive s_{i+1} form a vector of sums s, recording successive states of the random walk. Both have an arbitrary length n depending on the number of iterations.

An evolutionary random walk is symmetric and the probabilities associated with the plus and minus signs are again equal at $p = q = 0.5$. Note that for a given i_0, d_0 determines r_0, and r_0 determines d_0. Rate r_0 now has a distribution that reflects how values of the variable d_0 are related to i_0 on the scale of time i_0.

Initial steps of an evolutionary random walk for $d_0 = \boldsymbol{n}(0, 1)$ are illustrated in Figure 4.3a. The random walk here begins with a step to the left from time $t = 0$ to $t = 1$. The step has a step interval $i_0 = 1$, a step displacement d_0 of -1.165, and a step rate $r_0 = d_0 / i_0 = -1.165$. The second step from $t = 2$ to $t = 3$ has the same step interval, but now the displacement d_0 and rate $r_0 = 0.179$ are positive. This is followed by two steps to the left, a step to the right, and two steps to the left. Note that the step interval length is constant, but the displacement and the rate in each successive step vary because each is drawn from $d_0 = \boldsymbol{n}(0, 1)$.

4.2.3 Rates of Evolutionary Random Walks

Figure 4.3b shows the random walk of Figure 4.3a, drawn at the same scale with the same initial seed, but now each step displacement or difference is one fifth of that before, drawn from $\boldsymbol{n}(\mu, \sigma)$ with mean $\mu = 0$ and standard deviation $\sigma = 0.2$. The result appears to be stationary, but it is the same random walk as that in Figure 4.3a as we could show by adjusting the display of the horizontal axis.

Figure 4.3c shows the random walk of Figure 4.3a, drawn as before at the same scale and with the same initial seed, but now each step displacement or difference is five times that of Figure 4.3a, drawn from $\boldsymbol{n}(\mu, \sigma)$ with mean $\mu = 0$ and standard deviation $\sigma = 5$. In this case the result appears to be directional, moving rapidly to

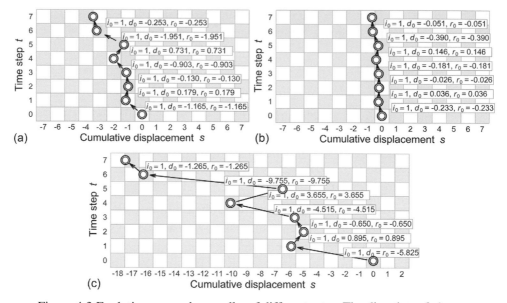

Figure 4.3 Evolutionary random walks of different rates. The direction of change (+ or −) is random. Step interval i_0, step distance d_0, and step rate r_0 are shown for each step in each time series. (a) Path of a seven-step random walk with step displacement d_0 drawn from a normal distribution of mean $\mu = 0$ and standard deviation $\sigma = 1$. Time series starts at the origin with coordinates ($t = 0$, $s = 0$) and ends at ($t = 7$, $s = -3.492$). (b) Path of the same seven-step random walk with standard deviation $\sigma = 0.2$. This series starts at the origin with coordinates ($t = 0$, $s = 0$) and ends at ($t = 7$, $s = -0.698$). (c) Path of the same seven-step random walk with standard deviation $\sigma = 5$. This series starts at the origin with coordinates ($t = 0$, $s = 0$) and ends at ($t = 7$, $s = -17.460$). Note that the step rate r_0 expressing the relationship of d_0 to i_0 is critically important for the realized tempo and mode of an evolutionary time series.

the left. Again, the random walk in Figure 4.3c is the same as that in Figure 4.3a as we could show by adjusting the display of the horizontal axis.

Figures 4.3a–c illustrate why quantitative analysis of evolutionary patterns is so important. The same generating process or mode, in this case the same random walk, can yield time series that look stationary, random, and directional just by varying their step rate. In the next chapter we shall learn to analyze time series to distinguish stationary, random, and directional evolutionary modes. We shall also learn to recover the step rates of the generating processes.

4.3 Brownian Diffusion

With the random walk of Figure 4.3a in mind, reproduced in Figure 4.4a, it is easy to imagine a second random walk starting from the same origin and following the

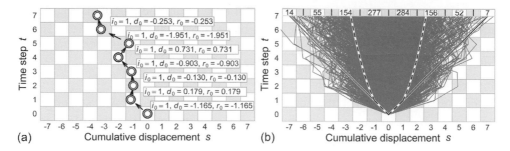

(a) Cumulative displacement *s*   (b) Cumulative displacement *s*

Figure 4.4 Evolutionary random walk and corresponding diffusion. For each time step the direction of change (+ or −) is random and the step displacement d_0 is drawn from a normal distribution with mean $\mu = 0$ and standard deviation $\sigma = 1$. (a) Path of the seven-step walk shown in Figure 4.3a, starting at the origin with coordinates ($t = 0$, $s = 0$) and ending at ($t = 7$, $s = -3.492$). The step interval i_0, step distance d_0, and step rate r_0 are shown for each step. (b) Paths of 1,000 seven-step random walks, each starting at the origin. Inset numbers in the top row show the number of walks reaching each of eight terminal bins. Here as in Figure 4.2, the variance of cumulative displacement increases in proportion to t (solid lines) and the standard deviation increases as $\sqrt{t}$ (dashed lines). Evolutionary random walks have more-variable step rates than simple random walks and the variance does not constrain the dispersion for walks as short as those shown here (the variance does constrain dispersion for longer walks).

same rules. What are the chances the second random walk will follow the same path as the first? The direction of each step, forward and to the left or to the right, is determined by chance. The two alternatives are equally probable, with $p = q = 0.5$. Thus the probability that the first step of the second random walk will take the same direction as the first step of the first random walk is 0.5, and this will be true for each of the six subsequent steps. Random determination of the signs of the steps by itself means that the probability that the second random walk will follow the path of the first is $0.5^7 = 0.0078$. This is actually the chance that a second random walk might follow the path of the first in Figure 4.2a where the displacements d_0 are constant. If we add the fact that each displacement d_0 in Figure 4.4a is determined by chance, then the probability that the second random walk will follow the path of the first is infinitesimal.

Figure 4.4b shows the traces of 1,000 random walks generated like that in Figure 4.4a. Each individual walk is different, but collectively they describe a coherent pattern. Here again the variance σ^2 increases in direct proportion to t (superimposed solid lines) and the standard deviation of cumulative displacements increases as the square root of t (or $t^{0.5}$; dashed lines). A 95% confidence interval for the cumulative displacement s is:

$$s_{(\alpha \leq 0.05)} = \pm 1.96 \cdot \sigma \cdot \sqrt{t} \qquad (4.5)$$

One way to visualize the developing pattern is to group the cumulative displacements after seven steps into bins. Bins comparable to those in Figure 4.2b are shown at the top of Figure 4.4b. The numbers in these bins (14–55–154–277–284–156–52–7) again differ only by chance and proportion from the corresponding row of Pascal's triangle (1–7–21–35–35–21–7–1).

Figure 4.4b illustrates the diffusive behavior of random walks studied in aggregate. If the trace of a one-dimensional random walk is a simple form of Brownian motion, then this aggregate diffusion is appropriately called Brownian diffusion. Both are named for the nineteenth-century Scottish botanist, paleobotanist, and microscopist Robert Brown, who published his observations in 1828. What Brown saw under his microscope were individual pollen grains moving in the two-dimensional pattern that Pearson later called a random walk. Sewall Wright (1931) called the aggregate pattern "random drift." Motoo Kimura (1964) preferred "diffusion."

4.4 Ornstein–Uhlenbeck Random Walks

Models of evolutionary time series include many alternatives to random walks, ranging from the simplest of functions such as the stationary model $y = 0$ or the directional model $y = t$, to more complex functions building on these. Each may or may not have added variance. One popular model borrowed from physics is the Ornstein–Uhlenbeck process, where successive s_i are drawn from repeated iteration of:

$$s_i + 1 = s_i - \alpha \cdot (s_i - \theta) + \boldsymbol{n}(\mu, \sigma) \qquad (4.6)$$

Here $\boldsymbol{n}(\mu, \sigma)$ is again a random deviate drawn from a normal distribution with mean $\mu = 0$ and finite standard deviation σ.

The Ornstein–Uhlenbeck model of Equation 4.6 has two parameters not present in Equation 4.4: a target or "optimum" value θ, and an intensity or "rate" parameter α. Note that the middle term $\alpha \cdot (s_i - \theta)$ drops out whenever $\alpha = 0$, or $s_i = \theta$, or both. When the middle term drops out, Equation 4.6 reduces to the evolutionary random walk of Equation 4.4. When $\alpha > 0$, then θ functions as an attractor, drawing s_{i+1} toward the target at each step by a factor depending on α and on the difference between s_i and θ in the previous step. When θ is not the initial value for a time series, then convergence is asymptotic and requires more than a few time steps.

Lande (1976) was the first to suggest the Ornstein–Uhlenbeck process as an evolutionary model. He used it to explore the efficacy (inefficacy) of random genetic drift moving a population from one adaptive zone to another. Application of Ornstein–Uhlenbeck to evolutionary studies has been extended by Hansen and Martins (1996), Hansen (1997), Butler and King (2004), and many others.

4.4.1 Ornstein–Uhlenbeck as a Stationary Process

The influence of the Ornstein–Uhlenbeck parameters α and θ on the random walk of Figure 4.3a is illustrated in Figure 4.5a, which shows how the original random walk is drawn back rapidly toward the initial-value target $\theta = 0$ at each step. The result is a more stationary time series than the original. Intensity α determines the balance between movement toward the target and random displacement due to $n(\mu, \sigma)$ at each step.

For comparison, Figure 4.5b shows the paths of 1,000 random walks with the original parameters ($\mu = 0$ and $\sigma = 1$), subject to the additional Ornstein–Uhlenbeck

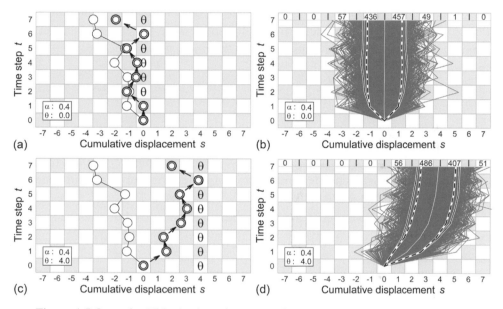

Figure 4.5 Ornstein–Uhlenbeck stationary random walk and corresponding diffusion. (a) Evolutionary random walk of Figures 4.3a and 4.4a with additional parameters θ and α. Here the target value θ is the initial value $\theta = 0$, and the constraining intensity of convergence α is set to $\alpha = 0.4$. Note that the constrained random walk remains closer to $\theta = 0$ than the unconstrained original (open circles). (b) Paths of 1,000 seven-step random walks starting at the origin with parameters $\theta = 0$ and $\alpha = 0.4$. The intensity of convergence constrains each random walk to a region near the target value. The mean (central white line) remains near $\theta = 0$. The standard deviation (dashed lines) and variance (solid lines) of the 1,000 paths stabilize in the first few steps of the Ornstein–Uhlenbeck process. (c) Random walk of Figures 4.3a and 4.4a with the target value θ set at $\theta = 4$. The intensity of convergence remains $\alpha = 0.4$. The constrained random walk is now much closer to $\theta = 4$ than the unconstrained original (open circles). (d) Paths of 1,000 random walks starting at the origin with $\theta = 4$ as the target value. The mean (central white line) now approaches the target asymptotically. The mean and variance again stabilize rapidly.

constraints provided by $\alpha = 0.4$ and $\theta = 0$. The intensity $\alpha = 0.4$ is sufficient to stabilize the constituent random walks within the first few time steps. When all 1,000 random walks are analyzed statistically, the standard deviation for steps 3–7 of the Ornstein–Uhlenbeck process is found to be stable and averages about 1.24 displacement units, and the variance for steps 3–7 is similarly stable and averages about 1.53 displacement units. Terminal values are normally distributed, but the distribution in bins at the top of Figure 4.5b is more tightly constrained than that of Figure 4.3b.

If the target θ is changed to a number greater or less than $\theta = 0$, then the distribution of random walks approaches the new value. This is illustrated in Figure 4.5c, where α remains 0.4 but θ is increased to $\theta = 4$. Figure 4.5d shows 1,000 random walks with $\alpha = 0.4$ and $\theta = 4$. The asymptotic convergence mentioned above is clearly visible, and the mean value remains less than the target through the first seven time steps. If the intensity α were increased (or decreased) from $\alpha = 0.4$, then the rate of transition from the original to any new value θ would increase (or decrease), and the variance of traces tracking the target θ would decrease (or increase). This is why α is an intensity as well as a rate.

4.4.2 Ornstein–Uhlenbeck as a Directional Process

Most applications of the Ornstein–Uhlenbeck process employ a constant-value target θ, but θ can also change through time. Figure 4.6a shows the influence of Ornstein–Uhlenbeck α and θ on the random walk of Figure 4.3a when θ changes as a function of time. The intensity α remains the same at $\alpha = 0.4$, but now the target is $\theta = t$. The resulting random walk is strongly influenced by the moving target and approaches it closely.

When we simulate 1,000 random walks as an Ornstein–Uhlenbeck process, the mean path (white line in Figure 4.6b) approaches the moving target but lags about one displacement unit behind it. Nevertheless, dispersion of paths about the mean path stabilizes rapidly as we saw in Figures 4.5b and 4.5d for Ornstein–Uhlenbeck time series with stationary targets. Here, as in those figures, the standard deviations and variances for the last five time steps average 1.24 and 1.53, respectively, even though the mean itself is moving.

4.4.3 Limitations of Ornstein–Uhlenbeck Diffusion

Lande (1976) used Ornstein–Uhlenbeck diffusion as a stationary process to show that the most time-consuming part of diffusion from one adaptive zone to another is movement out of the first zone against the force of natural selection. Random exploration of an adaptive landscape, and hence the chance of escaping an adaptive zone, is reduced exponentially in proportion to population size. In Lande's

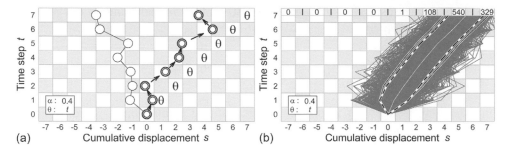

Figure 4.6 Ornstein–Uhlenbeck directional random walk and diffusion. (a) Evolutionary random walk of Figures 4.3a and 4.4a with additional parameters θ and α. Here target value θ is the time step $\theta = t$, and the constraining intensity of convergence is again $\alpha = 0.4$. Note that the constrained random walk remains closer to $\theta = t$ than the unconstrained original (open circles). (b) Paths of 1,000 seven-step random walks starting at the origin with parameters $\theta = t$ and $\alpha = 0.4$. The intensity of convergence constrains each random walk to a region near the target value. The mean value of the 1,000 paths (central white line) tracks the target θ but lags a displacement value behind it. Here as in Figures 4.5b and 4.5d, the standard deviation from the mean path (dashed lines) and the variance about the mean path (solid lines) stabilize in the first few steps of the Ornstein–Uhlenbeck process.

application the target θ is the optimum or peak of the resident adaptive zone, and the intensity α is the force of natural selection resisting movement away from this optimum. Lande noted that once the threshold or valley separating the resident adaptive zone from a new one was reached and crossed, movement into the new adaptive zone would be rapid because α would now be the force of natural selection favoring movement to the new θ representing the new optimum.

The attractive features of an Ornstein–Uhlenbeck model of diffusion are its rapid attainment of target values and its constrained variance. Ordinary Brownian diffusion is inefficient in reaching a target, and its variance expands linearly with time. However, the arbitrary and limiting feature of Ornstein–Uhlenbeck diffusion as an evolutionary model is the target θ, which is often conceived as the optimum of a new, empty, and stationary adaptive zone. The position of an optimum like this cannot be predicted but is only evident in hindsight, and it probably will not be recognized as a target if θ is moving through time. Addition of α as a force of natural selection distinguishes Ornstein–Uhlenbeck diffusion from a passive random drift model.

4.5 Note on Mean Absolute Deviations and Standard Deviations

Lower-case sigma in $d_0 = \boldsymbol{n}(\mu, \sigma)$ is approximately the rate of a random walk, but the step displacements d_0 drawn from $\boldsymbol{n}(\mu, \sigma)$ and the step rates $r_0 = d_0 / i_0$, have a mean absolute deviation *MAD* that is about 80% of the model standard deviation σ. The relationship of *MAD* to σ, from Geary (1935), is:

$$MAD = \sqrt{\frac{2}{\pi}} \cdot \sigma = 0.7979 \cdot \sigma \tag{4.7}$$

In modeling, if we want a sample of the resulting displacements or differences to have a realized standard deviation $s = 1$, then we have to draw d_0 from $d_0 = n\left(\mu, \sigma/\sqrt{(2/\pi)}\right)$.

4.6 Summary

1. A random walk is a pattern of change or movement in successive steps, with the change at each step governed by chance, independent of what came before.
2. Time t is the *independent* variable in random walks and in evolution. Other measures such as displacement d or cumulative displacement s are *dependent* variables because they change as a function of time.
3. An iterative process like a random walk generates change by a signed displacement or difference $+d_0$ or $-d_0$ in each time step i_0 of the process; i_0 is normally a constant "generation" time, but the step displacement d_0 can be a constant or a random variable. Time t is a count of accumulated i_0.
4. The ratio of d_0 to i_0 is the step rate of a process: $r_0 = d_0 / i_0$. The signs of successive d_0 change randomly, meaning the signs of r_0 also change randomly. Subscripts of i, d, and r flag the timescale involved: These are $\log_{10}$ values and the zero subscripts of i_0, d_0, and r_0 indicate that all involve change on a timescale of one step or generation.
5. An evolutionary random walk is a time series with successive s_i drawn from repeated iteration of $s_{i+1} = s_i \pm d_0$, where d_0 is a random variable drawn from a normal distribution with mean μ and standard deviation σ.
6. Individual random walks are by nature idiosyncratic and unpredictable, with a wide range of possible behaviors. Random walks aggregated as Brownian diffusion share some common characteristics: (1) terminal values are normally distributed with most random walks near their starting point and few in each tail, (2) the variance of the aggregate increases as a constant proportion of time, and (3) the standard deviation of the aggregate increases as the square root of time.
7. Ornstein–Uhlenbeck random walks include a target or "optimum" value θ, and an intensity or "rate" parameter α. These are useful for estimating the probability and waiting time for a biological population to leave a model adaptive zone by random diffusion or random drift.

5

Temporal Scaling and Evolutionary Mode

> We are now prepared for a closer analysis of the nature of chance fluctuations in a random walk. The results are startling. According to widespread beliefs a so-called law of averages should ensure that in a long coin-tossing game each player will be on the winning side for about half the time, and the lead will pass not infrequently from one player to the other ... Intuition leads to an erroneous picture of the probable effects of chance fluctuations.
>
> *William Feller,* An Introduction to Probability Theory and Its Applications, *Third Edition, 1968, p. 78*

William Feller's comments on random walks can be extended to evolutionary time series. Intuition is a poor substitute for analysis. Random walks may include sequences that appear to be stationary and sequences that appear to be directional that are sometimes just part of being random. What a pattern appears to show depends greatly on its scale and how it is drawn. In evolution, where the steps are generations, external constraints may be imposed by stabilizing or directional selection, converting what would otherwise be a random time series to one that is stationary or directional. Analysis is required to distinguish each mode of change from the other.

Sequence length is important for distinguishing stationary and directional time series from random series and from each other. This may seem obvious because in a coin-tossing experiment a long run of heads is less likely than a shorter run of heads, but how long is the run required to distinguish observations from random? We need a method for analyzing time series when these are complete that will also work when available information is sampled on scales of time greater than steps of the generating process and when incomplete sampling may occur anywhere in a time series.

The solution is *temporal scaling*, an exercise relating change to the scale of time. Temporal scaling involves: (1) the relationship of *differences* between values of a time series to the *intervals of time separating the values*, or, alternatively,

(2) the relationship of *rates of change* in the values of a time series to the *intervals of time separating the values*. As we shall see, temporal scaling of one is the complement of the other. We study time-series differences in relation to time to see whether and how these are dependent on time, and we study time-series rates in relation to time for the same reasons.

Many processes, including evolution, yield incremental change with each iteration of the process, and differences accumulate in relation to the number of iterations. The cumulative differences generally depend to some extent on the number of iterations. Temporal scaling is a way to test the expected dependence of cumulative differences on time by plotting empirical observations of one against the other: cumulative differences should increase with the passage of time. Time is the independent variable and cumulative difference is the dependent variable.

Rates are ratios of cumulative difference to time that summarize an expectation per iteration and are thus, in principle, independent of time. Temporal scaling is a way to test the expected independence of rates from time by plotting empirical observations of one against the other: Rates should remain stable with the passage of time. Time is again the independent variable, and rate is the dependent variable.

Temporal scaling of differences and temporal scaling of rates are most efficiently studied on proportional or logarithmic scales, and quantified in terms of slopes and intercepts. The temporal scaling slope of a time series tells us the evolutionary *mode* of the underlying process, whether stationary, random, or directional. The temporal scaling intercept tells us the evolutionary *tempo*, the step difference or, equivalently, the step rate on the time scale of the process.

The temporal scaling slope and intercept of a time series are, in principle, independent of each other. The slope does not determine the intercept, and the intercept does not determine the slope. The caveat of independence "in principal" is important because the differences and rates we are able to observe and quantify in evolutionary time series are sometimes affected by the physiological constraints of organic design and sometimes censored by limits on our powers of observation and measurement.

In the following text we consider temporal scaling and its quantification in more detail. We start by analyzing two time series familiar in commerce. Then we consider simple simulations where we know that a time series of interest is random, stationary, or directional. Finally we analyze a longer random time series where the statistical power is greater.

5.1 Temporal Scaling of Differences and Rates

For any particular difference d and its associated interval i, there is a rate $r = d/i$. This is true for step differences, intervals, and rates (Figure. 5.1), and it is true for longer differences, intervals, and rates. Further, for any sequential set or vector d of

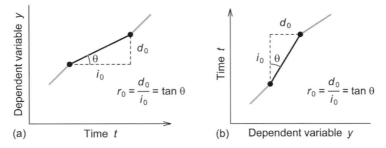

Figure 5.1 Geometric representations of step intervals i_0, differences d_0, and rates r_0. (a) Time t, the independent variable, is plotted on the horizontal axis, and the step rate r_0 is equal to the slope d_0/i_0 and tangent of angle θ measured from horizontal. (b) Time t, the independent variable, is plotted on the vertical axis, and the step rate r_0 is equal to the divergence d_0/i_0 and tangent of angle θ measured from vertical.

differences d, and the corresponding sequential set or vector $\boldsymbol{i}$ of intervals i, there is an associated sequential set or vector of rates $\boldsymbol{r}$ for which each constituent rate $r = d/i$. For a given set of intervals $\boldsymbol{i}$, the set of differences $\boldsymbol{d}$ is the inverse of the set of rates $\boldsymbol{r}$, and the set of rates $\boldsymbol{r}$ is the inverse of the set of differences $\boldsymbol{d}$.

$$\boldsymbol{r} = \boldsymbol{d}/\boldsymbol{i} \; ; \quad \boldsymbol{d} = \boldsymbol{r} \cdot \boldsymbol{i} \; ; \quad \text{and} \quad \boldsymbol{i} = \boldsymbol{d}/\boldsymbol{r} \quad\quad (5.1)$$

The rationale for studying evolutionary differences and rates in relation to their associated time scales is expressed in the "rescaled range" and "Hurst exponent" developed by Harold Hurst (1951). Hurst worked in Egypt where he was concerned with annual discharge of the Nile and the storage capacity of reservoirs. It is also expressed in the fractional dimension or fractal "roughness" of coastlines and other natural phenomena developed by Lewis Richardson (1961) and Benoit Mandelbrot (1967). The Hurst exponent of a random walk is $K = 0.5$ (from $s \propto t^{0.5}$ in Chapter 4), and the Mandelbrot fractional dimension of a random walk is $D = 1.5$ (halfway between a one-dimensional line and a two-dimensional plane). K and D are both proportional and describe relationships on logarithmic scales.

Equation 5.1 can be rewritten in logarithmic form:

$$\log \boldsymbol{r} = \log \boldsymbol{d} - \log \boldsymbol{i} \; ; \quad \log \boldsymbol{d} = \log \boldsymbol{r} + \log \boldsymbol{i} \; ; \quad \text{and} \quad \log \boldsymbol{i} = \log \boldsymbol{d} - \log \boldsymbol{r}$$

$$(5.2)$$

Here the set of differences or displacements $\boldsymbol{d}$ differs from the set of rates $\boldsymbol{r}$ by the set of intervals $\boldsymbol{i}$.

5.1.1 Base Rates and Net Rates

In Section 4.2 of the previous chapter we learned about *step* intervals, displacements, and rates (i_0, d_0, and r_0, where the subscript 0 is $\log_{10}$ of an interval length of one step or generation; see also Equation 4.2). Step intervals, step displacements, and step rates are the intervals and associated displacements and rates of the *generating process* of a time series.

Here we introduce the related concepts of *base* intervals, displacements, and rates. Base intervals, base displacements, and base rates are the time intervals and associated displacements and rates of our *finest sampling* of a time series. "Finest sampling" means that all successive samples we can observe are included, but this does not mean that all successive steps of the underlying generating process are included. Base intervals are the intervals between successive samples, and base differences and base rates are the displacements and rates associated with each base interval. Step intervals all have a uniform time length, whereas base intervals may differ greatly in length. When we know them, step intervals, displacements, and rates can be considered base intervals, displacements, and rates, but the reverse is not true. Base intervals, displacements, and rates cannot be step intervals, displacements, and rates when the steps are not sampled.

An additional distinction is important: that between *base* intervals, displacements, and rates, and *net* intervals, displacements, and rates. Base intervals are intervals between successive samples, while net intervals can span two or more base intervals. If samples are taken at times t_0, t_1, t_2, and t_3, then the intervals $t_1 - t_0$, $t_2 - t_1$, and $t_3 - t_2$ are base intervals, and the intervals $t_2 - t_0$, $t_3 - t_1$, and $t_3 - t_0$ are net intervals. The difference is important because base intervals are statistically independent of each other, while net intervals cannot be independent of the base intervals that compose them. Net intervals also often overlap, in which case they are not independent of each other.

For each base interval there is a base displacement and a base rate, and similarly for each net interval there is a net displacement and a net rate. When the intervals are statistically independent, as they are for base intervals, then the corresponding displacements and rates are independent as well. When the intervals lack independence, as they do when net intervals overlap with base intervals or net intervals overlap with each other, then the corresponding displacements and rates lack independence. Statistical independence is important for some purposes, while characterization of statistical dependence is important for others.

Finally, we can consider the number of intervals, displacements, and rates in a given time series. In the example here, with samples at t_0, t_1, t_2, and t_3 we found three base rates and three net rates. In general, for n successive samples the total number of intervals, displacements, and rates is given by:

$$N_{total} = \binom{n}{2} = \frac{n!}{2! \ (n-2)!} = \frac{n \cdot (n-1)}{2} \tag{5.3}$$

The number of base intervals and hence base displacements and base rates is simply:

$$N_{base} = n - 1 \qquad (5.4)$$

And the number of net intervals and hence net displacements and net rates is:

$$N_{net} = \frac{(n-1)\cdot(n-2)}{2} \qquad (5.5)$$

For $n = 4$, there are six total intervals, displacements, and rates; three are base intervals, displacements, and rates as we already saw, and three are net intervals, displacements, and rates. For the time series in Figure 4.3a, $n = 8$ and there are 28 total intervals, displacements, and rates; 7 base intervals, displacements, and rates; and 21 net intervals, displacements, and rates. The number of base intervals rises in linear proportion to the number of samples, but the number of net intervals rises as a quadratic proportion of the number of samples.

Examples are given here and in the following chapters of analyses that are longitudinal in the sense of tracing change through time (e.g., automobile production, stock exchange values) and analyses that are cross-sectional in the sense of comparing change across time (e.g., Olympic running records). Hendry and Kinnison (1999) distinguished such analyses as "synchronic" and "allochronic," but here longitudinal and cross-sectional are retained because of their familiarity.

5.1.2 Rates in Commerce: Automobile Production

The idea that rates of change might be related to the intervals of time used to calculate them is counterintuitive. This is because a rate involves division by time, and we normally expect such a quotient to be independent of time. An example of an ordinary application of rates is illustrated in Figure 5.2, based on information at the website ycharts.com (14 June 2016). United States' automakers produced 326,000 automobiles in January of 2015, 322,000 in February, 400,000 in March, and so forth, with very little incremental difference each month, reaching a cumulative difference, displacement, or total of 4,163,000 cars for the year. The year starts at zero, so each month adds a sequential contribution. Time is on the horizontal axis, the slopes are rates, and the rates for the months are little different from the rate for the year.

Total production divided by 12 gives a monthly rate of 0.347 million cars per month on a timescale of one year. This number is close to each of the 12 individual monthly rates on a timescale of a month, but all, together, are only a sample of the 78 production intervals, cumulative differences, and rates implicit in the production history. There are 13 nodes defining the beginnings and ends of the 12 production months (the first month starts at zero), so the number of samples is 13. The

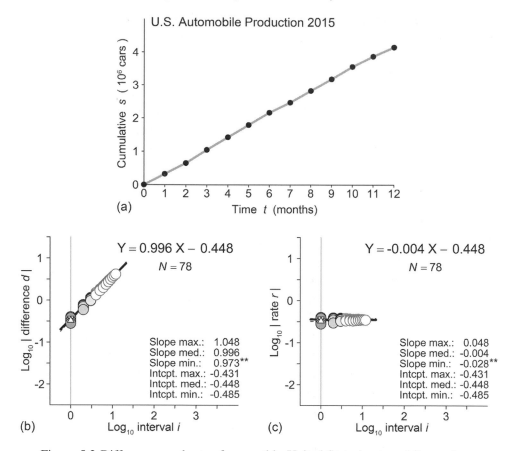

(a)

(b)

(c)

Figure 5.2 Differences and rates for monthly United States' automobile production in the year 2015. Time, the independent variable, is on the horizontal axis. (a) Gray lines connect cumulative total production for 12 successive months. Production ranged from a high of 400,000 cars in March (month 3) to a low of 279,000 cars in December (month 12). (b) LDI or log difference versus log interval analysis of the time series with circles representing changes or differences in production (in millions of cars) for corresponding intervals (in months). Solid line through the circles was computed using robust weighted linear modeling (see text). Bootstrapped confidence limits are virtually colinear. The slope of the scatter (0.996) and the confidence interval, which includes a slope of 1.000 but excludes slopes of 0.000 and 0.500 (double asterisk), together indicate that the mode of change is directional. The intercept of $10^{-0.448} = 0.356$ indicates a rate of production of 356,000 cars per month on a timescale of one month. (c) Log rate versus log interval (LRI) analysis of the time series with circles representing rates of production (in millions per month) for corresponding intervals (in months). Solid line was computed as before, and bootstrapped confidence limits are again virtually colinear. The slope of the scatter (−0.004) and the confidence interval, which includes a slope of 0.000 but excludes slopes of −1.000 and −0.500 (double asterisk), together indicate that the mode of change is directional. Here again the intercept of $10^{-0.448} = 0.356$ indicates a rate of production of about 356,000 cars per month on a time scale of one month. This rate is close to the median monthly production of 358,000 cars per month on a monthly time scale.

number of base intervals, differences, and rates is 12 (Equation 5.4). There are, in addition, 66 net intervals, differences, and rates (Equation 5.5). Thus full sampling yields 78 production differences and 78 production rates.

Each of the 78 production differences is plotted against its corresponding interval in Figure 5.2b. The relationships are proportional, so logarithmic axes are appropriate. Note that the slope of a line fit to the points (0.996) is almost exactly 1.0, which is the number expected for a perfectly smooth directional trend of increasing (or decreasing) monthly production through the year. The intercept in Figure 5.2b is -0.448, and $10^{-0.448} = 0.356$ million is an estimate of the monthly production of automobiles on the timescale of one month. Here, because the time series is directional, the monthly production on a timescale of one month (356,000 thousand) is close to the monthly median on a time scale of one year (358,000).

Each of the 78 production rates is plotted against its corresponding interval in Figure 5.2c. Again the relationships are proportional, so logarithmic axes are appropriate. Note that the slope of a line fit to the points (-0.004) is almost exactly zero, which is the number expected for a perfectly smooth trend of constant rate. The intercept in Figure 5.2c is again -0.448.

What conclusions can we draw from the automobile production information in Figure 5.2? First, the pattern of increasing production through time in Figure 5.2a is about as smooth a directional trend as one could find anywhere. Second, the slope of the distribution of differences on the LDI plot of Figure 5.2b is very close to 1.0, which is the hallmark of a directional trend. Third, the slope of the distribution of rates on the LRI plot of Figure 5.2c is very close to 0.0, which is another hallmark of a directional trend. Both slopes differ significantly, and greatly, from 0.5 and -0.5 expected for a random walk (double asterisk; see below). The LDI and LRI slopes of 0.996 and -0.004 are not independent. The difference between these slopes is always 1.0 because $\log d$ and $\log r$ are complementary reflections of $\log i$ (Equation 5.2). The intercepts, -0.448, are the same on both plots. If we know the slope and intercept of the difference versus interval scatter on an LDI plot, then we know the slope and intercept of the corresponding rate versus interval scatter on an LRI plot (and vice versa).

Automobile production is but one of many possible examples of more or less uniformly directional change. Time passes and the assembly process produces automobiles. Nature rarely works this way and, for reasons we shall explore, rates in nature are rarely independent of interval length.

5.1.3 Rates in Commerce: Stock Exchange Values

A second application of rates is illustrated in Figure 5.3a, based on information in Williamson (2017). The Dow Jones Industrial Average is a widely watched index

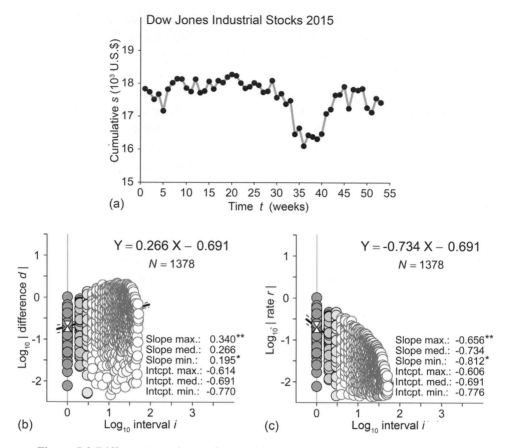

(a)

(b)

(c)

Figure 5.3 Differences and rates for weekly Dow Jones Industrial Average stock values in the year 2015. Time, the independent variable, is on the horizontal axis. (a) Gray lines connect average values for 53 successive weeks. Values range from a high of \$18.272 thousand in week 20 to a low of \$16.102 thousand in week 36. (b) LDI analysis of the time series, with circles representing changes or differences in value (in thousands of dollars) for corresponding intervals (in weeks). Solid line through the circles was computed using robust weighted linear modeling (see text). Dashed lines are bootstrapped confidence limits. The slope of the scatter (0.266) and the confidence interval, which excludes 0.000 (asterisk) and excludes both 0.500 and 1.000 (double asterisk), together indicate that the mode of change is random with a stationary component. The intercept of $10^{-0.691}$ = 0.204 indicates a rate of change of about \$204 per week on a weekly time scale. (c) LRI analysis of the time series with circles representing rates of change (in thousands of dollars per week) for corresponding intervals (in weeks). Solid line through the circles and confidence intervals were computed as before. The slope of the scatter (-0.734) and the confidence interval, which excludes -1.000 (asterisk) and excludes both 0.000 and -0.500 (double asterisk), together indicate that the mode is random with a stationary component. The intercept of $10^{-0.691}$ = 0.204 indicates again a rate of change of about \$204 per week on a weekly time scale.

in the United States' financial markets. This closed with a value of $17,832.99 at the end of the first week of 2015, and it closed with a value of $17,425.03 at the end of the 53rd week of 2015. The decline of $407.96 for the year yields an average decline of $7.85 per week on the timescale of a year. This might be all an investor wants to know, but the time series contains more information.

The time series of 53 values in Figure 5.3a spans 52 weeks and yields 52 base differences and 52 base rates (Equation 5.4), and 1,326 net differences and 1,326 net rates (Equation 5.5). Combined there are a total of 1,378 differences and a total of 1,378 rates. The time series in Figure 5.3a is different from that in Figure 5.2a. It is anything but smooth, exhibiting considerable roughness or volatility, and it runs more or less parallel to the time axis.

Each of the 1,378 Dow Jones differences is graphed against the corresponding interval length in Figure 5.3b. A line fit to the resulting scatter has a slope of 0.266, which is less easily interpreted than the slope in Figure 5.2b. The intercept in Figure 5.3b is -0.691, so $10^{-0.691} = \$0.204$ thousand is an estimate of the average weekly change in value on a timescale of one week. This weekly change of $204 on a timescale of one week is very different from the average weekly decline noted above of $7.85 per week on a timescale of a year.

Each of the 1,378 Dow Jones rates is graphed against the corresponding interval length in Figure 5.3c. A line fit to the resulting scatter has a slope of -0.734 that is again less easily interpreted than the slope in Figure 5.2c. The intercept in Figure 5.3c, like that in 5.3b, is -0.691.

The slope of the distribution of differences on the LDI plot of Figure 5.3b is significantly different from 1.000 expected for a directional time series and from 0.500 expected for a purely random time series (hence the double asterisk), and it is significantly different from 0.000 expected for a stationary time series (single asterisk). Lying between 0.500 and 0.000, the slope for the differences indicates a time series that we can call random with a stationary component.

Similarly, the slope of the distribution of rates on the LRI plot of Figure 5.3c is significantly different from 0.000 expected for a directional time series and from -0.500 expected for a purely random time series (double asterisk), and it is significantly different from -1.000 expected for a stationary time series (asterisk). Lying between -0.500 and -1.000, the slope for the rates indicates again a time series that we can call random with a stationary component.

5.2 Robust Linear Modeling

Time scale matters. The differences and the rates of a time series cannot both be independent of the interval of time over which they are calculated. If the *differences* are independent, as in a purely stationary time series, then the *rates* will be highly dependent. If the *rates* are independent, as in a purely directional time series like that in Figure 5.2, then the *differences* will be highly dependent. In a random time series like that in Figure 5.3, both the differences and the rates will be dependent on the timescale over which they were measured or calculated. Temporal scaling is an efficient way to identify and learn from these dependencies.

Given the importance of temporal scaling, there is a sense in which any method that documents the independence or dependence of differences and rates on their time scale is acceptable. The relationship can be illustrated visually by plotting differences or rates against their associated intervals and noting whether there is an increase or decrease as intervals get longer. A relationship, or absence of a relationship, can be quantified using ordinary least-squares regression to determine a slope and an intercept. Any method of exploration is better than employing differences or rates without investigating their dependence on interval length.

This said, it is also true that distributions of differences and rates graphed against their interval lengths are usually both skewed (asymmetrical) for a given timescale, and heteroscedastic (differently scattered) across a range of interval lengths. The most efficient and powerful way to deal with skewness and heteroscedasticity in a bivariate plot is to modify least-squares regression to make it "robust." Nonparametric options such as Theil–Sen regression are appropriate too but are limited by the sizes of matrices required to store what rapidly become very large numbers of slopes and intercepts.

The adjective *robust* in a statistical context generally means that a procedure is insensitive to small deviations from assumptions of symmetry and homoscedasticity that are commonly made in representing distributions of variation. The general approach, as in least-squares regression, is to find parameters that maximize the likelihood of our observations. The robust approach finds slopes and intercepts that maximize likelihood by minimizing mean absolute deviations of data from a model rather than minimizing mean square deviations of data from the model. These robust "M" or maximum likelihood estimates are widely used to estimate regression parameters (Huber 1981; Press et al. 1989).

The robust linear model or "rlm" script in the R "MASS" package (R Core Team 2017) was used to fit lines to the distributions of differences and rates in Figures 5.2 and 5.3. This is a weighted full-sampling procedure, here referred to as a robust weighted fit ("RWF"), which requires supplying a $3 \times N$ matrix of $\log_{10}$ intervals, $\log_{10}$ differences (or $\log_{10}$ rates), and prior weights for each case. Prior weights are calculated as the inverse of the interval length before logging. Initial values for the coefficients are found by least squares. The psi function is the default, psi.huber. Slope and intercept parameters are determined by iteration of reweighted least squares, with the maximum number of iterations set high at 200.

An alternative is a robust weighted median fit ("RWMF") where the "rlm" script is not applied to all observations as a full-sampling procedure but is applied to the median difference or median rate for each of the sampled time intervals. Simulations comparing the two using large samples suggest that RWMF yields a slightly higher slope than RWF, which is slightly closer to a known target value, and that RWMF and RWF yield virtually identical intercepts. For smaller samples with more volatile median differences and rates, like most samples analyzed in Chapters 7–9 here, RWF is sufficiently accurate.

Confidence intervals for the slope and intercept can be computed by bootstrapping (Efron and Tibshirani, 1986; 1993). This involves 1,000 successive resamplings of the rows of the $3 \times N$ matrix described above, each with its interval, difference or rate, and weight. The resampling is done with replacement, so individual rows can be sampled more than once. A new rlm slope and a new rlm intercept are computed for each successive resampling. Each slope and intercept is appended to a corresponding vector. The 1,000-row vectors of slopes and intercepts are then sorted, with the 26th, the median, and the 975th values highlighted. The median for each vector provides a new expected value for the slope or intercept, and the 26th and 975th values for each define their respective 95% confidence limits. Frequencies of entries in the sorted vectors define probability distributions for the slopes and intercepts

Bootstrapping 1,000 times is enough to yield a stable median and a representative confidence interval but not enough to ensure stability of the confidence limits. This instability shows up in the slight noncomplementarity of confidence limits for slopes in temporal scaling of differences and rates, and in the slight inequality of confidence limits for intercepts (compare, for example, the confidence limits of slopes and intercepts for differences and rates in Figures 5.3b and 5.3c).

5.3 Temporal Scaling: Modes of Change

Three simulations will help to demonstrate how temporal scaling can be used to distinguish and recognize random, stationary, and directional time series. The examples given are deliberately simple, and the time series involved are so short that several lack the statistical power required to fully constrain their interpretation. The purpose of the examples is to illustrate random, stationary, and directional modes of change, and how they can be recognized.

5.3.1 Temporal Scaling of a Random Time Series

The step intervals, displacements, and rates for the seven-step random-walk time series of Figure 4.3a are listed in that figure. These are also the base intervals, displacements, and rates for the random walk involved. The random walk of Figure 4.3a is illustrated again in Figure 5.4a, where seven black lines connect eight successive samples and define the base intervals. The six gray lines in Figure 5.4a connect successive samples after skipping one, and these define six net intervals. Five lines (not shown) can be drawn that connect successive samples after skipping two, etc. Finally, one line can be drawn that connects the first and last of the samples, after skipping six, and this defines the single longest net interval.

The seven black lines for the seven step intervals in Figure 5.4a correspond to seven successive step differences. The differences are plotted as gray circles on the log difference–log interval or LDI graph in Figure 5.4b. The interval lengths for the independent variable time are plotted as $\log_{10}$ values on the horizontal axis. For a step interval of 1, the log interval for each is $\log_{10}(1) = 0$. The corresponding differences in the cumulative-displacement dependent variable are plotted as $\log_{10}$ absolute values on the vertical axis. Random walks are symmetrical in the sense that a negative displacement for a given interval is equivalent to a positive displacement for that interval. Each is an independent realization of the same process. Hence differences in a negative direction can be combined as absolute values with differences in a positive direction. The base differences in Figure 5.4a range in absolute value from 0.130 to 1.951 (see Figure 4.3a), yielding $\log_{10}$ values that range from -0.886 to 0.290.

The six gray lines for the six successive two-step intervals in Figure 5.4a correspond to six successive two-step differences. These differences are plotted as medium-gray circles on the LDI graph of Figure 5.4b. For a step interval of 2 the log interval for each is $\log_{10}(2) = 0.301$. Corresponding differences range in absolute value from 0.049 to 2.204, yielding $\log_{10}$ values ranging from -1.319 to 0.343. Log differences for the longer intervals and hence longer time scales of Figure 5.4a are plotted as lighter gray or open circles in Figure 5.4b.

The model in Figure 5.4a is a seven-step random walk with each step difference d_0 drawn from a normal distribution with mean $\mu = 0$ and standard deviation $\sigma = 1$ — which we know because the model is a simulation that we constructed. When step differences are drawn randomly from a normal distribution with these parameters, then the observed intercept step differences and step rates are expected to average the mean absolute deviation, MAD, where $MAD = 0.7979 \cdot \sigma$ (Eq. 4.7). Solving for σ, $\sigma = MAD / 0.7979 = 1.2533$. Observed MAD step differences and step rates (intercepts) for a random walk will generally underestimate the model step difference and rate σ by a factor of $\sqrt{(2/\pi)}$. Stating this differently, to recover the model value of σ, the observed MAD intercept for step differences or step rates must be multiplied by $1/\sqrt{(2/\pi)}$.

The slopes, intercepts, and associated confidence intervals in Figures 5.4b and 5.4c are robust weighted fits (RWF) with bootstrapped confidence intervals as

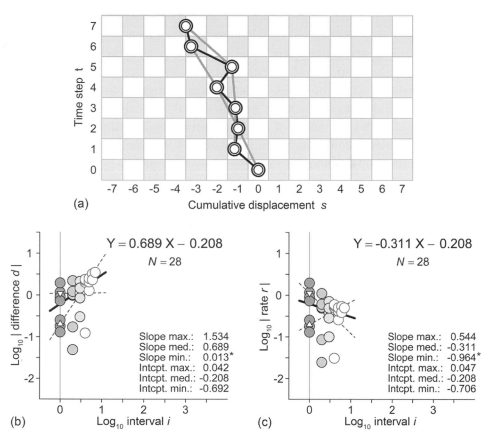

(a)

(b)   (c)

Figure 5.4 Differences and rates for a short simulated random walk. The independent variable, time, is on the vertical axis. (a) Time series of Figure 4.3a. Black lines connect eight states of the series, spanning seven intervals of step size 1 ($i_0 = 1$) yielding seven step differences d_0 and seven step rates r_0. Gray lines connect states spanning six successive intervals of step size 2 ($i_{0.30} = 2$) yielding six net differences $d_{0.30}$ and six net rates $r_{0.30}$. Overlapping lines for intervals spanning 3 through 7 steps are not shown. (b) Log difference versus log interval (LDI) plot for all differences and corresponding intervals separating all values of the time series in the first panel. Seven dark-gray circles represent differences corresponding to the seven black lines in the first panel. Six medium-gray circles represent differences corresponding to the six gray lines, etc. Solid line through the circles and dashed confidence limits for the slope were computed by robust weighted fit (RWF) and bootstrapping (see text). The slope of the scatter of differences (0.689) shows how these scale with time. Note that a 95% confidence interval for slopes includes 0.500 expected for temporal scaling of differences in a random walk. Asterisk indicates that the slope of 0.000 expected for a stationary time series is excluded. (c) Log rate versus log interval (LRI) plot for all rates and corresponding intervals separating all values of the time series in the first panel. Seven darker-gray circles represent rates corresponding to the seven black lines in the first panel. Six medium-gray circles are rates corresponding to the six gray lines, etc. Solid line through the circles and dashed confidence limits for the slope

described above. For the random walk in Figure 5.4a we expect the scatter of differences and intervals on an LDI plot to have a slope equal to the Hurst exponent of 0.500, and an intercept equal to the mean absolute deviation 0.798 (-0.098 on a $\log_{10}$ scale). The slope estimated in Figure 5.4b is 0.689 (solid line). A 95% confidence interval for the calculated slope ranges from 0.013 to 1.534 (dashed lines). This interval includes the value of 0.500 expected for a random walk, and it excludes the value of 0.000 expected for a stationary time series (asterisk). The intercept in Figure 5.4b is -0.208. A 95% confidence interval for the intercept ranges from -0.692 to 0.042 (open triangles in Figure 5.4b). This interval includes the expected value of -0.098.

We expect the scatter of rates and intervals on an LRI plot to have a slope equal to the complement of the Hurst exponent; that is, we expect a slope of -0.5. As before, we expect the LRI intercept to be equal to the mean absolute deviation, where $\log_{10} \sqrt{(2/\pi)} = -0.098$. The slope estimated in Figure 5.4c is -0.311 (solid line). A 95% confidence interval for the calculated slope ranges from -0.964 to 0.544 (dashed lines). This interval includes the expected value of -0.500, and it excludes the value of -1.000 expected for a stationary time series (asterisk). The intercept calculated in Figure 5.4c is -0.208. A 95% confidence interval for the intercept ranges from -0.706 to 0.047 (open triangles in Figure 5.4c), and this interval again includes the expected value of -0.098.

LDI and LRI temporal scaling observations in Figures 5.4b and 5.4c match expectation for a time series known to be a random walk. We can exclude the possibility that it is a stationary time series (asterisk), but we do not have the statistical power in this short time series to exclude that it is directional.

5.3.2 Temporal Scaling of a Stationary Time Series

A stationary time series is shown in Figure 5.5a. The series was constructed using the d_O values from Figures 4.3a and 5.4a, but this series is stationary because each d_O, whether positive or negative, is an individual and temporary displacement from the starting value. This is a special case of an Ornstein–Uhlenbeck process where $\theta = 0$ and $\alpha = 1$ (see Equation 4.6).

Seven black lines connect the eight successive samples and define seven step intervals. Lines connecting successive samples by skipping one (not shown) define

---

Figure 5.4 (*cont.*) were computed by RWF and bootstrapping (see text). The slope of the scatter of rates (-0.311) shows how these scale with time. Note that a 95% confidence interval for slopes includes -0.500 expected for temporal scaling of rates in a random walk. Asterisk indicates that the slope of -1.000 expected for a stationary time series is excluded.

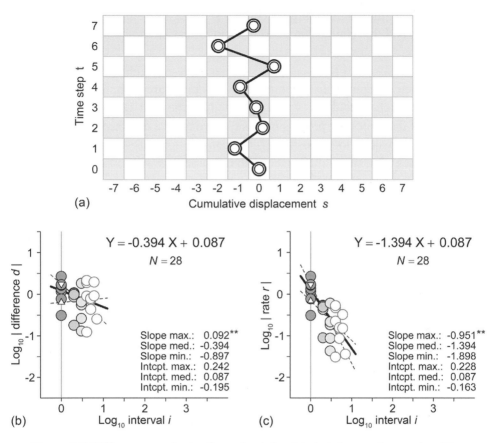

Figure 5.5 Differences and rates for a short simulated stationary time series. The independent variable, time, is on the vertical axis. (a) Time series with d_0 values from Figure 4.3a. The series here differs from that in Figures 4.3a and 5.2a because differences in successive steps are all displacements from the initial value of zero. Black lines connect eight states of the series, spanning seven intervals of step size 1 ($i_0 = 1$) yielding seven step differences d_0 and seven step rates r_0. Lines for intervals spanning 2–7 steps (not shown) contribute the remaining 21 net differences and rates. (b) LDI plot with slope, intercept, and confidence intervals computed like those in Figure 5.4b. The slope of the scatter of differences (−0.394) shows how these scale with time. Note that a 95% confidence interval for slopes includes 0.000 expected for temporal scaling of differences in a stationary series. Double asterisk indicates that the slopes of 0.500 and 1.000 expected for random and directional series are excluded. (c) LRI plot with slope, intercept, and confidence intervals computed like those in Figure 5.2c. The slope of the scatter of rates (−1.394) shows how these scale with time. Note that a 95% confidence interval for slopes includes the slope of −1.000 expected for the scaling of rates in a stationary time series. Double asterisk indicates that the slopes of 0.500 and 0.000 expected for random and directional series are excluded.

six net intervals, lines connecting successive samples by skipping two define five net intervals, and this can be extended to a single net interval connecting the first and last of the samples.

The seven black lines for the seven step intervals in Figure 5.5a correspond to successive step differences in the time series. As in Figure 5.4b, these and each of the 21 net differences and intervals are plotted on the LDI graph in Figure 5.5b. Symmetry of the stationary process means that a negative displacement for a given interval is equivalent to a positive displacement for that interval. Hence the absolute values of negative differences can be combined with positive differences. The base differences in Figure 5.5a range in absolute value from 0.130 to 1.951 (see Figure 4.3a), yielding $\log_{10}$ values that range from -0.886 to 0.290.

The slopes, intercepts, and associated confidence intervals in Figures 5.5b and 5.5c are robust weighted fits (RWF) with bootstrapped confidence intervals as described above. For a stationary time series we expect the scatter of differences and intervals on an LDI plot to have a slope equal to 0.000, and an intercept equal to the standard deviation, which is 1.000 here or 0.000 on a $\log_{10}$ scale. The slope estimated in Figure 5.5b is -0.394 (solid line). A 95% confidence interval for the calculated slope ranges from -0.897 to 0.092 (dashed lines) including the expected value of 0. The intercept calculated in Figure 5.5b is 0.087. A 95% confidence interval for the intercept ranges from -0.195 to 0.242 (open triangles in Figure 5.5b), including the expected value of 0.000.

For a stationary time series we expect the scatter of rates and intervals on an LRI plot to have a slope equal to -1.000. As before, we expect the LRI intercept to be equal to the LDI intercept, which is $\sigma = 1.000$ or, on a $\log_{10}$ scale, 0.000. The slope estimated in Figure 5.5c is -1.394 (solid line). A 95% confidence interval for the calculated slope ranges from -1.898 to -0.951 (dashed lines), and this interval includes the expected value of -1.000. The intercept calculated in Figure 5.5c is 0.087. A 95% confidence interval for the intercept ranges from -0.163 to 0.228 (open triangles in Figure 5.5c), and this interval again includes the expected value of 0.000.

LDI and LRI temporal scaling observations in Figures 5.5b and 5.5c match expectation for a time series known to be stationary, and here we have enough statistical power to exclude alternative interpretations of the time series as random or directional (double asterisks).

5.3.3 Temporal Scaling of a Directional Time Series

A directional time series is shown in Figure 5.6a. The series was constructed using the d_O values from Figures 4.3a and 5.4a, but this series is directional because signs of the successive d_O are no longer random but are constrained to be positive. Seven

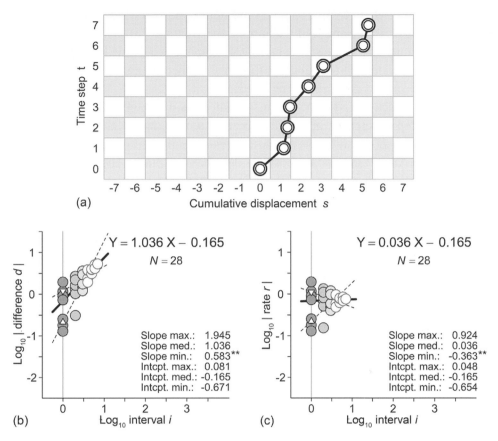

(a)

(b)

(c)

Figure 5.6 Differences and rates for a short simulated directional trend. The independent variable, time, is on the vertical axis. (a) Time series with d_0 values from Figure 4.3a. The series here differs from that in Figures 4.3a and 5.4a because signs of the successive d_0 values are not random but are constrained to be positive. Black lines connect eight successive states of the series, spanning seven successive intervals of step size 1 ($i_0 = 1$) yielding seven step differences d_0 and seven step rates r_0. Lines for intervals spanning 2–7 steps (not shown) contribute the remaining 21 net differences and rates. (b) LDI plot with slope, intercept, and confidence intervals computed like those in Figure 5.4b. The slope of the scatter of differences (1.036) shows how these scale with time. Note that a 95% confidence interval includes the slope of 1.000 expected for differences in a stationary series. Double asterisk indicates that the slopes of 0.000 and 0.500 expected for stationary and random series are excluded. (c) LRI plot with slope, intercept, and confidence intervals computed like those in Figure 5.4c. The slope of the scatter of rates (0.036) shows how these scale with time. Note that a 95% confidence interval includes the slope of 0.000 expected for rates in a stationary series. Double asterisk indicates that the slopes of -1.000 and -0.500 expected for stationary and random series are excluded.

black lines connect successive samples and define the seven step intervals. Lines connecting successive samples by skipping one (not shown) define six net intervals, lines connecting successive samples by skipping two define five net intervals, and this can be extended to a single net interval connecting the first and last of the samples.

The seven black lines for the seven step intervals in Figure 5.6a correspond to successive step differences in the time series. As in Figures 5.4b and 5.5b, these and each of the 21 net differences and intervals are plotted on the LDI graph in Figure 5.6b. All differences are positive. The base displacements in Figure 5.6a range in absolute value from 0.130 to 1.951 (see Figure 4.3a), yielding $\log_{10}$ values that range from -0.886 to 0.290.

The slopes, intercepts, and associated confidence intervals in Figures 5.6b and 5.6c are robust weighted fits (RWF) with bootstrapped confidence intervals as described above. For a directional time series we expect the scatter of differences and intervals on an LDI plot to have a slope equal to 1.000, and an intercept equal to the mean absolute deviation. *MAD* here is 0.798, or -0.098 on a $\log_{10}$ scale. The slope estimated in Figure 5.6b is 1.036 (solid line). A 95% confidence interval for the calculated slope ranges from 0.583 to 1.945 (dashed lines), and this interval includes the expected value of 1.000. The intercept calculated in Figure 5.6b is -0.165. A 95% confidence interval for the intercept ranges from -0.671 to 0.081 (open triangles), and this interval includes the expected value of -0.098.

For a directional time series we expect the scatter of rates and intervals on an LRI plot to have a slope equal to 0.000. As before, we expect the LRI intercept to be equal to the mean absolute deviation, which is 0.798 or -0.098 on a $\log_{10}$ scale. The slope estimated in Figure 5.6c is 0.036 (solid line). A 95% confidence interval for the calculated slope ranges from -0.363 to 0.924 (dashed lines), and this interval includes the expected value of 0.000. The intercept calculated in Figure 5.6c is -0.165. A 95% confidence interval for the intercept ranges from -0.654 to 0.048 (open triangles), and this interval again includes the expected value of -0.098.

LDI and LRI temporal scaling observations in Figure 5.6b and 5.6c match expectation for a time series known to be directional, and here again we have enough statistical power to exclude alternative interpretations of the time series as stationary or random (double asterisks).

5.3.4 Likelihood Interpretation of Temporal Scaling Slopes

We have seen that random, stationary, and directional time series differ in their LDI and LRI temporal scaling slopes. If we organize examples in the preceding sections in terms of increasing slope, the estimated LDI slopes are -0.394, 0.689,

and 1.036 for stationary, random, and directional time series, respectively, compared to expected LDI slopes of 0.000, 0.500, and 1.000 for these modes. Similarly, the estimated LRI slopes are -1.394, -0.311, and 0.036 for stationary,
random, and directional time series, respectively, compared to expected LRI slopes
of -1.000, -0.500, and 0.000 for these modes. Corresponding LDI and LRI
slopes differ by 1 because the underlying time series differences and time series
rates are complementary. Thus LDI and LRI plots are equally useful for recovering
temporal scaling slopes.

It is common to assign, or attempt to assign, a time series to a particular mode of
change: stationary, random, or directional. This is straightforward when confidence
limits include the slope expected for one hypothesis and exclude the slopes
expected for the other two, but this is not always the case. Sometimes, when
statistical power is small, no mode is excluded, and sometimes, when statistical
power is great, all three modes are excluded. It is still possible to compare modes
as hypotheses by comparing the probabilities or relative frequencies of the data at
hand corresponding to each hypothesis — which is a comparison in terms of
relative likelihood (Edwards, 1972; 1992).

Bootstrap frequency distributions for the slopes of differences and rates for each
of the random, stationary, and directional time series of Figures 5.4 through 5.6 are
shown in Figure 5.7. The frequencies are proportions of a bootstrapped sample of
1,000, meaning that a frequency of 64, for example, corresponds to a probability of
0.064. The likelihood of any hypothesis for a given set of data (observations or
simulation results) is proportional to the probability of the data given the hypothesis. The likelihood of a second hypothesis for the same set of data is similarly
proportional to the probability of the data given the second hypothesis. The
strategy of relative likelihood involves comparison of different hypotheses for
the same data because then the ratio of any two likelihoods is just the ratio of
their frequencies or probabilities, and the unknown constant of proportionality
drops out.

Note that calculations of relative likelihood and support depend on the same
probability distribution used to specify critical values in significance testing, and,
as a rule, likelihood support can be calculated in any situation permitting significance testing. For example, in Figure 5.7a, frequency associated with an LDI slope
of 0.0 for differences in the random time series of Figure 5.4 is 1, and the
probability is $1/1000 = 0.001$. The frequency for a slope of 0.5 in the time series
is 57 and the probability is $57/1000 = 0.057$, and the frequency for a slope of 1.0 is
18 and the probability is $18/1000 = 0.018$. If we work with the frequencies and
calculate their six possible ratios (three greater than 1 and three less than 1), we see
that the likelihood of the time series in Figure 5.4 being random based on
differences is $57/1 = 57.00$ times greater than the likelihood of it being stationary.

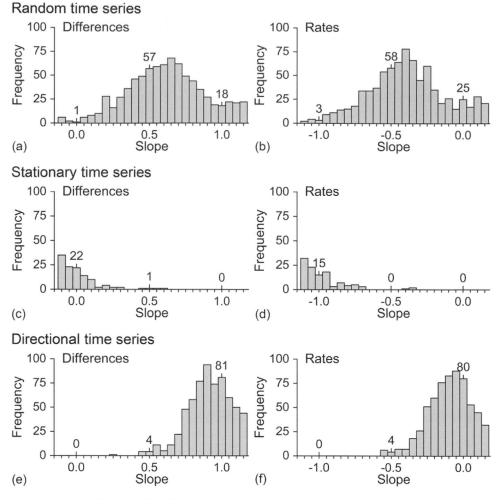

Figure 5.7 Relative likelihood approach to interpretation of the simulated time series in Figures 5.4, 5.5, and 5.6. Frequency associated with each temporal scaling slope represents the probability of that slope value (area of 1,000 units under each curve is the number of bootstrapped slopes). Slope values of 0.0, 0.5, and 1.0 for differences, or −1.0, −0.5, and 0.0 for rates, are associated with hypotheses of stationary, random, and directional change, respectively. The frequency associated with each slope value (e.g., 1, 57, and 18) yields a per-mil probability (e.g., 0.001, 0.057, 0.018), and ratios of the probabilities provide estimates of relative likelihood for comparison of the hypotheses. (a) The random time series simulated in Figure 5.4 is 0.057/0.018 = 3.17 times more likely to be random than directional, and 0.057/0.001 = 57 times more likely to be random than stationary. (b) Independently bootstrapped distribution for rates gives corresponding relative likelihoods of 0.058/0.025 = 2.32 and 0.058/0.003 = 19.33. (c) The stationary time series simulated in Figure 5.5 is 0.022/0.001 = 22 times more likely to be stationary than random, and incalculably (0.022/0.000) more likely to be stationary than directional. (d) Independently bootstrapped distribution

The likelihood of the time series being directional is 18/1 = 18.00 times greater than the likelihood of it being stationary. Finally, the likelihood of the time series being random is 57/18 = 3.17 times greater than the likelihood of it being directional. We cannot rule out any of the canonical hypotheses, stationary, random, or directional, in this way, but the hypothesis of random change has a greater relative likelihood than any other. Thus random is the best supported of the three possible modes.

The bin in Figure 5.7a that has the greatest frequency is not the bin corresponding to a slope of 0.5 expected for a random time series, but this is rather the bin corresponding to a slope of 0.65 lying between random and directional. The latter has an associated frequency of 68, meaning that an interpretation on the directional side of random has the maximum likelihood. This is consistent with our results for hypothesis testing in Figure 5.4b, where the slope of 0.000 expected for a stationary time series fell outside our calculated confidence interval.

The probabilities for slopes based on rates in Figure 5.7b yield relative likelihoods similar to those based on differences in Figure 5.7a. They differ slightly because the probability distribution in Figure 5.7b is based on an independent bootstrap analysis of LRI slopes for the Figure 5.4 time series. Here too the single best canonical hypothesis is that the time series is random. In this case the bin of greatest frequency is the bin corresponding to a slope of −0.40, and the interpretation with maximum likelihood is again on the directional side of random. This too is consistent with our results for hypothesis testing in Figure 5.4c, where the slope of −1.000 expected for a stationary time series fell outside our calculated confidence interval.

We can make similar analyses of relative likelihoods for the Figure 5.5 simulation of a stationary time series, based on the probability distributions in Figures 5.7c and 5.7d. In Figure 5.7c the greatest calculable relative likelihood for the canonical modes is 22/1 = 22.00 for stasis over random change. In Figure 5.7d the greatest relative likelihood is 15/0 for stationary over both random and directional change — which is incalculable. Relative likelihood yields the same interpretation as hypothesis testing. Maximum likelihood favors slopes on

---

Figure 5.7 (*cont.*) for rates gives incalculable relative likelihoods of 0.015/0.000 and 0.015/0.000. (e) The directional time series simulated in Figure 5.6 is 0.081/0.004 = 20.25 times more likely to be directional than random, and incalculably (0.081/0.000) more likely to be directional than stationary. (f) Independently bootstrapped distribution for rates gives corresponding relative likelihoods of 0.080/0.004 = 20 and incalculable (0.080/0.000). Maximum likelihoods lie between random and directional in panels a and b and e and f, and more negative than any of the graphed slopes in panels c and d.

the stationary side of random for both differences and rates due to and reflecting the small number of samples involved.

Likelihood yields another clear result for the directional time series simulated in Figure 5.6. The associated probability distributions are illustrated in Figures 5.7e and 5.7f. In Figure 5.7e the likelihood of the Figure 5.6 time series being directional is 81/4 = 20.25 times greater than the likelihood of it being random, and it is 81/0 or incalculably greater than the likelihood of it being stationary. The likelihood of the time series in Figure 5.7f being directional is 80/4 = 20 times greater than the likelihood of it being random, and again incalculably greater than the likelihood of it being stationary. Likelihood interpretation as directional is consistent with the confidence intervals for slopes calculated in Figures 5.6b and 5.6c, where both stationary and random hypotheses were ruled out.

There is value in analysis of time series from controlled simulations, as in Figures 5.4 through 5.6 because we know the mode of the process being studied in advance. The same can be said for artificial selection experiments on plants and animals in a laboratory setting. Nature is different because: (1) the underlying mode or modes of change are not known, (2) modes are nonexclusive and may compete simultaneously, and (3) there is added uncertainty because modes change over time.

5.4 Temporal Scaling: Rates of Change

The slopes of temporal scaling plots tell us about modes of change, as we have just seen. Now we investigate the intercepts that result from temporal scaling and tell us about rates of change.

Our previous simulations were deliberately simple, involving few steps. To investigate rates of change it is helpful to consider longer simulations providing greater statistical power. A 1,600-step random walk is illustrated in black in Figure 5.8a, superimposed on a background of a thousand 1,600-step random walks. Each of the thousand random walks in the background, shown in gray, has the same rate parameters as the time series shown in black. The displacement or difference at each step is a random deviate drawn from a normal distribution $n(0, 1)$. Centering the distribution at zero means we can expect one half of the rates to be positive and one half to be negative, but the signs themselves are determined by chance.

The background sample of 1,000 random walks in Figure 5.8a enables us to visualize some important statistical properties of random walks. First, the variance σ^2 increases in proportion to the number of time steps t (open circles in Figure 5.8a), which for $\sigma = 1$ is the long-term statistical limit for all random walks. Second, the standard deviation σ increases as $\sqrt{t}$ (open squares in

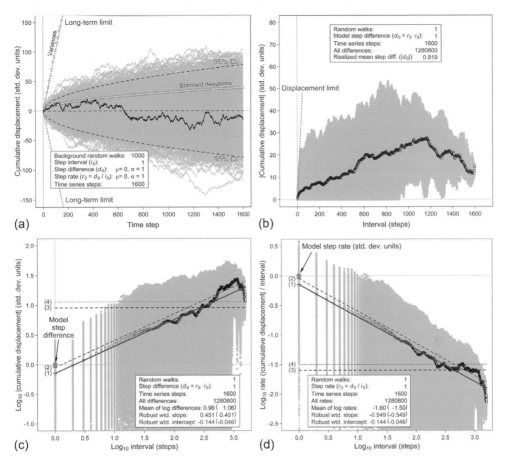

Figure 5.8 Full sampling of the differences and rates of a simulated random walk to recover its mode and tempo. (a) Random walk of 1,600 steps (solid black line) with successive step differences d_0 drawn from a normal distribution with $\mu = 0$ and $\sigma = 1$. This is superimposed on the paths of 1,000 random walks with the same parameters (gray lines). Open circles trace the variances and open squares the standard deviations for successive steps. Parabolic dashed lines enclose a 95% confidence envelope for random walk positions and the Brownian diffusion associated with these. (b) 1,600 one-step differences + 1599 two-step differences + ... + one 1,600-step difference = 1,280,800 cumulative differences for the random walk in the previous panel. Each difference is plotted as an absolute value against its interval length. Open squares show median differences for each interval length. (c) Cumulative differences and intervals in the previous panel plotted on proportional (logarithmic) scales. The slope of the robust weighted fit (RWF) to the full sample of points is 0.451 – which is close to 0.500 expected for a random walk. The overall mean difference for all intervals is $10^{0.96} = 9.12$ standard deviation units (intercept labeled '3'), greatly overestimating the intercept mean difference for one-step intervals of $10^{-0.144} = 0.718$ standard deviation units (intercept '1'; or $10^{-0.046} = 0.899$ for intercept '2,' reflecting addition of 0.098 on a $\log_{10}$ scale or division by $\sqrt{(2/\pi)}$ to convert a mean deviation in the

Figure 5.8a). And third, $\pm 1.96 \cdot \sigma \cdot \sqrt{t}$ defines 95% confidence limits (parabolic dashed lines in Figure 5.8a) within which we expect to find 95% of all of the random walks. This 95% confidence interval is the range within which Brownian diffusion is reasonably probable.

If we focus on the single random walk illustrated in black in Figure 5.8a, there are 1,601 successive samples starting at $t_0 = 0$, and we can calculate 1,600 displacements or differences from one sample to the next. The displacements all have an interval length of one step. By chance, some will be positive and some will be negative. These are plotted as absolute values against their interval length on the displacement versus interval graph of Figure 5.8b. Continuing as we did in Figure 5.4, we can calculate 1,599 two-step displacements or differences from time steps zero to two, one to three, etc. that we can continue until we calculate a single 1,600-step difference from time step zero to 1,600. The total number of differences, 1,280,800, is given by Equation 5.3. All are plotted as absolute values against their interval length on the displacement versus interval graph of Figure 5.8b. The displacement limit shown by a dashed line corresponds to the long-term limit shown in Figure 5.8a. Open squares in black trace the median displacements.

The $\log_{10}$ values of the cumulative displacements and intervals in Figure 5.8b are shown on an LDI plot in Figure 5.8c, and the corresponding $\log_{10}$ rates are shown on an LRI plot in Figure 5.8d. Median displacements and rates for each interval are shown by open squares. RWF lines fit to the LDI and LRI scatters have slopes of 0.451 and -0.549, respectively, which are close to the slopes of 0.500 and -0.500 expected for random walks. The slopes for the LDI and LRI plots also differ from each other by the expected value of 1.000.

Here we are less interested in the slopes than in the intercepts. The intercepts on the LDI and LRI plots are both -0.144, showing, as expected, that both yield the same information. However, the realized value of the estimates for each intercept is clearly less than the model standard deviation of $\sigma = 1$, for which $\log_{10} \sigma = 0.000$.

Figure 5.8 (*cont.*) result to the standard deviation of the model). (d) Rates and intervals for differences in the previous panel plotted on proportional (logarithmic) scales. The slope of the robust weighted fit (RWF) to the full sample of points is -0.549, which is close to -0.500 expected for a random walk. The overall mean rate for all intervals of $10^{-1.60} = 0.025$ standard deviation units (intercept '3') greatly underestimates the tempo given by the intercept mean rate for one-step intervals of $10^{-0.144} = 0.718$ standard deviation units (intercept '1'; or $10^{-0.046} = 0.899$ for intercept '2,' reflecting addition of 0.098 on a $\log_{10}$ scale or division by $\sqrt{(2/\pi)}$ to convert a mean absolute deviation in the result to the standard deviation of the model). Three-dimensional representations of panels c and d are shown in Figure 5.9.

What is being estimated is the mean absolute deviation rather than the standard deviation. Remembering from Equation 4.7 that $MAD = \sqrt{(2/\pi)} \cdot \sigma$; then $\sigma = MAD / \sqrt{(2/\pi)}$, $\log_{10} \sigma = \log_{10} MAD - \log_{10} \sqrt{(2/\pi)}$, and $\log_{10} \sigma = \log_{10} MAD + 0.098$. Our intercept of -0.144 thus corresponds to an estimate for the model σ of -0.046. The estimate is still slightly less than $\log_{10} \sigma = 0$, but it is closer to the model.

The distribution of differences in Figure 5.8c and the distribution of rates in Figure 5.8d are easy to visualize as they are being drawn on a computer screen, but these are more difficult to visualize as static maps. Figure 5.8c is redrawn as a three-dimensional frequency plot in Figure 5.9a, and Figure 5.8d is redrawn as a three-dimensional frequency plot in Figure 5.9b. $\log_{10}$ interval length is the independent variable on the flat maps in Figures 5.8c and 5.8d, and on the three-dimensional frequency representations in Figures 5.9a and 5.9b. $\log_{10}$ cumulative displacement or difference and $\log_{10}$ rate are the corresponding dependent variables. The most salient characteristic of each difference distribution and each rate distribution is its skewness, with a short tail on the side of larger differences and higher rates, and a long tail on the side of smaller differences and lower rates. The distributions are also heteroscedastic. Distributions of differences and distributions of rates associated with shorter interval lengths are both larger and broader, with greater variance, and those associated with longer interval lengths are both smaller and narrower, and have lesser variance.

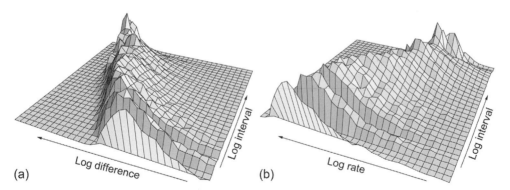

(a)

(b)

Figure 5.9 Three-dimensional representations of the LDI and LRI distributions analyzed in Figures 5.8c and 5.8d. (a) Distribution of $\log_{10}$ differences in relation to $\log_{10}$ interval lengths. (b) Distribution of $\log_{10}$ rates in relation to $\log_{10}$ interval lengths. Differences and rates depend on the interval over which each is measured or calculated. Vertical axis is frequency, and foreground peaks are at 184 differences or rates, respectively, in each of the two plots. Note the steep shoulder on the large-difference and high-rate sides of the distributions, and the long tails on the small-difference and low-rate sides of each. The distributions flatten to a single difference or rate for the longest interval sampled.

5.5 Fractal Geometry of Evolutionary Time Series

The Ornstein–Uhlenbeck equation, Equation 4.6 in Chapter 4, provides a general model for a stationary time series (when $\alpha = 1$ and θ is the intercept), and for a random walk (when $\alpha = 0$ and θ drops out). The Ornstein–Uhlenbeck equation will also generate a directional time series when $\alpha = 1$ and θ is proportional to the step number or time. These are the canonical modes of change with time: stationary, random, and directional.

We have seen that the temporal scaling of differences in a time series can range from 0.0 to 0.5 to 1.0, and these nodal values correspond to stationary, random, and directional modes of change. Similarly, we have seen that the temporal scaling of rates in a time series can range from -1.0 to -0.5 to 0.0, and these too correspond to stationary, random, and directional modes. Stationary, random, and directional time series each have a characteristic geometry that can be related to their temporal scaling slopes. The geometries are illustrated in Figure 5.10.

A stationary time series is one of great "roughness," where the lines representing successive steps alternate so frequently that at some scale they fill a two-dimensional Euclidean area. A directional time series, on the other hand, is "smooth." It is not necessarily straight, but it is often nearly so. A directional time series comes closest to our normal concept of a one-dimensional Euclidean line. A random time series is neither rough like a stationary series nor smooth like a directional series but something in between. It can be described as having a fractional or "fractal" dimension D that is halfway between the dimension of an area and the dimension of a line, that is $D = 1.5$.

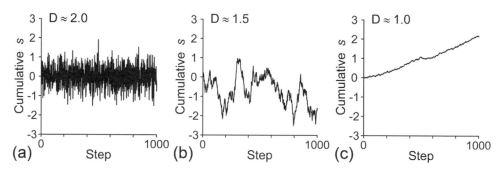

Figure 5.10 Fractal dimension D for (a) a stationary time series; (b) a random time series; and (c) a directional time series. The stationary time series in (a) is "rough" and has lines sufficiently packed to fill a two-dimensional area ($D \approx 2.0$). The directional time series is "smooth" and corresponds to our conventional concept of a one-dimensional line ($D \approx 1.0$). The random time series is halfway between a line and a plane in a proportional sense and has a "fractional" or fractal dimension $D \approx 1.5$.

Hurst (1951) recognized that the differences in a random time series have a temporal scaling slope of 0.5, and Mandelbrot (1967; 1983) introduced the idea of fractional dimensions and recognized that random time series have a fractional dimension $D = 1.5$. Notice that there is a simple relationship between the slopes $m_d = 0.0$, 0.5, and 1.0 for the temporal scaling of differences in stationary, random, and directional time series, and the corresponding D values of 2.0, 1.5, and 1.0. The sum $D + m_d = 2.0$, meaning that $D = 2 - m_d$ and $m_d = 2 - D$.

Similarly, there is a simple relationship between the slopes $m_r = -1.0$, -0.5, and 0.0 for the temporal scaling of rates in stationary, random, and directional time series, and the corresponding D values of 2.0, 1.5, and 1.0. Here the sum $D + m_r = 1.0$, meaning that $D = 1 - m_r$ and $m_r = 1 - D$. We can take this a step farther by substituting $D = 1 - m_r$ into $D = 2 - m_d$. This tells us that $1 - m_r = 2 - m_d$, and finally $m_d - m_r = 1$, $m_d = m_r + 1$ and $m_r = m_d - 1$. We have seen these relationships many times earlier in the chapter when comparing temporal scaling slopes for differences and rates of the same time series.

5.6 Cross-Sectional Rates

The time series examined above, concerning automobile production, industrial stock prices, and simulated random, stationary, and directional time series, have all been longitudinal. That is, all involved sequential or successive samples of the same process analyzed in relation to time. Temporal scaling is also useful for comparing cross-sectional samples of a process on different scales of time. Here all the intervals, differences, and rates are net intervals, differences, and rates.

We can end this chapter by posing a cross-sectional question. Who is the faster runner, a person running 10 meters per second for a distance of 100 meters, or a person running 8 meters per second for a distance of 800 meters? The first is a rate on a timescale of 10 seconds, and the second is a rate on a timescale of 100 seconds. Olympic records for men and women running sprints, dashes, and longer distances are listed in Table 5.1. The records are shown graphically as points in Figure 5.11a. Time is on the *x*-axis, and the average rate a person ran to achieve a given record is the slope of a line connecting the point to the origin. The points are individual records that do not form a sequential or longitudinal time series. No runner ran at the record-setting rate of a shorter race to set the record for a longer race. These are cross-sectional data that nevertheless tell us about running and human performance.

Everyone knows from experience that a person can run faster for a short distance than for a long distance, so no one will be surprised that rates for Olympic-record running performances decrease with distance as shown in the LRI plot in Figure 5.11b. The decreases in rate are proportional to distance, so the records shown decrease linearly when plotted on logarithmic axes.

Table 5.1 *Olympic outdoor running records for men and women in 2016*

Distance (meters)	Men's time (seconds)	Women's time (seconds)	Men's rate (m/s)	Women's rate (m/s)
100	9.63	10.62	10.38	9.42
200	19.30	21.34	10.36	9.37
400	43.03	48.25	9.30	8.29
800	100.91	113.43	7.93	7.05
1,500	212.07	233.96	7.07	6.41
5,000	777.82	866.17	6.43	5.77
10,000	1,621.17	1,757.45	6.17	5.69

Source: www.iaaf.org/records/by-category/olympic-games-records

We can ask for running, as we did previously for other processes, whether what we see is stationary, random, or directional. To start, if successive records in Figure 5.11a were time series, their combined trajectory would be relatively straight and smooth as expected for a directional process. We can confirm directionality in the cross-sectional records by temporal scaling on an LRI plot, with running rates plotted against intervals of time in their denominators. The temporal scaling slopes for men and women calculated in Figure 5.11c are −0.113 and −0.112, respectively, which are both close to 0.000 expected for rates of a directional process. If longitudinal records were available for successive stages in the 10-kilometer record run, we would expect these to have a similar temporal scaling slope. This slope would probably be even closer to 0.000 because disciplined pacing would make progress in running more like the progress in production of automobiles shown in Figure 5.2.

Running in a race is normally directional, but we can also image a race that might yield a random or even stationary mode of change. What if a runner in a long race became dehydrated, overheated, and confused, continuing only in fits and starts? What if a runner fell or pulled up lame and could not finish the race? These are not expected, to be sure, but either could happen. In such cases, analyses of rates on different scales of time would be helpful in quantifying and recognizing the unexpectedly random or stationary modes of change.

The intercepts of lines fit to the Olympic records for men and women in Figure 5.11c are 1.140 and 1.097, respectively, which correspond, following exponentiation, to rates of $10^{1.140} = 13.8$ meters per second and $10^{1.097} = 12.5$ meters per second, both on a time scale of 1 second. Intercepts bring us back to the question posed above: Who is the faster runner, the person running 10 meters per second for a distance of 100 meters, or the person running 8 meters per second for a distance of 800 meters?

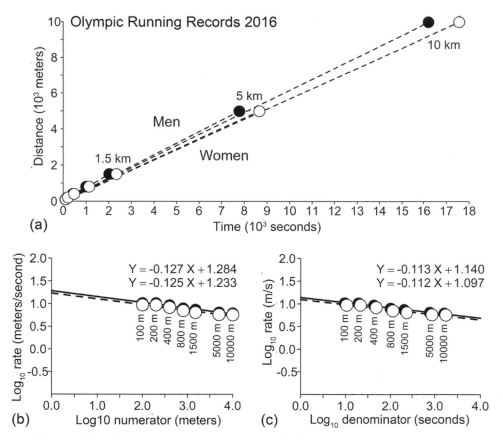

(a)

(b)

(c)

Figure 5.11 Olympic outdoor running records for sprints, dashes, and longer distances (Table 5.3). These are cross-sectional records rather than longitudinal time series. (a) Distances run as a function of time for Olympic men (filled circles) and women (open circles). The average rate for the distance run is given by the slope of a line connecting the record to the origin. (b) LRI plot showing how rates for men and women decrease with the numerator or distance run. (c) LRI plot showing how rates for men and women decrease with the denominator or time run. Lines fit to the rates for men and women in panel c have slopes of −0.113 and −0.112, respectively, which are close to the LRI slope of 0.000 expected for a directional process. Source for Olympic records: International Association of Athletics Federations.

One way to compare the rates is to calculate the intercept expected for each of the two competitors (Fig. 5.12). The intercept for the person running 10 meters per second for a distance of 100 meters will be $y_{100} = \log_{10}(100/10) - -0.113 \cdot \log_{10}(10)$, which is $1 + 0.113 = 1.113$. The intercept for the person running 8 meters per second for a distance of 800 meters will be $y_{800} = \log_{10}(800/100) - -0.113 \cdot \log_{10}(100)$ which is $0.903 + 0.226 = 1.129$. Temporal scaling in relation to Olympic records indicates that the intercept for the person running 800 meters is higher,

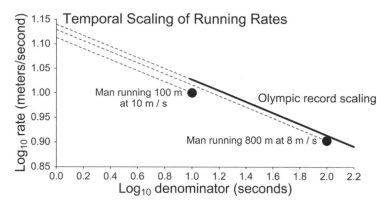

Figure 5.12 Running rates compared in the context of the temporal scaling of Olympic records shown in Figure 5.11c. Note that a person running 800 meters at a rate of 8 meters per second appears faster on all scales of time than a person running 100 meters at 10 meters per second. Rates over different distances on different scales of time cannot be compared directly.

meaning he or she appears to be faster. The comparison could be made on any common scale of time, and the person running 8 meters per second for 800 meters will appear to have the higher rate.

Differences and rates can be compared on many scales of time, but valid comparison requires that they be compared on the same scale of time. The only way to be certain about the faster runner would be to ask the competitors to run in a race of the same distance, or ask them to run for the same amount of time. Then of course one runner might be faster for one distance or time, and the other might be faster for a different distance or time.

Rates like these must have the same numerator or the same denominator if they are to be compared, and that numerator or denominator is integral to the comparison.

5.7 Summary

1. Step rates were introduced in the previous chapter. Base intervals, differences, and rates represent sequential nonoverlapping sampling of a time series where again differences and rates are statistically independent of each other but not independent of constituent step rates.
2. Net intervals, differences, and rates are those that sample differences and rates for multiple and often overlapping step or base intervals. Net differences and rates are statistically dependent on constituent step and base differences and rates, and often dependent on other net differences and rates.

3. Temporal scaling is a method for analysis of: (1) the quantitative dependence of *differences* between samples in a time series on the amount of time separating the samples and (2) the quantitative dependence of *rates* of change in a time series on the amount of time separating the samples (the denominator of the rate).

4. Robust linear modeling provides efficient recovery of temporal scaling slopes and intercepts.

5. Temporal scaling of differences and temporal scaling of rates are complementary and interchangeable in the sense that they recover complementary slopes and lead to the same inference of time-series *mode* The two are interchangeable too in the sense that they recover the same step-rate intercept reflecting time-series *tempo*.

6. Random time series have differences that scale with a slope at or near 0.500 on a log difference versus log interval or LDI plot. The corresponding rates scale with a slope at or near -0.500 on a log rate versus log interval or LRI plot.

7. Stationary time series have differences that scale with a slope at or near 0.000 on an LDI plot. The corresponding rates scale with a slope at or near -1.000 on an LRI plot.

8. Directional time series have differences that scale with a slope at or near 1.000 on an LDI plot. The corresponding rates scale with a slope at or near 0.000 on an LRI plot.

9. Differences in a stationary time series can be independent of their associated intervals, and rates in a directional time series can be independent of their associated intervals, but differences and rates cannot both be independent of interval length.

10. A stationary time series is one of great "roughness," filling an area with fractal dimension $D \approx 2.0$. A random time series has intermediate roughness and fractal dimension $D \approx 1.5$. A directional time series is relatively smooth and linear, with fractal dimension $D \approx 1.0$.

11. Cross-sectional information about differences and rates can be analyzed using LDI and LRI plots like those used for longitudinal time series. In this case all intervals, differences, and rates are net intervals, differences, and rates.

12. Rates must have the same numerator or the same denominator if they are to be compared, and that numerator or denominator is integral to the comparison.

6

Directional Selection, Stabilizing Selection, and Random Drift

In previous chapters we learned that biological populations are variable, and many of the phenotypic characteristics we see in a population of organisms are normally (log normally) distributed. This normality is both a manifestation and an indication of the additive nature of variation determined by many genes. Normal curves have two parameters: one, mean μ, specifies the location of the curve, and the other, standard deviation σ, indicates the dispersion of the population forming the curve. Differences we see in organisms separated by hundreds, thousands, and millions of years represent an accumulation of changes on a generational timescale. Here we consider how artificial selection in the laboratory and natural selection in the field produce evolutionary change through time.

6.1 Directional Selection

Directional selection can take many forms involving simple truncation or more complex selection gradients.

6.1.1 Truncation Selection in Theory

Truncation selection is as old as plant and animal breeding, and it undoubtedly occurs in nature as well. It is widely known to be the most efficient form of directional selection (Crow and Kimura, 1979). Truncation is efficient in the sense that it yields the greatest change in an average phenotype for a given cost in selective mortality. "Selective mortality" does not necessarily refer to the death of individuals but to loss of their genetic contribution in subsequent generations for a full spectrum of possible reasons that may of course include the death of individuals.

The theory is straightforward, involving a normally distributed parental population, with some proportion of the population selected to establish the next

110

generation. Sewall Wright (1969) called the selected proportion p. The selected portion can lie to the right or left of a truncation or abscission point t, where t is measured as a negative or positive distance from the population mean μ. If we substitute $x = t/\sigma$ for t, then the resulting truncation point x is expressed in standard deviation units. The equation for a standard normal curve yields the ordinate z for any x (Pearson and Hartley, 1966):

$$z = \frac{1}{\sqrt{2\pi}} e^{-0.5x^2} \tag{6.1}$$

Following Wright, the proportion p lying to the right of x is given by:

$$p = \frac{1}{\sqrt{2\pi}} \int_{t/\sigma}^{\infty} e^{-0.5x^2} dx \tag{6.2}$$

If z is the ordinate at truncation point x, and h^2 is the heritability, then the expected response is a change in the mean from μ to $\mu + \Delta\mu$, where:

$$\Delta\mu = h^2 \cdot \sigma \cdot (z/p) \tag{6.3}$$

The corresponding selection intensity is:

$$\Delta\mu/h^2 = \sigma \cdot (z/p) \tag{6.4}$$

Heritability h^2 lies in the range $0 \leq h^2 \leq 1$, but h^2 is rarely known with any certainty. Thus the relationship of the portion selected, p, and the response to selection, $\Delta\mu$, is rarely so clearly determined as Equations 6.3 and 6.4 might seem to imply. We are primarily interested in the response to selection, $\Delta\mu$, in any generation, which is then the step rate of change for the generation.

Lande (1976) gives $\Delta\mu$ in his equation 11 for "weak truncation selection," where weak selection requires the truncation point x to be so far in one tail or the other of the normal curve that the selected proportion p is effectively $p = 1$. Adding p, the Lande equation becomes (in Wright's notation):

$$\Delta\mu = \pm \frac{h^2 \cdot \sigma}{p \cdot \sqrt{2\pi}} e^{-0.5x^2} \tag{6.5}$$

where the sign of the response is opposite the sign of x. The normal curve is bilaterally symmetrical. Thus negative selection, retaining phenotypes to the left of the truncation point and removing those to the right, is the mirror image of positive selection.

The geometry of truncation for an example of positive selection is illustrated in Figure 6.1a, which is drawn to scale for a truncation point $x = -1.282$ (chosen as

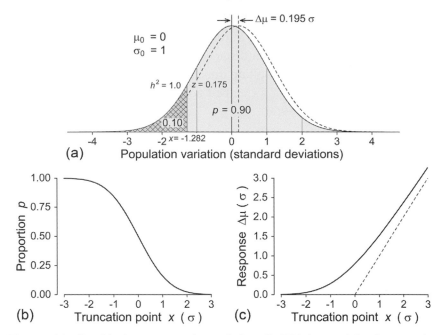

Figure 6.1 Graphical representation of Sewall Wright model of truncation selection. (a) Variables x and z are on the horizontal and vertical axes, respectively. For full heritability ($h^2 = 1$), removal of the cross-hatched area is expected to move the mean for founders of the next generation by $\Delta\mu = 0.195$ standard deviation units. Values of x, p, and $\Delta\mu$ for a standard normal curve with $\mu = 0$ and $\sigma = 1$ are found in table 5.6 of Wright (1969). (b) Proportion of area p to the right of truncation point x. (c) Response $\Delta\mu$ as a function of the position of truncation point x. In the right tail the response $\Delta\mu$ is asymptotic to $\Delta\mu = x$ (dashed line).

a value given by Wright, 1969, in his table 5.6). The truncation point separates the distribution into an area $1 - p$ lying to the left of the truncation point that is not selected (i.e., removed), and an area p lying to the right of the truncation point that is selected (i.e., retained). Here $1 - p$ (cross-hatched) represents 0.10 of the whole, and p represents 0.90 of the whole. The ordinate at $x = -1.282$ is $z = 0.175$. With full heritability, truncation at $x = -1.282$ can be expected to increase the mean by 0.195 standard deviation units in the next generation. However, the response will be proportionally less for any heritability falling short of full (Equation 6.3). In practical terms, achieving the same response for a lesser heritability requires moving the truncation point to the right to reduce the proportion selected.

The proportion p selected in an experiment like that in Figure 6.1a declines as the truncation point moves to the right (Figure. 6.1b). Proportion p starts high as a full proportion when the truncation point is in the left tail of a normal curve. It then declines rapidly to an inflection point corresponding to the mean of the normal curve. Proportion p continues to decline as the truncation point moves farther to the right, but p declines more slowly as it approaches minimal values in the right tail of the normal curve. In contrast, the response $\Delta\mu$ increases continuously as the truncation point moves from the left to the right tail of the normal curve (Figure 6.1c).

Seven parameters in Equations 6.1–6.4 and Figure 6.1 (μ, σ, h^2, p, x, z, and $\Delta\mu$), are required to characterize selection involving truncation, and an eighth parameter, the step interval i_0 or generation time, is required to quantify change in a time series of successive generations. The response, $\Delta\mu$, describing how the mean μ changes from one generation to the next, is measured in standard deviation units. This is why means, standard deviations, and generation times are essential for quantifying evolutionary change.

It is important to note a parameter that does not appear in Equations 6.1–6.4, nor in Figure 6.1: population size. For any population with a reasonable chance of survival, and for anything less than extreme truncation, the response to selection depends on the truncation point and the resulting mass proportion p of the population, but the response is independent of population size per se. Truncation experiments in the laboratory are carried out using relatively small populations, not because there is an advantage in terms of response but to avoid the cost of housing and processing larger populations.

6.1.2 Truncation Selection in a Scaled Example

Forms of selection can be illustrated with a personal story. Twenty or twenty-five years into my teaching career I began to be impressed that university students are taller now than they were when I started teaching, and I realized that I had been teaching long enough to have taught some of the parents of my students. Anthropologists often speak of a "secular trend" of increasing human stature, and Gordon Townsend Bowles (1932) wrote a classic Harvard thesis on the subject (more on this later). Putting these ideas together, I started to wonder what the change I seemed to be seeing in young-adult stature might require in terms of selection.

To investigate I estimated the stature of my generation when we were young adults. My estimate for mean stature in my generation was $e^{5.140} = 170.7$ cm (≈ 5 feet, 7 inches). In Chapter 2 we saw that linear measurements such as stature in organisms with determinate growth, such as humans, commonly have a coefficient of variation ($s/\bar{x}$) of about 0.050, which is equivalent to a standard deviation of 0.050 units on a natural-logarithmic scale. The resulting model distribution representing stature in my generation is shown as the face of the normal solid of Figure 6.2b and Figure 6.2c. The abscissa is the horizontal or x-axis, the y-axis is the depth ordinate of the solid (here uniformly distributed), and the z-axis is the

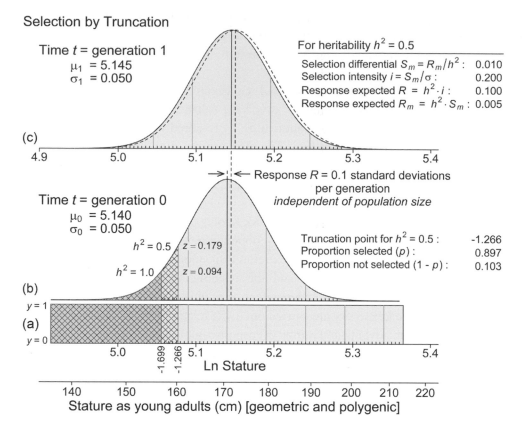

Figure 6.2 Truncation selection illustrated by a hypothetical one-generation change in human stature. (a) Plan view of normal solid x and y axes in generation 0. (b) Face view of normal-solid x and z axes in generation 0. (c) Face view of normal-solid x and z axes in generation 1. A mean for ln stature of $\mu_0 = 5.140$ (170.7 cm) in generation 0 is modified by truncation to yield a mean of $\mu_1 = 5.145$ (171.6 cm) in subsequent generation 1. Standard deviation in both is $\sigma = 0.050$. Truncation at $\mu_0 - 1.699 \cdot \sigma$ yields an expected response to selection of $R = 0.100$ standard deviation units when heritability $h^2 = 1$. Truncation at $\mu_0 - 1.266 \cdot \sigma$ is required to yield the same response when heritability $h^2 = 0.5$. Note that mass-selection response in a viable population is independent of population size.

vertical axis as it was in Figure 6.1. It is not necessary to represent the frequency distribution as a normal solid in this example, but the representation will be useful later: x is the measured variable in the views of Figure 6.2a and b, while y in panel a and z in panel b are the related frequencies.

Some care is required to relate x-dimension external measurements on a natural-logarithmic scale to x-dimension location and dispersion on an internal standard-deviation scale. The former is calibrated in logarithmic (ln) units on a scale external to the normal and uniform distributions, and the latter in standard deviation units on a scale internal to the distributions.

In this example R, the response to selection, is 0.100 or one tenth of a standard deviation unit (Figure 6.2c; chosen for reasons that will become clear in later chapters). R in the terminology of Falconer (1981), is equivalent to $\Delta\mu$ in the notation of Wright (1969; Figure. 6.1). R and $\Delta\mu$ are the difference, in standard deviation units between the mean of the parental generation and the realized mean for offspring of the selected parents.

R can also be expressed in measurement units. The response in measurement units is called R_m to distinguish it from R in standard deviation units. Then $R_m = s \cdot R$, and for $s = 0.050$ and $R = 0.100$, we have $R_m = 0.005$ cm.

R is the difference between the mean of the parental generation and the *realized* mean for offspring of the selected parents. The realized response depends on the heritability h^2 of the trait being studied. To achieve $R = 0.100$ with $h^2 = 0.5$ requires a selection intensity i of 0.200 standard deviation units ($i = R/h^2$, and $0.200 = 0.100/0.5$). With a heritability of 0.5, the mean must move twice as far to achieve the same result because half of the gain will be lost in the next generation. R is then the product of the heritability and the intensity ($R = h^2 \cdot i$, and here $0.100 = 0.5 \cdot 0.200$). The selection differential S_m is the log-measurement equivalent of the selection intensity ($S_m = R_m/h^2$, and $S_m = i \cdot s$). In the example in Figure 6.2, $S_m = 0.005/0.5 = 0.010$ and $S_m = 0.200 \cdot 0.050 = 0.010$.

Darker cross-hatching in Figure 6.2a and Figure 6.2b shows truncation of the parental generation (generation 0) required to move the mean by 0.1 standard deviation units for founders of the following generation under an assumption of full heritability ($h^2 = 1$). Here the truncation point is 1.699 standard deviations to the left of the mean (at 5.055 on the ln measurement scale). The derived generation in the upper panel (generation 1) was founded by offspring of the selected parents. The mean is affected by selection, but we assume that natural fecundity will both replace the nonselected individuals, maintaining population size, and at the same time restore preselection population variability.

A more realistic example of truncation, with $h^2 = 0.5$, is shown by the combination of darker and lighter cross-hatching in the lower panel of Figure 6.2. Now the truncation point is only 1.266 standard deviations to the left of the mean (at 5.077 on the ln measurement scale), and the proportion of the population selected ($p = 0.897$) is still almost nine times as great as the proportion not selected (0.103). The response is maintained at 0.1 standard deviations per generation on a timescale of one generation, but with a heritability of $h^2 = 0.5$ the cost is greater in terms of selective deaths. Here again the response is a mass response, independent of population size.

Truncation selection is widely practiced in experimental field and laboratory settings, as we shall see in the next chapter, but some form of gradient selection may be more important in nature.

6.1.3 Gradient Selection in a Scaled Example

The simplest form of gradient selection applied to a normal distribution of population variation involves sampling under a normal curve and a linear gradient during a sweep across the normal distribution. The geometry is illustrated in Figure 6.3a and Figure 6.3.b, where the abscissa of the normal solid is again the x-axis, there is an ordinate on the y axis, and there is a second ordinate on the z axis. The response to selection can be estimated by calculating a weighted mean for all products $y \cdot z$ resulting from $y = f_1(x)$ and $z = f_2(x)$ for arbitrarily small increments of x spanning the range of observed variation. The first function, yielding y, has the form:

$$y = a + gx \qquad (6.6)$$

where a is a small positive intercept and g is the gradient or slope of the line. The second function, yielding z, is that in Equation 6.1 for a standard normal curve. Figure 6.3a is a graph of y versus x, and Figure 6.3b is a graph of z versus x.

The volume representing the product of areas under both curves is given by:

$$p_g = \frac{1}{\sqrt{2\pi}} \int_{-5}^{+5} e^{-0.5x^2} \cdot (a + g \cdot (x+5)) \; dx \qquad (6.7)$$

The integral is evaluated from $x = \bar{x} - 5 \cdot s$ to $x = \bar{x} + 5 \cdot s$, which includes virtually all of the volume under a normal solid. Note that for $g = 0$ the contribution of y to the volume p_g is the constant a, which is uniform and does not affect the mean on the x-axis.

For a given gradient g the response to selection R can be calculated by integrating Equation 6.7 in arbitrarily small increments and calculating a new weighted $\bar{x}_g$, weighted by the volume p_g for each increment. This can be repeated for increasing values of g until the desired response is achieved. For the response $R = \bar{x}_g - \bar{x} = 0.1 \cdot s$ shown in Figure 6.3, the target will be:

$$\bar{x}_g = \bar{x} + R/h^2 \qquad (6.8)$$

For $a = 0.1$ and $h^2 = 1.0$, the target is $\bar{x}_g = \bar{x} + 0.100$, the response is $R = 0.100$, and the required gradient is $g = 0.020$. For $a = 0.1$ and $h^2 = 0.6$, the target is $\bar{x}_g = \bar{x} + 0.167$, and the required gradient is $g = 0.100$. There is no solution for $h^2 = 0.5$ because the response is asymptotic and cannot reach the required value of $R = 0.200$ (Fig. 6.4).

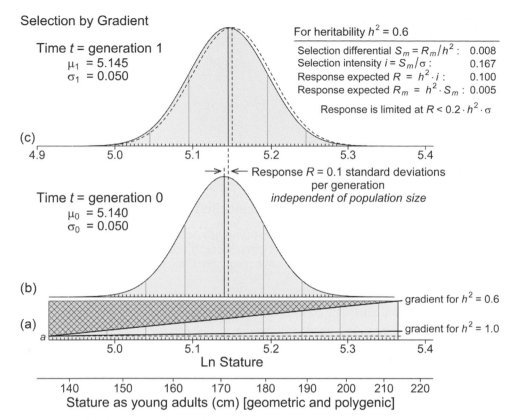

Figure 6.3 Gradient selection illustrated by a hypothetical one-generation change in human stature. (a) Plan view of normal-solid x and y axes in generation 0. (b) Face view of normal-solid x and z axes in generation 0. (c) Face view of normal-solid x and z axes in generation 1. A mean for ln stature of $\mu_0 = 5.140$ (170.7 cm) in generation 0 is modified by gradient selection to yield a mean of $\mu_1 = 5.145$ (171.6 cm) in subsequent generation 1. Standard deviation in both is $\sigma = 0.050$, and the arbitrary gradient intercept is $a = 0.1$. A linear selection gradient of $g = 0.020$ is sufficient to yield the observed response $R = 0.100$ standard deviation units when heritability $h^2 = 1.0$. However, a selection gradient of $g = 0.100$ is required to yield the same response when $h^2 = 0.6$. Note that this response cannot be achieved when $h^2 \leq 0.5$ (Figure 6.4). Note too that mass-selection response in a viable population is independent of population size. Figure is modified from Gingerich (2014), reproduced by permission of Springer Nature Publishing

If, as mentioned above, truncation is the most efficient form of directional selection, then a linear selection gradient like that in Figure 6.3 is one of the most limited and least efficient. Gradient selection does retain the advantage of operating on a mass proportion of the population. Also possible are selection gradients more complicated than linear that can be combined with truncation in ways not explored

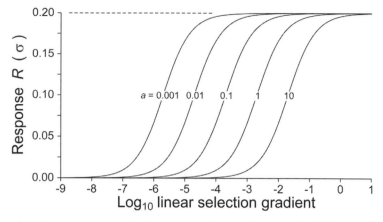

Figure 6.4 Response curves for linear selection gradients with different values of the arbitrary gradient intercept a. For a given value of a, low gradients yield very little response, and higher gradients have responses limited by $R \leq 0.200 \cdot s$.

here. Selection gradients that combine truncation with curvature will logically be intermediate in promoting directional change.

6.2 Stabilizing Selection

What happens when we combine truncation selection in one tail of a normal curve, like that described in Section 6.1.1, with similar truncation in the opposite tail? What happens when we combine positive selection by gradient, like that described in Section 6.1.2, with similar negative selection by gradient? In instances where selection in one tail exactly balances that in the other tail, selection differentials add to zero, selection intensities add to zero, and the expected response is little or no change from one generation to the next (Figure 6.5). Balanced selection in both tails is commonly called "stabilizing selection" because it yields no change.

Assigning stabilizing selection a name distinct from directional selection makes it sound like a different process is involved, but both are manifestations of the same process. The stabilizing selection example outlined here has selection balanced in the left and right tails, with selection differentials adding to zero. This is a special case in a spectrum ranging from (a) purely directional selection in a positive direction with no opposing selection in a negative direction to (b) purely directional selection in a negative direction with no opposing selection in a positive direction. Populations in nature necessarily lie somewhere on this spectrum.

The examples chosen for illustration in Figures 6.2 and 6.3 are special cases too. The example of directional selection in Figure 6.2 has a positive selection differential of $S_m = +0.010$ and a negative selection differential of $S_m = -0.000$.

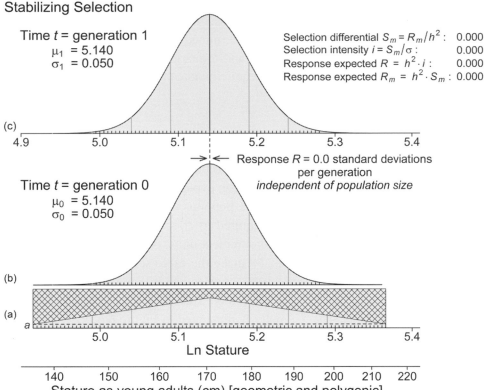

Stabilizing Selection

Time t = generation 1
μ_1 = 5.140
σ_1 = 0.050

Selection differential $S_m = R_m/h^2$:   0.000
Selection intensity $i = S_m/\sigma$:   0.000
Response expected $R = h^2 \cdot i$:   0.000
Response expected $R_m = h^2 \cdot S_m$: 0.000

(c)

4.9   5.0   5.1   5.2   5.3   5.4

→ ← Response R = 0.0 standard deviations
per generation
independent of population size

Time t = generation 0
μ_0 = 5.140
σ_0 = 0.050

(b)

(a)

a

5.0   5.1   5.2   5.3   5.4
Ln Stature

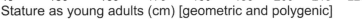

140   150   160   170   180   190   200   210   220
Stature as young adults (cm) [geometric and polygenic]

Figure 6.5 Gradient selection illustrated by a hypothetical one-generation change
in human stature. (a) Plan view of normal-solid x and y axes in generation 0. (b)
Face view of normal-solid x and z axes in generation 0. (c) Face view of normal-
solid x and z axes in generation 1. Balanced gradient selection in both tails will
yield a selection differential and a selection intensity of zero, and no change in
subsequent generation 1. Unbalanced selection (not illustrated here) would yield a
response and change. As before, mass-selection response in a viable population is
independent of population size.

Together these yield a net differential S_m = +0.010. If the negative selection
differential in Figure 6.2 was modeled as S_m = −0.005 instead of −0.000, then
positive and negative would add together to yield a net selection differential of
S_m = +0.005, and the expected response R_m would be one half of that shown in the
figure. If the negative differential in Figure 6.3 was modeled as S_m = −0.007
instead of −0.000, then positive and negative would add to yield a net selection
differential of S_m = +0.001 and the expected response R_m would be one eighth of
that shown in the figure.

In Section 4.4.1 above we discussed an Ornstein–Uhlenbeck model of a stationary stochastic process with a constant target 2. The process is stationary, but not

because selection in each step is balanced. In the Ornstein–Uhlenbeck model, a time series tracks θ by application of successive directional corrections, with the direction of the response in each step determined by the difference, positive or negative, between the mean at the time and the target. Thus stabilizing selection through multiple generations does not require maintenance of balanced selection in each generation.

6.3 Random Drift

Random drift is the principal alternative to evolutionary change or stasis controlled by directional or stabilizing selection. We can explore this using the same example of change in human stature developed in previous sections of this chapter. Here we are asking again what change is required to effect a positive response in the mean of $R_m = 0.005$ in natural-log measurement units, or equivalently, a response of $R = 0.1$ standard deviation units. To make the exercise realistic, we assume a heritability of $h^2 = 0.5$, meaning we require a selection differential of $S_m = 0.010$ in natural-log measurement units or, equivalently, a selection intensity of $i = 0.200$ in standard deviation units. The geometry is illustrated in Figure 6.6.

In this example there is no selection by truncation or selection by gradient to help us move the mean. We have to rely on a sample drawn at random from the normal distribution of generation 0 to found the population of subsequent generation 1. The mean of the sample $\bar{x}$ will be the mean μ_1 of the new population. We assume, as before (but with less certainty), that fecundity will maintain population variation.

One sample of 10 individuals is represented by vertical bars in Figure 6.6a. This was drawn at random from the normal distribution in Figure 6.6b. Most of the variation in Figure 6.6b is found near mean $\mu_0 = 5.140$, so it is not surprising that the sample of 10 is also clustered near this mean. However, the mean of our random sample of 10 individuals is $\bar{x} = 5.136$, which is less than the original mean μ_0 and a change in the wrong direction from our target of $\mu_0 + S_m = 5.150$ (dashed vertical line in Figure 6.6c). This first sample is a warning that it might not be easy to move a mean in the desired direction by random drift.

We can begin to understand the nature of the problem by asking how much of the normal distribution in Figure 6.6b lies at or to the right of our target of $\mu_0 + S_m = 5.150$ required to effect a response of at least $\mu_0 + R_m = 5.145$ when $h^2 = 0.5$. A table of the normal probability function (e.g., Pearson and Hartley, 1966, table 4) shows that 0.421 of the area under a normal curve lies 0.200 standard deviations or more from the mean — we can confirm this by drawing numbers at random from the distribution in Figure 6.6b. For 10,000 samples of size $N_e = 1$, a total of 4,211 were equal or greater than 5.150, and the proportion at or to the right of our

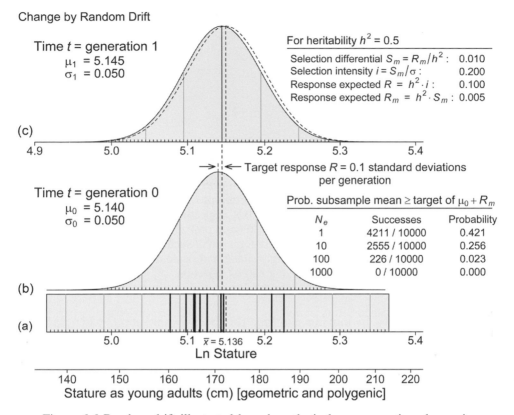

Change by Random Drift

Time t = generation 1
μ_1 = 5.145
σ_1 = 0.050

For heritability $h^2 = 0.5$

Selection differential $S_m = R_m/h^2$:   0.010
Selection intensity $i = S_m/\sigma$:   0.200
Response expected $R = h^2 \cdot i$:   0.100
Response expected $R_m = h^2 \cdot S_m$: 0.005

(c)

4.9   5.0   5.1   5.2   5.3   5.4

→ ← Target response R = 0.1 standard deviations
per generation

Time t = generation 0
μ_0 = 5.140
σ_0 = 0.050

Prob. subsample mean ≥ target of $\mu_0 + R_m$

N_e	Successes	Probability
1	4211 / 10000	0.421
10	2555 / 10000	0.256
100	226 / 10000	0.023
1000	0 / 10000	0.000

(b)

(a)

5.0   5.1 $\bar{x}$ = 5.136 5.2   5.3   5.4
Ln Stature

140   150   160   170   180   190   200   210   220
Stature as young adults (cm) [geometric and polygenic]

Figure 6.6 Random drift illustrated by a hypothetical one-generation change in human stature. (a) Plan view of normal-solid x and y axes in generation 0. (b) Face view of normal-solid x and z axes in generation 0. (c) Face view of normal-solid x and z axes in generation 1. A mean for ln stature of $\mu_0 = 5.140$ (170.7 cm) in generation 0 can be modified by random drift to yield a mean of $\mu_1 = 5.145$ (171.6 cm) in subsequent generation 1. However, as the inset table in panel b shows, the probability of achieving the change by random drift decreases rapidly as sample size N_e increases (see Figure. 6.7). The probability for a run of such random-drift responses in successive generations is vanishingly small, and random drift has little power in the face of directional or stabilizing natural selection.

target is 0.421. For samples of size $N_e = 1$ the probability of success in reaching our target by random drift is 0.421 (and the probability of failure to reach the target is 0.579). However, it is not realistic to imagine founding a population in generation 1 with a sample of one individual.

If we repeat the exercise by drawing 10,000 samples of size $N_e = 10$, $N_e = 100$, $N_e = 1,000$, our success in drawing a sample with a mean greater than or equal to our target of $\mu_0 + S_m = 5.150$ declines from 2,555 successes for $N_e = 10$, to 226 for $N_e = 100$, and to 0 for $N_e = 1,000$. Random drift is acutely sensitive to sample size,

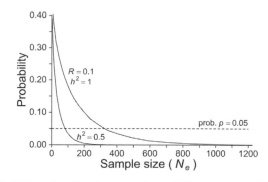

Figure 6.7 Probability of achieving a random-drift response of $R = 0.1$ standard deviation in a generation for samples of different sizes. The two curves shown represent heritabilities of $h^2 = 1$ and $h^2 = 0.5$, respectively. Note that for $h^2 = 0.5$ any sample size greater than about $N_e = 80$ has a probability of achieving $R = 0.1$ that is less than $p = 0.05$. The probability for a run of such random-drift responses in successive generations is vanishingly small, and random drift has little power in the face of directional or stabilizing natural selection.

and it is ineffective in moving a mean, for even a generation, when samples approach the effective size of most populations breeding in nature. Part of the problem is that random samples are inherently self-balancing because half move the mean in one direction and half move the mean in the opposite direction.

The critical dependence of random drift on sample size is illustrated in Figure 6.7, where the probability of reaching a target response of $R = 0.1$ standard deviation units is plotted against sample size. For heritability $h^2 = 1$, a sample of size 320 will reach the target by drift alone with a probability of 0.05. For a more realistic value of $h^2 = 0.5$, a sample of size 80 will reach the target by drift alone with a probability of 0.05. Stated differently, we can expect a sample of 80 to reach the target in no more than one generation out of 20. The probability of reaching the target two generations in a row would be $0.05^2 = 0.0025$, and the probability of reaching it three generations in a row would be $0.05^3 = 0.000125$. Subdividing a larger breeding population into samples of 80 is not effective either because it is still true that only one in 20 will reach the target, and fully half will change in the opposite direction. Random drift pales to insignicance in comparison to directional and stabilizing selection and their power to move and stabilize populations in terms of mass proportion, independent of population size, generation after generation.

6.4 Summary

1. The response to artificial or natural selection can be expressed in terms of a selection differential in measurement units, or a selection intensity in standard

deviation units. Heritability determines the proportional relationship of response to selection in both cases.

2. Truncation is the most efficient form of directional selection. The response is independent of population size and the response curve increases indefinitely.

3. Gradient selection is a much less efficient form of directional selection. The response is independent of population size, but the response curve for a linear gradient is limited at $R < 0.2 \cdot h^2 \cdot \sigma$. More complicated gradients are possible, which can be combined with truncation in ways not explored here.

4. Stabilizing selection requires a balance of truncation or gradient selection (or both). Stability in an evolutionary time series can also be achieved through a succession of directional corrections, with the response in each generation determined by the difference between the mean at the time and some optimal value.

5. Random drift is acutely sensitive to sample size and inherently self-balancing. Consequently, random drift has much less power to move a mean or to constrain it than any form of mass selection.

7

Phenotypic Change in Experimental Lineages

Why are mice the size they are? Or rats, or men? The ease with which
artificial selection can change almost any character proves that natural
selection could do so too...

Douglas S. Falconer, Proceedings of the International
Conference on Quantitative Genetics, *1977, p. 19*

"Artificial" selection and preferential breeding of plants and animals that have
desirable characteristics are surely as old as the history of human cultivation and
domestication. These are practiced for a reason: because they work, producing
economically, behaviorally, or aesthetically advantageous results. Artificial selection does not replace natural selection but is almost always subordinate and hence
antagonistic to it. When we select breeding stock we do so based on one or a
few desirable characteristics, but an organism is always more complex. Underlying
genomes are integrated in the sense that (1) they are often pleiotropic, with
individual genes affecting multiple phenotypic traits and (2) the traits themselves
are often polygenic, with multiple genes contributing to an individual trait. Genetic
integration rarely prohibits change, but it is conservative and inhibitory. Integration
probably always acts to make change slower.

There are three ways to study evolutionary change: one is experimental, another
is observational, and the third is inferential. Here we start by examining what
we can do in the field and laboratory to change the phenotypes of experimental
lineages. The next chapter, following this one, will consider phenotypic change we
can observe or reliably infer in evolutionary field studies. A subsequent chapter
will consider phenotypic change we can observe or reliably infer in the fossil
record. Field and fossil studies are observational when we see direct evidence
of change through time, and they are inferential when we cannot see everything
and have to piece together evidence from different sources. In all three of these
chapters we will quantify change in terms of rate to enable comparisons of
different organisms and different scales of time.

Selection experiments commonly involve four taxa: (1) domestic maize or corn of the genus and species *Zea mays*, (2) fruit or vinegar flies of the genus *Drosophila*, (3) domestic chickens of the genus and species *Gallus gallus*, and (4) house mice of the genus and species *Mus musculus*. Examples of each are analyzed here. Each experiment was carried out and change recorded on a timescale of generations.

Selection experiments are long-term studies with major investments of time and energy. Early investigations were sufficiently novel that reasonably complete statistics were published. Later studies tended to illustrate results graphically with less complete statistics. Here we are interested to quantify the effectiveness of selection in each experiment, and the rates of phenotypic (morphological, or in one case behavioral) change that were achieved in each generation.

We often refer to experimental selection as "laboratory selection," which fits a narrow expectation of working within four walls. Experiments on fruit flies and mice usually take place within the confines of a typical laboratory. However, fields and farms are also places to work and places to carry out experimental studies. These are the laboratories for experimental studies of maize and chickens. All rates calculated in this chapter can be found at www.cambridge.org/evolution.

7.1 Experiments with Maize or Corn, *Zea mays*

Two kinds of experiments with maize or American corn are considered here. The first two sets of experiments were carried out to increase the number of longitudinal seed-kernel rows found on an ear of maize. A third set of experiments, actually one grand experiment, was initiated in an attempt to increase and decrease both protein and oil in the maize seed kernel. The latter grand experiment is notable for the sustained response achieved and for the longevity of the study itself.

7.1.1 Rows of Seed Kernels

Johann Friedrich Theodor Müller, or Fritz Müller, is best known as a biologist for developing the concept of Müllerian mimicry. Earlier, while working as an advisor to farmers in the Itajaí Valley of Brazil, Müller made a study of artificial selection in maize. This was carried out as an exercise in the years 1867, 1868, and 1869 to see whether it was possible to augment the number of rows of seed kernels on maize cobs or ears. Müller started with a sample of 100 ears of maize, which had rows of kernels ranging in number from 8 to 16. He did not publish statistics for the starting sample, but the mean number of rows in the starting set was seemingly 12 or close to 12. One ear in Müller's initial sample had 18 kernel rows.

Müller carried out three experiments in 1867, one with seed from a 14-row ear, another with seed from a 16-row ear, and a third with seed from an 18-row ear. When planted, the seeds yielded a large number of descendent ears of maize, with means of 12.610 for seed from the 14-row ear, 14.076 for seed from the 16-row ear, and 14.906 for seed from the 18-row ear. Thus the response observed in the first generation was proportional to selection intensity as reflected in the choice of seed increasingly different from the starting mean.

To calculate a rate of change we first recognize that adding a row to an 8-row ear is different from adding a row to a 16-row ear. Proportional differences such as these are appropriately analyzed in terms of logarithms. The natural logarithm (ln) of 8 is 2.079, of 10 is 2.303, of 12 is 2.485, etc. Müller reported the percentages of 8-row ears, 10-row ears, etc., at the end of each of his nine experiments, and we can use these percentages as weights to calculate the means and variances for the samples resulting from each experiment. The variances are closely similar, and an overall standard deviation (0.126 in ln units) can be calculated as the square root of the pooled variances. Since the first experiment started with 12 rows (2.485) and ended with 12.610 rows (2.527; calculated from Müller's proportions for ln rows), the response was 0.042 ln units, which is 0.336 in standard deviation units. Division by the number of generations (one in the first experiment) yields a rate of 0.336 standard deviations per generation on a timescale of one generation. The remaining one-generation rates are 1.202 and 1.654 standard deviations per generation, reflecting their more intense selection.

Müller's three experiments in 1868 yielded responses of 1.248, 1.586, and 1.978 in standard deviation units, for rates of change of 0.624, 0.793, and 0.989 standard deviations per generation on a timescale of two generations. Finally, the three experiments in 1869 yielded responses of 2.003, 2.103, and 2.303 for rates of 0.668, 0.701, and 0.768 standard deviations per generation on a time scale of three generations. These results are summarized in Table 7.1. All of Müller's rates were high, reflecting the intensity of his selection. The experiments were short, so it is not clear how long one might expect to sustain such a response.

Müller published his results 16 years after completing the selection experiments. He was inspired to publish by contradictory interpretations of inheritance voiced by Francis Galton and August Weismann (Müller, 1886). Müller achieved remarkable responses to selection, and recognized too that his experiments supported Galton's view of regression to the mean (what Galton called "regression to mediocrity").

A second set of experiments was carried out by Hugo Marie de Vries, a professor of botany at the University of Amsterdam. De Vries is best known for introducing the term "mutation" as a genetic mechanism for the origin of species, and also, in 1900, for being one of three botanists credited with rediscovering

Table 7.1 *Rates of response in two early experiments to increase the number of kernal rows in maize*

Müller (1886)

Int.	Rate	Int.	Rate	Int.	Rate
1	0.336	2	0.624	3	0.668
1	1.202	2	0.793	3	0.701
1	1.654	2	0.989	3	0.768

de Vries (1909)

Int.	Rate	Int.	Rate	Int.	Rate	Int.	Rate	Int.	Rate	Int.	Rate	Int.	Rate
1	0.150	2	0.246	3	0.285	4	0.293	5	0.369	6	0.311	7	0.393
1	0.343	2	0.353	3	0.341	4	0.424	5	0.344	6	0.434		
1	0.362	2	0.339	3	0.451	4	0.344	5	0.452				
1	0.317	2	0.495	3	0.338	4	0.475						
1	0.674	2	0.348	3	0.527								
1	0.023	2	0.454										
1	0.885												

Rates are in haldanes, standard deviations per generation; *int.* is the corresponding interval length in generations

Mendelian inheritance. In the years from 1886 through 1899 de Vries repeated and amplified Müller's selection experiments on maize. In his first experiment, from 1886 through 1896, de Vries founded each new generation with seed selected from maize ears with a high number of rows. Then in a second experiment, from 1897 through 1899, he reversed the direction of selection. The only data presented, in de Vries (1909: figure 18), are those for the first seven years of positive selection. It is not clear whether de Vries was aware of Müller's investigation when he carried out his experiments. De Vries, like Müller, waited a long time to publish his results.

De Vries started his experiment in 1886 with seed from maize ears having a mean of 13 kernel rows. From this initial sample he selected seed for the 1887 generation from one or more ears with 16 kernel rows. Successive generations were founded with new seed selected from one or more ears with 20, 20, 24, 22, 22, and finally, in 1874, from one or more ears with 22 kernel rows. The whole experiment was illustrated in a scaled "fan" diagram with time on the vertical axis and the number of kernel rows on the horizontal axis. The number of kernel rows for seed used to found each new generation was plotted as a point at the beginning of each year. This point formed the origin of the fan, and then a horizontal line spanning the range of derived kernel-row variation was drawn at the end of the year. Relevant statistics were represented by a solid middle line connecting the origin to the mean of the range, and by two dashed lines connecting the origin to adjacent quartiles on the range. Assuming the variation to be normally distributed, we can calculate the standard deviation for each sample as the average of the two quartile distances from the mean divided by 0.6745. The variance is then the square of the standard deviation. The samples for 1891 and 1892 have unusual variances, so here the median of variances was used to calculate the pooled standard deviation (0.157 ln units).

In this calculation, and whenever means and standard deviations are calculated directly from arithmetic measurements, it is necessary to use the natural logarithm of the arithmetic mean ($\ln \bar{A}$) to represent the desired geometric mean ($\bar{G}$). It is well known that $\ln \bar{A} \geq \bar{G}$, and overestimation increases as the coefficient of variation for a sample of arithmetic measurements increases (this coefficient of variation is equivalent to the standard deviation of a natural-logged geometric sample). However the overestimate is small: $\ln \bar{A} - \bar{G} < 0.025$ for standard deviations and differences commonly used to calculate rates.

Step rates for de Vries' observations in successive generations, calculated as shown above for Müller's data, are 0.885, 0.023, 0.674, 0.317, 0.362, 0.343, and 0.150 standard deviations per generation on a timescale of one generation. De Vries' observations can also be analyzed through multiple generations to yield net rates on timescales of two through seven generations. The compound net rates are listed with de Vries' step rates in Table 7.1. Comparison of rates for de Vries' experiment with those for Müller's experiments shows that Müller's rates were a little higher than de Vries' rates, reflecting Müller's more extreme selection. Change was positive, reflecting the direction of selection in both sets of experiments.

7.1.2 Chemical Composition of Maize Kernels

Selection experiments of a different kind, involving the chemical composition of maize kernels, were started in 1896 by Cyril G. Hopkins at the University of Illinois Agriculture Experiment Station. The experiment, known as the Illinois Long-Term Selection Experiment, continues today. The initial objective was to determine whether the chemical composition of the kernel could be changed by selection. Once change was demonstrated the objective became a more general exploration of the limits of selection.

In the first publication on the Illinois experiment, Hopkins (1899) provided percentages of ash, protein, fat, and carbohydrate for each ear in a baseline 1896 set of 163 maize ears. These included seed to found all of the Illinois selection lineages, including the Illinois High Protein or IHP lineage analyzed here. The sample size, mean, standard deviation, and coefficient of variation for the 1896 protein percentages published by Hopkins are $N = 163$, $\bar{x} = 10.9245$, $s = 1.0493$, and $V = 0.0960$. Analysis requires that percentages be normalized to an arithmetic scale, which can be done using the following inverse-cosine angular transformation (adapted from Wright, 1968, p. 258):

$$\theta = \frac{1}{\pi} \cos^{-1} \left[1 - 2 \cdot (0.01 \cdot x) \right] \tag{7.1}$$

Multiplication of x by 0.01 in Equation 7.1 is necessary to change any original percentage to a 0–1 scale before transformation. The θ resulting from transformation is also expressed on a 0 to 1 scale. This arccosine transformation is similar to the arcsine transformation of Sokal and Rohlf (1981, p. 427), but the latter is expressed on a 0–1.571 scale. Statistics for the transformed sample of Hopkins' protein percentages are $N = 163$, $\bar{x} = 0.2142$, $s = 0.0107$, and $V = 0.0502$. Note that the sample standard deviation is now a substantially smaller proportion of the mean than it appeared to be before transformation ($V = 0.0502$ compared to $V = 0.0960$).

When θ is measured in radians, the standard deviation σ expected from an underlying binomial distribution is independent of x and purely a function of sample size N. The binomial σ associated with θ in Equation 7.1 is:

$$\sigma = \frac{2}{\pi} \cdot \sqrt{\frac{1}{4N}} \tag{7.2}$$

If we substitute $N = 163$ for the transformed sample of Hopkins' protein percentages, the binomial σ is 0.0249, which is more than twice the observed $s = 0.0107$. This is a cautionary tale about using a binomial σ as a substitute for s in rate calculations: In empirical research an expectation is no substitute for an observation.

To calculate rates in standard deviation units, we have to transform sample standard deviations from percentage units to the same arithmetic scale as θ. Transformed standard deviation s_{ab} can be estimated as one half the difference between transformed values of $\bar{x}-s$ and $\bar{x}+s$, as follows:

$$s_{ab} = \frac{s_b - s_a}{2}, \text{where } s_a = \frac{1}{\pi} \cos^{-1}[1 - 0.02{\cdot}(\bar{x}-s)] \text{ and}$$

$$s_b = \frac{1}{\pi} \cos^{-1}[1 - 0.02{\cdot}(\bar{x}+s)] \qquad (7.3)$$

Application to Hopkins' sample mean and standard deviation, $\bar{x} = 10.9245$ and $s = 1.0493$, yields the same $s_{ab} = 0.0107$ that was calculated from the sample of transformed percentages.

John W. Dudley and Robert J. Lambert published a review of the first 100 generations of selection for oil and protein in the Illinois Long-Term Selection Experiment, starting with Hopkins' first sample measured in 1896 and continuing to their own 1999 results (experiments were halted from 1942 to 1944). Dudley and Lambert (2004) provided the number of ears selected for planting (generally 12), empirical means, and standard deviations for each generation. Here the Illinois High Protein or IHP results are used to calculate rates.

The Illinois High Protein time series and resulting differences between generations and rates per generation are summarized in Figure 7.1. Generation number is plotted on the X axis in Figure 7.1a, and the normalized percentage of protein measured in each generation is plotted on the Y axis. Vertical lines associated with each mean value span a range of two standard deviations (mean ± 1 std. dev.). The initial mean value for protein in 1896 was 0.215 and the mean value for protein after 100 generations of selection in 1999 was 0.362. This is an increase of 0.147 units. The average standard deviation for all 101 samples from 1896 through 1999 was 0.0144. Thus the increase in protein was almost exactly 10 standard deviation units, for a long-term average rate of increase of 0.1 standard deviations per year (on a timescale of 100 years). The time series for protein in Figure 7.1a climbs steadily overall, but it also has many pauses and reversals.

Dudley and Lambert (2004) reported selection differentials, S, calculated as the difference between (1) the mean for the seed selected to found the next generation

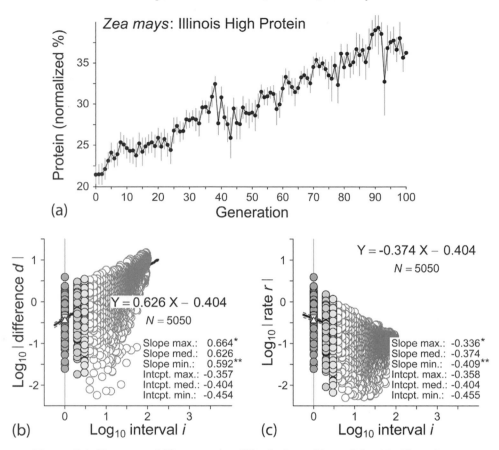

Figure 7.1 *Zea mays* 100-generation Illinois Long-Term Selection Experiment selecting for high protein content. (a) Realized gain during the experiment. Solid circles are means; vertical lines span ±1 standard deviation. (b) LDI log-difference-interval plot for 5,050 differences between all samples in the experiment. (c) LRI log-rate-interval plot for 5,050 rates of change between all samples in the experiment. Median slopes of 0.626 and −0.374 in (b) and (c) lie between, and are significantly different from, expectations for directional and random time series (between 1.000 and 0.500, and between 0.000 and −0.500, respectively). Median intercept of −0.404 is the same in (b) and (c), indicating a step rate h_0 of $10^{-0.404} = 0.394$ standard deviations per generation.

and (2) the mean for the population from which the seed was drawn. Normalized, the Dudley–Lambert selection differentials average about $S = 1.219$ standard deviations per generation. S provides an upper limit for the expected response. Simplifying slightly, response $R = h^2 \cdot S$, where h^2 is the heritability. As a rule heritability $h^2 < 1$, and $R = S$ is an equality that can only be achieved when heritability is complete, that is when heritability $h^2 = 1$.

Differences d between generations for intervals ranging from 1 to 100 generations are shown in the log-difference-interval or LDI graph of Figure 7.1b, where 5,050 differences are plotted. Similarly, rates of change r between generations for intervals ranging from 1 to 100 generations are shown in the log-rate-interval or LRI graph of Figure 7.1c, where the corresponding 5,050 rates are plotted. The LDI and LRI plots both have intercept medians of -0.404, indicating a median rate $10^{-0.404} = 0.39$ standard deviations per generation on a timescale of one generation. A bootstrapped 95% confidence interval for this median ranges from about 0.35 to 0.45 standard deviations per generation.

The maximum of the 100 one-generation differences (the maximum rate on a one-generation timescale) in Figure 7.1 is 0.590 standard deviation units on a log scale ($10^{0.590} = 3.888$ standard deviations), and the minimum of the 100 one-generation differences (the minimum rate on a one-generation time scale) is -1.607 standard deviation units on a log scale ($10^{-1.607} = 0.025$ standard deviations). Note that all of the one-generation rates of Müller (1886) and de Vries (1909) in Table 7.1 fall within or very nearly within the range of one-generation rates found in the much longer Illinois High Protein study.

The median slopes for the LDI and LRI plots, 0.626 and -0.374 respectively, always differ by 1.000 because each is a complement of the other. A bootstrapped two-tailed 95% confidence interval for the LDI slope ranges from 0.592 to 0.664. The former exceeds expectation for the slope of a purely stationary time series (0.000), and also exceeds expectation for the slope of a random walk (0.500) — excluding both interpretations. The latter (0.664) excludes expectation for a purely directional time series (1.000). A bootstrapped two-tailed 95% confidence interval for the LRI slope ranges from -0.409 to -0.336. The former exceeds expectation for the slope of a purely stationary time series (-1.000), and also exceeds expectation for the slope of a random walk (-0.500) — excluding both interpretations. The latter (-0.336) excludes expectation for a purely directional time series (0.000).

LDI and LRI slopes for the 100-generation Illinois High Protein time series as a whole both tell us the same thing: The time series is directional with a substantial random component, reflecting both the longer-term efficacy and the shorter-term unpredictability of experimental selection.

7.2 Experiments with Fruit Flies, *Drosophila* Species

The fruit fly *Drosophila melanogaster* is a laboratory animal of European origin that is widely used in selection and other experiments. *Drosophila pseudoobscura* is a North American species more commonly studied in the field.

7.2.1 *Eye Facets in* **Drosophila melanogaster**

An early selection experiment on *Drosophila melanogaster* involved the number of eye facets in white bar-eye fruit flies. Charles Zeleny carried out a 42-generation study at the University of Illinois, starting in 1914. He selected for a high facet number in one lineage and a low facet number in a parallel lineage (Zeleny, 1922). One complication of this study is Zeleny's expression of all of his results on what he called a "factorial" scale. He recognized that "the geometric or logarithmic mean and the dispersion of the logarithms of the facet numbers should form the basis of the biological analysis" (Zeleny, 1922, p. 13), but then effectively reinvented logarithms.

The zero point of Zeleny's factoral scale was the mean eye facet number in his unselected population of white bar-eye fruit flies (58.8 in females and 111.4 in males). Each factorial unit above the mean was incremented by a factor of 1.1 times the previous facet number, starting with the mean. Each factorial unit below the mean was decremented by a factor of $1/1.1 = 0.909$ times the previous facet number, again starting with the mean. The natural log value of eye facet number is a linear function of the factorial value. Thus log values are easily recovered from factoral values using the linear relation $L = a \cdot F + b$, where a and b are 0.0953 and 4.0741 respectively for females, and 0.0953 and 4.7131 respectively for males. This works directly for converting means. Standard deviations can be converted as one-half the difference between the conversions of $F = (\bar{x}-s)$ and $F = (\bar{x}+s)$, following the strategy of Equation 7.3 above.

Analyses of Zeleny's selection results for eye facet number (in his tables 12–15) are shown in Figure 7.2. The female low selection, female high selection, male low selection, and male high selection lineages are each shown separately in Figure 7.2a. High and low selection lineages for each sex diverged rapidly from a common parental population and then seemingly stabilized for much of the experiment.

Pedigrees show that female and male high-selection lines were bred from one to seven brother–sister pairs (harmonic mean 3.06, with $2 \cdot 3.06 = 6.12$ founding individuals). The female and male high-selection lines can be considered replicates of a single high-selection experiment, and rates for both are combined in the log-rate-interval or LRI graph of Figure 7.2b, where the corresponding 1,678 rates are plotted. The LRI plot has a median intercept of -0.355, indicating a median rate $10^{-0.355} = 0.442$ standard deviations per generation on a timescale of one generation. A bootstrapped 95% confidence interval for this median ranges from about $10^{-0.418} = 0.382$ to $10^{-0.299} = 0.502$ standard deviations per generation. The 77 one-generation step rates in Figure 7.2b range from a minimum of -2.402 standard deviation units on a log scale ($10^{-2.402} = 0.004$ standard deviations per generation)

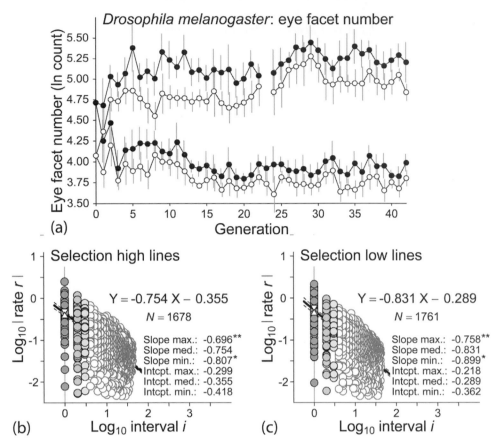

Figure 7.2 *Drosophila melanogaster* 42-generation experiments of Zeleny (1922) selecting for high and low eye facet numbers. (a) Realized gain and loss during each experiment. Open (female) and closed (male) circles are means; vertical lines span ±1 standard deviation. (b) LRI log-rate-interval plot for 1,678 rates of change in male and female high selection lines. (c) LRI plot for 1,761 rates of change in male and female low selection lines. Median slopes of −0.754 and −0.831 in (b) and (c) lie between, and are significantly different from, expectations for random and stationary time series (−0.500 and −1.000, respectively). Median intercepts are −0.355 and −0.289 in (b) and (c), indicating step rates h_0 of $10^{-0.355} = 0.442$ and $10^{-0.289} = 0.514$ standard deviations per generation.

to a maximum of 0.396 standard deviation units on a log scale ($10^{0.396} = 2.489$ standard deviations per generation).

The median temporal scaling slope in Figure 7.2b is −0.754, with a 95% confidence interval ranging from −0.807 to −0.696. The former excludes −1.000 expected for stasis, and the latter excludes both −0.500 and 0.000 expected for a random time series and for sustained directional change. Thus the high-selection time series as a whole appears to be stationary with a random component.

Pedigrees show that female and male low-selection lines were bred from one to eight brother–sister pairs (harmonic mean 2.56, with 2 · 2.56 = 5.12 founding individuals). The female and male low-selection lines can be considered replicates of a single low-selection experiment, and rates for both are combined in the log-rate-interval or LRI graph of Figure 7.2c, where the corresponding 1,761 rates are plotted. The LRI plot has a median intercept of −0.289, indicating a median rate $10^{-0.289} = 0.514$ standard deviations per generation on a timescale of one generation. A bootstrapped 95% confidence interval for the median ranges from about $10^{-0.362} = 0.435$ to $10^{-0.218} = 0.605$ standard deviations per generation. The 82 one-generation step rates in Figure 7.2c range from a minimum of −2.026 standard deviation units on a log scale ($10^{-2.026} = 0.009$ standard deviations per generation) to a maximum of 0.331 standard deviation units on a log scale ($10^{0.331} = 2.143$ standard deviations per generation).

The median temporal scaling slope in Figure 7.2c is −0.831, with a 95% confidence interval ranging from −0.899 to −0.758. The former excludes −1.000 expected for stasis, and the latter excludes both −0.500 and 0.000 expected for a random time series and for sustained directional change. Thus the low-selection time series as a whole, like the high-selection series, appears to be stationary with a random component.

7.2.2 *Phototaxis in* **Drosophila pseudoobscura**

Theodosius Dobzhansky and Boris Spassky conducted fruit-fly selection experiments on behavior rather than morphology. Their research was carried out at Rockefeller University in New York City. The study of interest here involved phototaxis, or movement in relation to light: movement toward light in the positive case, and away from it in the negative case. Dobzhansky and Spassky (1967) used a connected series of tubes reminiscent of a Galtonian quincunx that required the flies to make a series of choices, positive or negative, in relation to light. After a sequence of 15 such choices a fly found itself in one of 16 terminal tubes. Fifteen negative decisions would place a fly in tube number 1, and 15 positive decisions would place a fly in tube number 16. Flies making similar numbers of negative and positive decisions were found in tubes 8 or 9.

The initial experimental run started with 300 cross-bred flies. At the end of the run the flies were sorted by sex and scored. Scores were determined by tube number, weighted by the number of flies in each. Female flies in the initial run had a mean score of 8.70, with a standard deviation in scores of 2.74. Male flies had a mean score of 8.64 with a standard deviation of 3.02. Both means are close to the number 8.50 expected for a sample of flies making equal numbers of positive and negative choices. Subsequent runs involved a new cohort of 300 flies bred

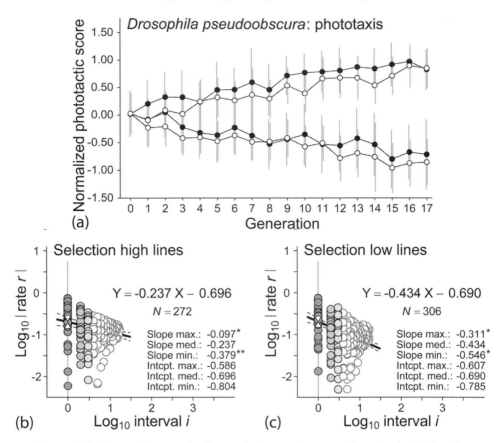

(a)

(b)

(c)

Figure 7.3 *Drosophila pseudoobscura* 17-generation experiments of Dobzhansky and Spassky (1967) selecting for high and low phototactic response. (a) Realized gain and loss during each experiment. Open (female) and closed (male) circles are means; vertical lines span ±1 standard deviation. (b) Log-rate-interval plot for 272 rates of change in male and female high-selection lines. (c) LRI plot for 306 rates of change in male and female low-selection lines. Median slopes of −0.237 and −0.434 in (b) and (c) lie between expectations for directional time series and random time series. High-selection lines (b) are significantly different from directional, random, and stationary, but low-selection lines (c) cannot be distinguished from random. Median intercepts are −0.696 and −0.690 in (b) and (c), indicating step rates h_0 averaging $10^{-0.696} = 0.201$ and $10^{-0.690} = 0.204$ standard deviations per generation.

from the 25 most negative males and 25 most negative females in the low-line experiment, and a new cohort of 300 flies bred from the 25 most positive males and 25 most positive females in the high-line experiment. The Dobzhansky–Spassky phototaxis selection results, rescaled using Equations 7.4 and 7.5, are shown in Figure 7.3a.

Dobzhansky and Spassky (1967) noted that the variance decreased and the distributions of variation became more skewed in each of their experimental lineages, but they made no attempt to correct this in presenting their results. Decreasing variance and skewing are inevitable in an experiment where the sample variances and number of generations are high relative to the range of choices available. We can correct this using an arcsin transformation to normalize the original scores x after conversion to proportions p on a -1 to $+1$ scale, where a and b define the original arithmetic range of scores:

$$x^* = f(x) = \sin^{-1}p, \text{ where } p = \left(x - \frac{a+b}{2}\right) \Big/ \max . \left(x - \frac{a+b}{2}\right) \quad (7.4)$$

In the Dobzhansky–Spassky experiments $a = 1$ and $b = 16$, so $p = (x - 8.5) / (16 - 8.5)$. When $x = a = 1$, then $p = -1$ and $x^* = -1.57$. When $x = b = 16$, then $p = +1$ and $x^* = 1.57$. Division of x* values by 1.57 completes the transformation to a -1 to $+1$ scale.

Following the logic of Equation 7.3, we can normalize the standard deviations as:

$$s^* = \frac{x_b^* - x_a^*}{2}, \text{ where } x_a^* = f_{Eq.7.4}(x - s) \text{ and } x_b^* = f_{Eq.7.4}(x + s) \quad (7.5)$$

In simple terms, we are averaging the results of first substituting $x - s$ for x in Equation 7.4, then substituting $x + s$ for x in Equation 7.4.

The female and male high-selection lines can be considered replicates of a single high-selection experiment, and rates for both are combined in the log-rate-interval or LRI graph of Figure 7.3b, where the corresponding 272 rates are plotted. The LRI plot has a median intercept of -0.696, indicating a median rate $10^{-0.696} = 0.201$ standard deviations per generation on a timescale of one generation. A bootstrapped 95% confidence interval for the median ranges from about $10^{-0.804} = 0.157$ to $10^{-0.586} = 0.259$ standard deviations per generation. The 29 one-generation step rates in Figure 7.3b range from a minimum of -2.225 standard deviation units on a log scale ($10^{-2.225} = 0.006$ standard deviations per generation) to a maximum of -0.131 standard deviation units on a log scale ($10^{-0.131} = 0.740$ standard deviations per generation).

The median temporal scaling slope in Figure 7.3b is -0.237, with a 95% confidence interval ranging from -0.379 to -0.097. The former excludes -1.000 expected for stasis and -0.500 for a random time series, and the latter excludes 0.000 expected for sustained directional change. Thus the high-selection time series as a whole appears to be directional with a random component.

The female and male low-selection lines can be considered replicates of a single low-selection experiment, and rates for both are combined in the log-rate-interval or LRI graph of Figure 7.3c, where the corresponding 306 rates are plotted. The LRI plot has a median intercept of -0.690, indicating a median rate $10^{-0.690} = 0.204$ standard deviations per generation on a timescale of one generation. A bootstrapped 95% confidence interval for the median ranges from about $10^{-0.785} = 0.164$ to $10^{-0.607} = 0.247$ standard deviations per generation. The 34 one-generation step rates in Figure 7.3c range from a minimum of -1.874 standard deviation units on a log scale ($10^{-1.874} = 0.006$ standard deviations) to a maximum of -0.129 standard deviation units on a log scale ($10^{-0.129} = 0.743$ standard deviations). These numbers are closely comparable to those for the high-selection line.

The median temporal scaling slope in Figure 7.3c is -0.434, with a 95% confidence interval ranging from -0.546 to -0.311. The former excludes -1.000 expected for stasis, and the latter excludes 0.000 expected for sustained directional change, but neither excludes -0.500 expected for a random time series. Thus the low-selection time series is more random than the high-selection series.

7.3 Experiments with Domestic Chickens, *Gallus gallus*

7.3.1 Shank Length

I. Michael Lerner and Everett R. Dempster carried out a classic selection experiment on domestic chickens, selecting for increased size as measured by shank length (tarsometatarsus length) on live birds (Lerner and Dempster, 1951; Lerner, 1954). This research was started in 1938 and continued through 1952, at the University of California, Berkeley, and the affiliated California Agricultural Experiment Station (now part of the University of California, Davis).

Lerner and Dempster provided counts of the birds selected, mean shank length, and phenotypic variance for each generation in the selected line. They also provided counts of the birds involved and the mean shank length for each generation in the control population. The median number of birds in the selected line was 88 and the median number of birds in the control population was 346. When variances were not reported (e.g., for shank length in the control population), the mean of known shank length variances was substituted. Lerner and Dempster's results for 14 generations of selection are plotted and analyzed in Figure 7.4a.

Mean shank length at the beginning of the study was 9.69 cm, which increased by 13%, to 10.99 cm, during 14 generations of selection (Fig. 7.4a). Rates of change for the selected line are shown on the log-rate-interval or LRI graph of Figure 7.4b. The LRI plot has a median intercept of -0.513, indicating a median rate $10^{-0.513} = 0.307$ standard deviations per generation on a timescale of one generation. A bootstrapped 95% confidence interval for the median ranges from

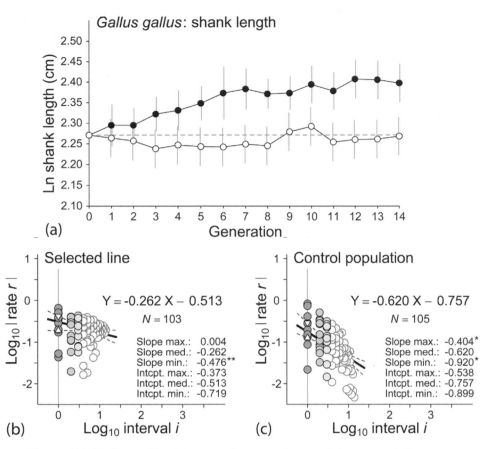

Figure 7.4 *Gallus gallus* 14-generation experiment of Lerner and Demptster (1951) selecting for longer leg shank length. Supplemental information was provided by Lerner (1954). (a) Realized gain for shank length in the selected lineage (solid circles) compared to the control population (open circles). (b) Log-rate-interval plot for 103 rates of change in the selected line. Median slope of −0.262 is significantly different from expectation for stasis and random change, but not from that for directional change. (c) LRI plot for 105 rates of change in the control population. Median slope of −0.620 is significantly different from expectation for stasis or directional change, but not from that for random change. Median intercepts are −0.513 and −0.757 in (b) and (c), indicating step rates h_0 of $10^{-0.513} = 0.307$ and $10^{-0.757} = 0.175$ standard deviations per generation.

about $10^{-0.719} = 0.191$ to $10^{-0.373} = 0.424$ standard deviations per generation. The 13 one-generation step rates in Figure 7.4b range from a minimum of −1.362 standard deviation units on a log scale ($10^{-1.362} = 0.043$ standard deviations per generation) to a maximum of −0.181 standard deviation units on a log scale ($10^{-0.181} = 0.659$ standard deviations per generation).

The median temporal scaling slope in Figure 7.4b is −0.262, with a 95% confidence interval ranging from −0.476 to +0.004. The more negative slope excludes both −1.000 expected for stasis and −0.500 expected for a random time series. Thus the selected lineage as a whole is interpreted as directional.

In the control population shank length decreased slightly from 9.69 to 9.66 cm (Fig. 7.4a). Rates of change for the control population are shown on the log-rate-interval or LRI graph of Figure 7.4c. The LRI plot has a median intercept of −0.757, indicating a median rate $10^{-0.757} = 0.175$ standard deviations per generation on a timescale of one generation. A bootstrapped 95% confidence interval for the median ranges from about $10^{-0.899} = 0.126$ to $10^{-0.538} = 0.290$ standard deviations per generation. The 14 one-generation step rates in Figure 7.4c range from a minimum of −1.663 standard deviation units on a log scale ($10^{-1.663} = 0.022$ standard deviations per generation) to a maximum of −0.083 standard deviation units on a log scale ($10^{-0.083} = 0.826$ standard deviations per generation).

The median temporal scaling slope in Figure 7.4c is −0.620, with a 95% confidence interval ranging from −0.920 to −0.404. The more negative slope excludes −1.000 expected for stasis. The more positive slope excludes 0.000 expected for a directional time series. Neither excludes −0.500 expected for random change. Thus the control population as a whole is interpreted as changing randomly.

7.3.2 Egg Production

Another experimental study of artificial selection in single-comb white leghorn chickens was initiated in 1950 by Robert S. Gowe and colleagues at the Central Experimental Farm, which is part of the Ottawa Research and Development Centre, Department of Agriculture, Canada. The principal publications of interest here are by Gowe et al. (1959) and Gowe and Fairfull (1985). The Gowe study continued for 30 years and 30 generations.

Gowe and colleagues established an artificial selection line of chickens in 1950 that they called the "Ottawa strain" or "strain 3." Simultaneously they established a control line, which was sometimes called "strain 5." A second selection line, called "New" or "strain 4," was added in 1951. The two selection lineages were maintained as random-breeding populations averaging about 30 breeding males and 275 breeding females. These yielded about 1,950 female progeny each year that were then moved to hen houses and tested for egg production. The control strain generally involved a similar number of breeding males and females, and their progeny were tested similarly. Breeding males and females in the selection lines were chosen each year based on their contribution to egg production. Production was calculated as the mean number of eggs laid by each hen in a 273-day production year, starting from the time progeny were moved into the hen houses.

Gowe and Fairfull (1985) provided an illustration (their figure 2) of egg production for each strain for each generation and this has been digitized. Gowe and Fairfull provided lists of variance components (their tables 4 and 6) for each selected strain for each generation from which minimum phenotypic variances for each generation can be derived. Variances were not provided for the control strain, so the variance for each control-line generation was estimated as the mean for the two selection strains in that generation.

A plot of egg production for the selection and control lines is shown in Figure 7.5a. Production numbers are logged because the value of adding an egg is related to the number produced before. Mean egg production at the beginning of the strain-3 experiment was 66.9 eggs per hen per production interval, which increased by 72%, to 115.0 eggs after 30 generations of selection. Comparable numbers for the strain-4 experiment are 63.7, increasing by 88% to 119.8. Mean egg production at the beginning of the control lineage was 70.0 eggs per hen per production interval that increased by 17%, to 81.6 eggs after 30 generations of selection.

Gradual increase in egg production in the control lineage can be attributed to improvements in nutrition, housing, lighting, and vaccinations over the course of the 30-year experiment (Gowe and Fairfull, 1985: p. 131). These improvements no doubt influenced the selection lines as well, which is why it is important to have controls. The sharp drop in production during generations 19 and 20 in Figure 7.5a was associated with mortality due to Marek's disease. This affected both selected strains and the control strain, and consequently generations 19 and 20 were omitted from all of the rate analyses here.

Rates of change for the selected lines are shown on the log-rate-interval or LRI graph of Figure 7.5b. The LRI plot has a median intercept of -1.021, indicating a median rate $10^{-1.021} = 0.095$ standard deviations per generation on a timescale of one generation. A bootstrapped 95% confidence interval for the median ranges from about $10^{-1.102} = 0.079$ to $10^{-0.943} = 0.114$ standard deviations per generation. The 51 one-generation step rates in Figure 7.5b range from a minimum of -2.325 standard deviation units on a log scale ($10^{-2.325} = 0.005$ standard deviations per generation) to a maximum of -0.323 standard deviation units on a log scale ($10^{-0.323} = 0.475$ standard deviations per generation).

The median temporal scaling slope in Figure 7.5b is -0.232, with a 95% confidence interval ranging from -0.313 to -0.154. The more negative confidence limit excludes both -1.000 expected for stasis and -0.500 expected for a random time series, and the more positive limit excludes 0.000 expected for a directional time series. Thus the selected lineages taken together are interpreted as directional with a random component.

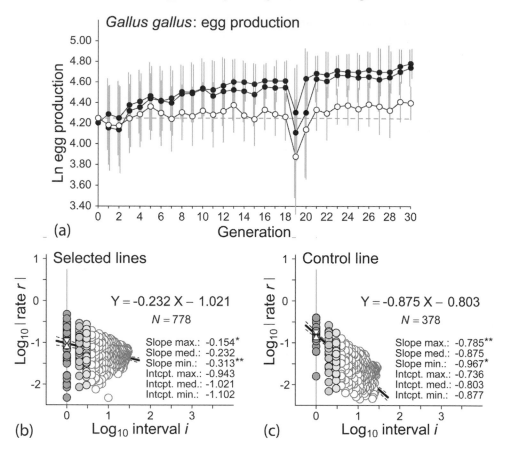

Figure 7.5 *Gallus gallus* 30-generation experiment of Gowe and Fairfull (1985) selecting for increased egg production. (a) Natural logarithm of the realized gain in production for each of the two selection lineages (solid circles) compared to production in the control lineage (open circles). Production was calculated as the mean number of eggs laid by a hen in a 273-day production interval. (b) Log-rate-interval plot for 778 rates of change in the selection lines. Median slope of -0.232 for the selection lines lies between expectation for directional change and random change, and is significantly different from both. (c) LRI plot for 378 rates of change in the control line. Median slope of -0.875 for the control line lies between expectation for stasis and random change, and is significantly different from both. Median intercepts are -1.021 and -0.803 in (b) and (c), indicating step rates h_0 of $10^{-1.021} = 0.095$ and $10^{-0.803} = 0.157$ standard deviations per generation. Experiment was interrupted by Marek's disease in generation 19.

Rates of change for the control lineage are shown on the log-rate-interval or LRI graph of Figure 7.5c. The LRI plot has a median intercept of -0.803, indicating a median rate $10^{-0.803} = 0.157$ standard deviations per generation on a timescale of one generation. A bootstrapped 95% confidence interval for the median ranges from about $10^{-0.877} = 0.133$ to $10^{-0.736} = 0.184$ standard deviations per gcneration. The 26 one-generation step rates in Figure 7.5c range from a minimum of -1.798 standard deviation units on a log scale ($10^{-1.798}$ = 0.016 standard deviations per generation) to a maximum of -0.399 standard deviation units on a log scale ($10^{-0.399}$ = 0.399 standard deviations per generation).

The median temporal scaling slope in Figure 7.5c is -0.875, with a 95% confidence interval ranging from -0.967 to -0.785. The more negative limit excludes -1.000 expected for stasis. The more positive limit excludes both 0.000 expected for a directional time series and -0.500 expected for random change. Thus the control population as a whole is interpreted as stationary with a random component.

7.4 Experiments with Laboratory Mice, *Mus musculus*

The final investigation considered here is a long-term replicated selection experiment involving the body weight of the laboratory mouse *Mus musculus*. The study was initiated and carried out by Douglas S. Falconer at the Institute of Animal Genetics in Edinburgh, Scotland. The research was initiated in 1960, starting with four years and 14 generations of background study of the cross-bred strain of mice later used as the founder stock for the selection experiment itself. In the background study large samples of male and female mice were found to have mean six-week body weights of $\bar{x} = 22.24$ grams (sexes averaged; 3.102 on a natural log scale). Within-generation variances for six-week weight were reported as 14.34 g^2 for males and 7.15 g^2 for females, yielding an average variance of $s^2 = 10.75$ g^2 and an average within-generation standard deviation of $s = 3.278$ g (0.147 on a natural log scale).

The replicated selection experiment itself started in 1964 when the founder stock was subdivided into six replication lines, labeled A–F. Each replication sample was then divided into a large selection line, a control, and a small selection lineage, for a total of 18 parallel mouse lines. All 18 of the original lines were maintained for an additional 23 generations. Each line was bred from eight single-pair matings in each generation, with the matings designed to minimize inbreeding. All progeny were weaned at three weeks of age and weighed at six weeks. Selection of males and females to found the next generation was made at the time of weighing. Details are given in Falconer (1973).

Falconer published the sex-averaged mean weights for each generation in each replicate lineage in the form of a diagram (his figure 6), and in 1992 he was generous enough to provide copies of the handwritten tabulations from which the diagram was drawn. Falconer's weights, logged, were used to construct Figure 7.6a. The weights were then combined with the within-generation standard deviation derived above ($s = 0.147$) to calculate rates in standard deviations

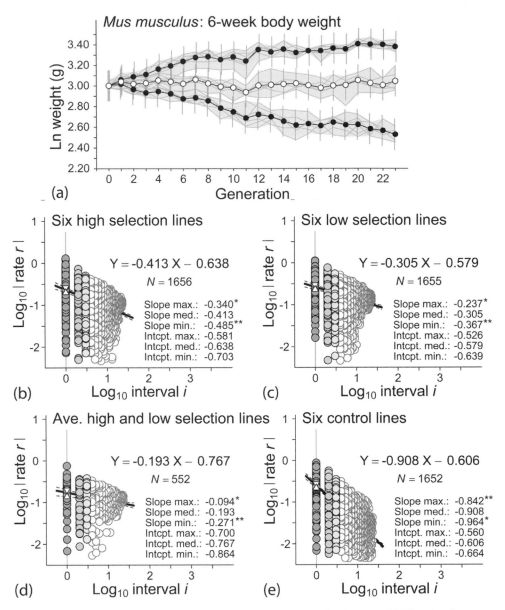

Figure 7.6 *Mus musculus* 23-generation experiment of Falconer (1973) selecting for large and small body weight. (a) Natural logarithm of the realized gain in weight for six replicated lineages selected to be larger (upper gray envelope), six replicated lineages selected to be smaller (lower gray envelope), and six replicated lineages not subject to experimental selection (middle gray envelope). Running mean values for the high- and low-selection lines are shown with solid circles, and means for the control lines are shown with open circles. (b) Log-rate-interval plot for 1656 rates of change in the high-selection lines. Median slope of −0.413 lies between expectation for directional change and random change, and is significantly different from both. (c) LRI plot for 1655 rates of change in the

per generation for each of the 18 replicate lines on timescales ranging from one to 23 generations.

The average weight of the mice used to found each set of replication lines was 22.24 *g*. Twenty-three generations of selection for large size yielded an increase of 47% to a mean weight of 32.59 *g* for all six of the large lines. Twenty-three generations of selection for small size yielded a decrease of 38% to a mean weight of 13.87 *g* for the small lines. During this time the control increased by 5% to a mean weight of 23.29 *g* for the six control lines. Expressed in standard deviations, these changes are +3.157, −2.553, and +0.320, respectively, which correspond to an average rate of +0.137 standard deviations per generation for the large selection lines, −0.111 standard deviations per generation for the small selection lines, and +0.014 standard deviations per generation for the control lines, all on a timescale of 23 generations.

Rates of change for the six replicate lineages selected to be large are shown on the log-rate-interval or LRI graph of Figure 7.6b. The LRI plot has a median intercept of −0.638, indicating a median rate $10^{-0.638} = 0.230$ standard deviations per generation on a timescale of one generation. A bootstrapped 95% confidence interval for the median intercept ranges from about $10^{-0.703} = 0.198$ to $10^{-0.581} = 0.262$ standard deviations per generation. The 138 one-generation step rates in Figure 7.6b range from a minimum of −2.140 standard deviation units on a log scale ($10^{-2.140} = 0.007$ standard deviations per generation) to a maximum of 0.112 standard deviation units on a log scale ($10^{0.112} = 1.294$ standard deviations per generation).

The median temporal scaling slope in Figure 7.6b is −0.413, with a 95% confidence interval ranging from −0.485 to −0.340. The more negative confidence limit excludes both −1.000 expected for stasis and −0.500 expected for a random time series, and the more positive limit excludes 0.000 expected for a directional time series. Thus the large-selection lineages taken together are interpreted as directional with a random component.

Figure 7.6 *(cont.)* low-selection lines. Median slope of −0.305 lies between expectation for directional change and random change, and is significantly different from both. (d) LRI plot for 552 rates of change in the averaged high- and low-selection lines. Median slope of −0.193 lies between expectation for directional change and random change, and is significantly different from both. (e) LRI plot for 1652 rates of change in the control lines. Median slope of −0.908 lies between expectation for random change and a stationary time series, and is significantly different from both. Significant difference from random indicates that control lines were subject to stabilizing natural selection. Median intercepts are −0.638 and −0.579 in (b) and (c), indicating step rates h_0 of $10^{-0.638} = 0.230$ and $10^{-0.579} = 0.264$ standard deviations per generation.

Rates of change for the six replicate lineages selected to be small are shown on the log-rate-interval or LRI graph of Figure 7.6c. The LRI plot has a median intercept of -0.579, indicating a median rate $10^{-0.579} = 0.264$ standard deviations per generation on a timescale of one generation. A bootstrapped 95% confidence interval for the median intercept ranges from about $10^{-0.639} = 0.230$ to $10^{-0.526} = 0.298$ standard deviations per generation. The 138 one-generation step rates in Figure 7.6c range from a minimum of -1.788 standard deviation units on a log scale ($10^{-1.788} = 0.016$ standard deviations per generation) to a maximum of 0.047 standard deviation units on a log scale ($10^{0.047} = 1.114$ standard deviations per generation).

The median temporal scaling slope in Figure 7.6c is -0.305, with a 95% confidence interval ranging from -0.367 to -0.237. Here as in Figure 7.6b the more negative confidence limit excludes both -1.000 expected for stasis and -0.500 expected for a random time series, and the more positive limit excludes 0.000 expected for a directional time series. Thus the small-selection lineages taken together are interpreted as directional with a random component.

If we average the six replicate lineages selected to be large, and average the six replicate lineages selected to be small, the results are shown in Figure 7.6a as black lines and filled circles superimposed on the gray envelopes enclosing high- and low-replicated selection lines. Each of the averaged black lines is smoothed relative to the underlying individual lineages. The distribution of rates for the averaged high- and low-selection lines is shown on the LRI graph of Figure 7.6d. The median temporal scaling slope is now -0.193, with a 95% confidence interval ranging from -0.271 to -0.094. Interpretation of the selection lineages combined this way is still directional with a random component, but averaging enhances the directionality and clarifies the response to selection.

Finally, rates of change for the six replicate control lineages are shown on the LRI graph of Figure 7.6e. The LRI plot has a median intercept of -0.606, indicating a median rate $10^{-0.606} = 0.248$ standard deviations per generation on a timescale of one generation. A bootstrapped 95% confidence interval for the median intercept ranges from about $10^{-0.664} = 0.217$ to $10^{-0.560} = 0.275$ standard deviations per generation. The 138 one-generation step rates in Figure 7.6e range from a minimum of -2.164 standard deviation units on a log scale ($10^{-2.164} = 0.007$ standard deviations per generation) to a maximum of -0.058 standard deviation units on a log scale ($10^{-0.058} = 0.875$ standard deviations per generation).

The median temporal scaling slope in Figure 7.6e is -0.908, with a 95% confidence interval ranging from -0.964 to -0.842. The more negative confidence limit for the control lineages excludes -1.000 expected for stasis, and the more positive limit excludes both -0.500 expected for a random time series and 0.000

expected for a directional time series. Thus the replicated control lineages taken together are stationary with a random component.

7.5 Discussion

Experimental lineages are important for four reasons. First, they provide direct evidence of amounts of change, and hence rates of change, as populations evolve from one generation to the next. Second, temporal scaling of rates in experimental control lineages shows that the control lineages are stationary and not random. Third, the generation-to-generation step rates observed in stationary and directional lineages are virtually the same, indicating that the alternative modes of stasis and directional change depend on the signs of the rates and not on differences in their magnitude. Fourth, the number of founders is known in each generation of a selection experiment, enabling calculation of an expected step rate and comparison to what we actually observe.

7.5.1 Step Rates in Experimental Lineages

There are 14 experimental lineages or sets of lineages analyzed in Table 7.1 and in Figures 7.1 through 7.6. The 14 experimental laboratory studies reviewed in this chapter yielded 418 high-line step rates, 254 low-line step rates, and 178 control step rates for a total of 672 independent selection step rates and 850 total step rates on a timescale of one generation. In addition, there were 9,156 high-line, 3,468 low-line, and 1,957 control net rates on longer time scales compounded from the step rates, making a grand total of 15,431 rates. The 14 studies range in size from that by Müller (1886), yielding three step rates to the parallel studies by Falconer (1973) that each yielded 138 step rates (Table 7.2).

It is interesting to compare the temporal scaling of differences and intervals on a log-difference-interval or LDI plot, with temporal scaling of log-transformed rates and intervals on a log-rate-interval or LRI plot. Both are shown side by side in Figure 7.7. Histograms for the untransformed high-line and low-line step differences and rates recorded here (not shown) are strongly skewed in a positive direction with a very long right tail. Logarithmic transformation does not remove the skew but reduces it and reverses it to negative skew, as shown in Figures 7.7a and 7.7b.

The histogram of Figure 7.7a shows the distribution of $\log_{10}$ values for the step differences recorded here. The 672 independent one-generation step differences have a median $\log_{10}$ value of -0.477, which corresponds to a median difference of 0.33 standard deviations per generation on a timescale of one generation. In

Table 7.2 *Generation-to-generation step rates for the experimental lineages analyzed here*

Study	Taxon	Trait	Source	Step rates (h_0)					
				N	Minimum	Q1	Median	Q3	Maximum
1	*Zea mays*	Kernel rows	Müller (1886)	3	0.336	—	1.202	—	1.654
2	*Zea mays*	Kernel rows	de Vries (1909)	7	0.023	0.234	0.343	0.518	0.885
3	*Zea mays*	Protein	Dudley and Lambert (2004)	100	0.025	0.302	0.556	1.070	3.888
4	*Drosophila melanogaster*	High eye-facet no.	Zeleny (1922)	77	0.004	0.287	0.474	0.708	2.491
5	*Drosophila melanogaster*	Low eye-facet no.	Zeleny (1922)	82	0.009	0.251	0.538	0.862	2.141
6	*Drosophila pseudoobscura*	High phototaxis	Dobzhansky and Spassky (1967)	29	0.006	0.157	0.249	0.496	0.739
7	*Drosophila pseudoobscura*	Low phototaxis	Dobzhansky and Spassky (1967)	34	0.013	0.175	0.223	0.334	0.743
8	*Gallus gallus*	Shank length	Lerner and Dempster (1951)	13	0.044	0.174	0.356	0.513	0.658
9	*Gallus gallus*	Control	Lerner and Dempster (1951)	14	0.022	0.098	0.152	0.261	0.826
10	*Gallus gallus*	Egg production	Gowe and Fairfull (1985)	51	0.005	0.058	0.107	0.192	0.475
11	*Gallus gallus*	Control	Gowe and Fairfull (1985)	26	0.016	0.126	0.178	0.218	0.400
12	*Mus musculus*	High body weight	Falconer (1973)	138	0.007	0.119	0.277	0.471	1.294
13	*Mus musculus*	Low body weight	Falconer (1973)	138	0.016	0.176	0.311	0.499	1.114
14	*Mus musculus*	Control	Falconer (1973)	138	0.007	0.132	0.250	0.404	0.875

All studies are high-line studies, increasing in trait value, except for those labeled "low" or "control." Rates are in haldanes (h_0), phenotypic standard deviations per generation on a timescale of one generation. *N* is number of step rates. One half of the step rates in each experiment lie between quartile limit values given in columns *Q1* and *Q3*.

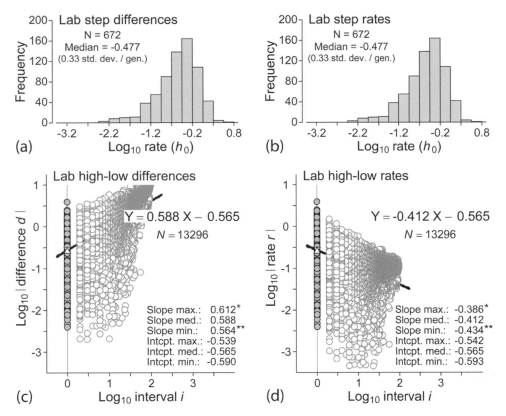

Figure 7.7 High-line and low-line selection differences and rates for all experimental studies in this chapter. (a) Histogram of high- and low-line step differences (column of shaded circles in panel c below). (b) Histogram of high- and low-line step rates (column of shaded circles in panel d below). (c) LDI log-difference-interval plot for all high- and low-line differences. (d) LRI log-rate-interval plot for all high- and low-line rates. Differences d are in standard deviation units, rates r are in standard deviation units per generation, and interval lengths i are in generations. Confidence intervals were bootstrapped with 1,000 replications. Note that the median slopes 0.588 in panel c and -0.412 in panel d are complementary, differing by one unit. Median intercepts in both panels of -0.565 indicate a step rate h_0 of $10^{-0.565} = 0.272$ standard deviations per generation.

Figure 7.7c, this is the distribution of differences for interval lengths of one generation, and the distribution of shaded circles that have zero value on the $\log_{10}$ x-axis. The histogram of Figure 7.7b shows the distribution of $\log_{10}$ values for the step rates recorded here. It shows the distribution of rates with interval lengths of one generation from Figure 7.7d. This is again the distribution of shaded circles with a $\log_{10}$ value of zero on the x-axis. Differences on a one-generation timescale are equivalent to rates on a one-generation timescale, or, inverting this statement, rates on a one-generation timescale are equivalent to differences on a one-generation timescale.

The LDI temporal scaling of high-line and low-line laboratory selection differences shown in Figure 7.7c is complementary to the LRI temporal scaling of high- and low-line laboratory selection rates shown in Figure 7.7d. In the LDI plot the differences are plotted against the corresponding number of generations over which each difference was measured. In the LRI plot the rates are the differences divided by their corresponding intervals, with these quotients plotted against the corresponding number of generations over which each difference was measured and each rate calculated.

The LDI distribution in Figure 7.7c gives us the same information about the underlying set of multiple time series as the LRI distribution in Figure 7.7d does. The median slope of 0.588 for the LDI distribution and the median slope of -0.412 for the LRI distribution differ from each other by one unit, reflecting their complementarity. The two slopes lie between the expectation for a directional time series and the expectation for a random time series (between expectations of 1.000 and 0.500, respectively, in the LDI case, and between expectations of 0.000 and -0.500, respectively, in the LRI case). Both slopes are significantly different from purely directional selection and from purely random change, and both slopes are even farther from expectation for a stationary time series (0.000 for LDI and -1.000 for LRI).

Temporal scaling slopes that are significantly different from expectation for directional change and from expectation for random change tell us two things: (1) Experimental selection has a directional effect that goes beyond random change, and (2) this effect retains a large random component, which may be due to chance alone and/or some combination of directional change and stasis.

The median LDI and LRI intercepts in Figures 7.7c and 7.7d provide the same information because both medians have the same value (-0.565). The 95% confidence intervals differ slightly because each was bootstrapped independently. A median intercept of -0.565 for an LDI distribution of high- and low-line differences points to a median rate of $h_0 = 0.27$ standard deviations per generation on a timescale of one generation — which is a little lower than the median step rate of $h_0 = 0.33$.

We can divide the full sample of step rates in Figure 7.8a into its components, which include: (1) high-line selection ($N = 418$), (2) low-line selection ($N = 254$), and (3) control rates ($N = 178$). All are left-skewed on a logarithmic scale. A histogram of positive or high-selection step rates (Figure 7.8b) is similar to that for negative or low-selection rates (Figure 7.8c). Medians for the two are, respectively, $h_0 = 0.32$ and $h_0 = 0.35$ standard deviations per generation. The median step rate for the control lineages (Figure 7.8d) is lower at $h_0 = 0.23$ standard deviations per generation.

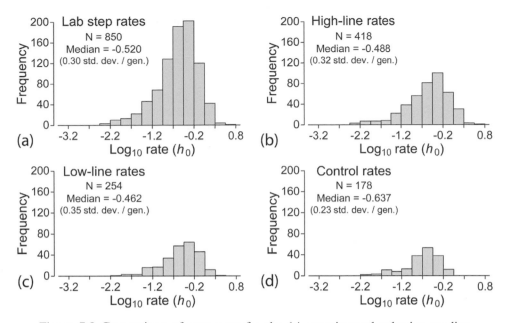

Figure 7.8 Comparison of step rates for the 14 experimental selection studies listed in Table 7.2. Rates, on a $\log_{10}$ scale, are in haldanes (h_0; standard deviations per generation on a timescale of one generation). (a) Histogram showing the distribution of all 850 experimental step rates. (b) High-selection-line step rates. (c) Low-selection-line step rates. (d) Control-line step rates. Note that the full sample of step rates has a median $h_0 = 0.30$ standard deviations per generation on a time scale of one generation. The high and low selection lines have a slightly higher median rate, while the control lines have a slightly lower median rate.

Step rates for all 14 studies are shown graphically in Figure 7.9, enabling comparison of rates for selection lineages (high, *H*; or low, *L*) and control lineages (*C*) in parallel studies. Parallel studies in Figure 7.9 are linked by arrows. In the first comparison of a selection lineage and a control lineage, that for studies 8 and 9 of shank length in chickens, high-line selection clearly has a higher median rate than the control ($h_0 = 0.356$ versus $h_0 = 0.152$ standard deviations per generation). However, in the next such comparison, that for studies 10 and 11 of egg production in chickens, the control median is higher than the high-line selection median ($h_0 = 0.178$ versus $h_0 = 0.107$ standard deviations per generation). In the final comparison, that for studies 12, 13, and 14 of body weight in mice, the high- and low-line selection medians are slightly higher than the control median ($h_0 = 0.277$ and $h_0 = 0.311$ versus $h_0 = 0.250$ standard deviations per generation). There is no systematic difference and it is doubtful that there is any significant difference in the step rates of control and selection lineages.

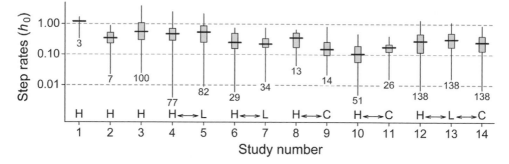

Figure 7.9 Comparison of step rates for the 14 experimental selection studies listed in Table 7.2. Rates, on a $\log_{10}$ scale, are in haldanes (h_0; standard deviations per generation on a timescale of one generation). Box and whisker plots show the median rate (horizontal lines), median quartiles (gray rectangles), and the full range of step rates (vertical lines) found in each study. Inset numbers are the sample size (N). Note that all of the median quartiles for $N > 3$ fall in the range of 0.1 to 1.0 standard deviations. The median for all step rates combined is $h_0 = 0.30$ (Fig. 7.8). Letters H, L, and C, represent high-line, low-line, and control experiments. Arrows connect related studies.

7.5.2 Temporal Scaling of Rates in Control Lineages

Three of the studies reported here have control lineages: the 14-generation Lerner and Dempster (1951) study of shank length in chickens; the 30-generation Gowe and Fairfull (1985) study of egg production in chickens; and the 23-generation Falconer (1973) study of body weight in mice. In the first of these (Figure 7.4), the median temporal scaling slope for the control line, -0.620, is more stationary than expected for a random time series (-0.500), but the difference from random is not statistically significant (possibly due to the short time series, relatively few rates, and hence relatively low statistical power). In the Gowe and Fairfull study (Figure 7.5) the time series is longer, there are more rates on a wider range of timescales, and the median temporal scaling slope for the control line, -0.875, is more stationary than random, with the difference being statistically significant. Finally, in the Falconer study (Figure 7.6) the temporal scaling slope for all six of the control lineages considered together, -0.908, is close to expectation for a stationary time series and, again, significantly different from random.

In many experimental settings "controls" are null models set up to represent random behavior as a baseline for evaluation of experimental results. However, here the two control experiments that involve the longest lineages can be shown to be significantly more stationary than random. The null model of randomness can be rejected in both experiments in favor of a stationary time series with a random component. The stationarity is coming from another form of selection, stabilizing natural selection. To move the population mean in an experimental setting, the directional selection applied must be strong enough to overcome whatever stabilizing natural selection is imposed by the experimental environment. The latter never disappears but is always present as a component in any response to directional selection.

7.5.3 Step Rates in Stationary and Directional Lineages

The comparison of stationary control and directional experimental-selection lineages is most robust in the Falconer (1973) mouse study. Here the median intercept for temporal scaling of the control lineages is −0.606 (Figure 7.6e). This lies within a 95% confidence interval for the temporal scaling intercept of the high- selection lineages, that is, within the limits of −0.703 and −0.582 (Figure. 7.6b). It also lies within a 95% confidence interval for the temporal scaling intercept of the low-selection lineages, with limits from −0.636 to −0.528 (Figure 7.6b). Stationary lineages responding to stabilizing selection differ from directional lineages responding to directional selection, not by the value of the step rates per se but by the pattern of change in their signs. The signs of the step rates reverse more frequently than random in stationary lineages and reverse less frequently than random in directional lineages.

7.5.4 Step Rates Expected for Change at Random

A final advantage of experimental-selection studies is that the number of individuals founding each new generation is known. This enables us to calculate the standard error (*SE*) of the expected mean phenotype from the observed phenotypic standard deviation. Next, following Equation 4.5, we can calculate the mean absolute deviation (*MAD*) of the expectation. The *MAD* value is the "null" step rate expected for an experimental population evolving at random that can then be compared to the step rate that was actually calculated for each experiment.

Expected step rates (MAD) for random change in the experimental lineages analyzed here are listed in Table 7.3, where they are compared to step rates actually observed in the experiments (OBS). The observed values are median step rates taken from Table 7.1, and exponentiated median intercepts taken from Figures 7.1–7.6. Observed step rates in Table 7.3 are consistently greater than corresponding null step rates. The median quotient of the two is 9.1. At minimum, observed step rates exceed the null expectation by a factor of 2.7, and, at maximum, the two

Table 7.3 *Expected number of generation-to-generation step rates for random change in the experimental lineages analyzed here*

Study	Taxon	Trait	Source	N^*	s	SE	MAD	OBS	Obs/Exp
1	*Zea mays*	Kernel rows	Müller (1886)	1	0.126	0.126	0.101	1.202	12.0
2	*Zea mays*	Kernel rows	de Vries (1909)	1 (?)	0.157	0.157	0.125	0.343	2.7
3	*Zea mays*	Protein	Dudley and Lambert (2004)	12	0.017	0.005	0.004	0.394	100.8
4	*Drosophila melanogaster*	High eye-facet no.	Zeleny (1922)	6.1	0.202	0.082	0.065	0.442	6.8
5	*Drosophila melanogaster*	Low facet no.	Zeleny (1922)	5.1	0.141	0.063	0.050	0.514	10.3
6	*Drosophila pseudoobscura*	High phototaxis	Dobzhansky and Spassky (1967)	50	0.383	0.054	0.043	0.201	4.7
7	*Drosophila pseudoobscura*	Low phototaxis	Dobzhansky and Spassky (1967)	50	0.488	0.069	0.055	0.204	3.7
8	*Gallus gallus*	Shank length	Lerner and Dempster (1951)	88	0.050	0.005	0.004	0.307	72.8
9	*Gallus gallus*	Control	Lerner and Dempster (1951)	346	0.172	0.009	0.007	0.175	23.7
10	*Gallus gallus*	Egg production	Gowe and Fairfull (1985)	305	0.225	0.013	0.010	0.095	9.3
11	*Gallus gallus*	Control	Gowe and Fairfull (1985)	320	0.223	0.012	0.010	0.157	15.8
12	*Mus musculus*	High body weight	Falconer (1973)	16	0.147	0.037	0.029	0.230	7.8
13	*Mus musculus*	Low body weight	Falconer (1973)	16	0.147	0.037	0.029	0.264	9.0
14	*Mus musculus*	Control	Falconer (1973)	16	0.147	0.037	0.029	0.248	8.4

Abbreviations: N^*, average number of individuals founding each new generation; s, phenotypic standard deviation in natural log units; SE, standard error ($s/\sqrt{N^*}$); MAD, mean absolute deviation ($\sqrt{(2/\pi)} \cdot SE$), which is the expected step rate; OBS, observed median step rate. Note that observed step rates exceed corresponding step rates expected for random change by factors ranging from 2.7 to 100.8.

differ by a factor of 100.8. This shows that experimental selection has a measurable and substantial effect on the magnitude of a step rate that goes beyond its influence on the sign of the step rate.

7.5.5 Experimental Selection and Natural Selection

The median step rates found in the high- and low-line experimental lineages studied here are fast: on the order of $h_0 = 0.32$ to $h_0 = 0.35$ standard deviations per generation on a timescale of one generation. Ten generations of selection at this rate, with no reversal, could in theory move the mean of a normally distributed population by more than three standard deviations or by about half the width of the normal curve itself. Twenty generations of selection like this could in theory move the population by virtually the entire width of a normal curve. In many situations there will, however, be countervailing stabilizing selection, which is why longer-term studies are valuable.

One response to selection at this rate is to imagine that experimental selection is artificial in some way: Yes, such rates can be achieved in an experimental setting, but natural selection could never move a population so fast in a natural environment. We have not yet compiled step rates of change in nature, but in the meantime it is worth recalling Douglas Falconer's admonition in the epigraph at the beginning of this chapter: "The ease with which artificial selection can change almost any character proves that natural selection could do so too."

7.6 Summary

1. Analysis of 14 selection experiments in 8 published studies yields 672 step rates quantifying high- and low-line selection from one generation to the next. The median rate for this sample is $h_0 = 0.33$ standard deviations per generation on a timescale of one generation.
2. High-line experiments selecting for increases in trait value yield virtually the same rates as low-line experiments selecting for decreases in trait value, with median h_0 values of 0.32 and 0.35, respectively. Both are near the selection median of $h_0 = 0.33$.
3. Control lineages not subject to experimental selection have a median $h_0 = 0.23$. When paired with corresponding selection experiments, control lineage rates may be more or less than the selection lineage rates. It is doubtful that there is any significant difference between the step rates of control and selection lineages.
4. Control lineages with a substantial number of generations have temporal scaling slopes showing that they are stationary rather than random, a result that is

attributable to stabilizing natural selection. To move the population mean, directional selection must be strong enough to overcome any stabilizing natural selection imposed by the environment.

5. What separates stationary lineages responding to stabilizing selection from directional lineages responding to directional selection is not the values of the step rates per se but the pattern of change in their signs. The signs of the step rates reverse more frequently in stationary lineages, and the signs reverse less frequently in directional lineages.

6. Step rates observed in selection experiments (high lines, low lines, and control lines) exceed "null" rates expected for purely random change by factors ranging from 2 to 100.

8

Phenotypic Change Documented in Field Studies

> Our thinking must not exclude the possibility of animals attaining
> extremely rapid rates of evolution.
> *Richard F. Johnston and Robert K. Selander,* Science, *1964, p. 550*

Field observations come directly from nature and are thus unquestionably natural. Field studies give us a way to test whether rates of biotic change generated in experimental selection settings are "artificial" and misleading in any way. Compromises are required because we do not have the same control over organisms and environments that we have in an experimental setting, and the time involved in terms of generations is sometimes known with less precision.

Field studies differ greatly in terms of the organisms studied, traits measured, analyses reported, and time recorded. Some studies are designed and organized in advance, while others are opportunistic, taking advantage of historical events such as extreme weather, environmental anomalies, or island colonization to calibrate time and calculate rates. Field studies are generally observational, but they may also be experimental and involve manipulation of organisms in the field or manipulation of the environment.

Some field studies are longitudinal and compare traits in an evolving population at successive times. Other studies are cross-sectional and compare traits across populations at the same time. The common time for comparison in this case is generally the present. Field studies often build on museum collections for baseline information and extend this with active field research. Museum collections are especially important when studying sequential change spanning a long interval of time. Field studies also depend on museums to archive and preserve specimens and other documentation of field research.

This chapter combines independent case studies carried out by many investigators. To impose some organization we will first consider observational studies that are longitudinal with respect to time, and then observational studies that are

Box 8.1

The protocol for analyzing evolutionary rates developed in previous chapters is outlined here to encourage more widely comparable quantification of rates. This can be implemented in a spreadsheet or, better, scripted in a computer language. At minimum five quantities are required, which are efficiently arrayed in a matrix $\mathbf{M}_{i,j}$. Each row i represents one of two or more independent observations in a time series, and $j = 5$ is the minimum number of quantities required. For each row:

1. $\mathbf{M}_{i,1}$ is a time term t_i reflecting the time of observation i.
2. $\mathbf{M}_{i,2}$ is the number n_i of cases included in observation i.
3. $\mathbf{M}_{i,3}$ is the mean $\bar{x}_i$ for the sample included in observation i, where the mean is ideally the mean of natural-log (ln) transformed measurements. When raw measurements are not available the natural logarithm of the more commonly published arithmetic mean is acceptable as a compromise (ideally with one-half of the squared coefficient of variation subtracted).
4. $\mathbf{M}_{i,4}$ is the standard deviation s_i for the sample included in observation i, where the standard deviation is ideally the standard deviation of ln-transformed cases. When raw measurements are not available the coefficient of variation, a quotient of the more commonly published standard deviation of raw measurements divided by the mean of the raw measurements, is acceptable as a compromise. A compromise standard deviation can also be recovered from a standard error based on raw measurements by multiplying the standard error by the square root of the number n_i of corresponding cases.
5. $\mathbf{M}_{i,5}$ is the generation time g_i required to convert $\mathbf{M}_{i,1}$ to generations. This is usually a constant, but it may also be a function of the mean $\mathbf{M}_{i,3}$.

For the ith or nth combination of rows j and k in $\mathbf{M}$ we can now calculate:

1. $I_n = (\mathbf{M}_{k,1} - \mathbf{M}_{j,1}) / \bar{g}$ or $(t_k - t_j) / \bar{g}$, which is the temporal interval separating observations j and k expressed in terms of the average number of generations involved ($\bar{g}$ is mean generation time). In practice, any interval I_n as long or longer than a reproductive cycle but shorter than $\bar{g}$ is an interval $I_n = 1$. What happens between reproductive cycles effects change on a generational time scale even though the time involved may not be a full generation.
2. $d_n = \mathbf{M}_{k,3} - \mathbf{M}_{j,3}$ or $\bar{x}_k - \bar{x}_j$, which is the difference between the means for observations j and k in natural-log units.
3. $v_j = \mathbf{M}_{j,4}{}^2$ or $s_j{}^2$ and $v_k = \mathbf{M}_{k,4}{}^2$ or $s_k{}^2$, which are the variances or squared standard deviations for observations j and k.
4. $s_n = \sqrt{\left[(n_j - 1) \cdot v_j + (n_k - 1) \cdot v_k\right] / (n_j + n_k - 2)}$, which is the pooled standard deviation for comparison of observations j and k. Values for n_j and n_k are the sample sizes in $\mathbf{M}_{j,2}$ and $\mathbf{M}_{k,2}$.
5. $D_n = d_n / s_n$, which is the difference between the means for observations j and k in standard deviation units.

Box 8.1 (cont.)

6. R_n = abs $(D_n) / I_n$, which is the rate of change for observations j and k expressed in haldanes h, or the rate of change expressed in standard deviations per generation. The appropriate subscript for h is given by $\log_{10} I_n$.

Additional quantities:

7. The quantities $\log_{10} I_n$, $\log_{10}$ abs(D_n), and $\log_{10} R_n$ are important for temporal scaling of differences and rates.
8. An '*sbn*' code is useful for classifying rates, where: *sbn* = 1 for independent step rates on a time scale of one generation; *sbn* = 2 for independent base rates on a time scale greater than one generation; and *sbn* = 3 for all net rates combining two or more step rates or base rates.
9. A weight W_n for weighted regression, where $W_n = 1 / I_n$.

cross-sectional with respect to time. Finally, we will consider experimental studies that are both longitudinal and cross-sectional. All, taken together, provide a representative overview of the rates of microevolutionary change we can observe in nature. The protocol involved in calculating rates, adapted from Chapter 7, is outlined in Box 8.1.

A broad and representative range of field studies is included here, but the chapter is not an exhaustive review. Some important studies could not be included because they lack basic information, including sample sizes and standard deviations or standard errors, and some studies have undoubtedly been overlooked. Readers are encouraged to analyze the studies overlooked and any new time series to improve our understanding of evolutionary change in nature. All rates calculated in this chapter can be found at www.cambridge.org/evolution.

8.1 Observational Studies: Longitudinal with Respect to Time

Classification of a study as longitudinal or cross-sectional requires some thought because it affects how we evaluate the time involved. Longitudinal studies may be short or long in terms of time, involving a single generation or many generations. All of the laboratory studies in the previous chapter were longitudinal. Cross-sectional studies start from a single divergence point and are always reasonably long because some time must pass for noticeable differences to accumulate. Also, divergence times are rarely known with sufficient precision to enable comparison on very short scales of time.

Colonization studies epitomize both the difficulty and the importance of distinguishing longitudinal studies from cross-sectional studies. Colonization usually involves dispersal of a local subset of a large stable ancestral population into a new and different habitat. It is reasonable to expect, but rarely possible to prove, that the ancestral population remained stable, while a descendent population or populations evolved longitudinally. If evolutionary change is measured by analyzing differences between ancestral and descendent populations, then the interval of time involved is simply the one-way time since divergence. However, if evolutionary change is measured by analyzing differences between two or more descendent sister populations, then the interval of time involved is the two-way sum of divergence times for each pair of descendent populations.

8.1.1 Rhode Island House Sparrows Killed in a Storm

A classic early study of natural selection is an analysis of house sparrows (*Passer domesticus*) carried out by Hermon C. Bumpus while a professor of comparative zoology at Brown University in Rhode Island, in the eastern United States. The sparrows were caught following a severe winter storm and brought to Bumpus's laboratory, where some of the birds survived and some perished. Bumpus studied both groups as a test of selection, and concluded that the birds that survived did so because they possessed advantageous characteristics, while the birds that perished were eliminated because they were unfit (Bumpus, 1899). Bumpus published an extensive series of measurements on all of the birds, and the measurements are probably more valuable than his conclusions. Male sparrows are the sex with the most birds and the most survivors identified as adults. Measurements of nine characteristics for 35 adult males that survived and 24 that perished are listed in Bumpus's table I. The combined sample of 59 adult male birds is reanalyzed here.

We can compare means and standard deviations for the entire sample of birds with those for birds that survived the winter storm. If the entire sample is representative of sparrows living in Rhode Island in 1898 when they died, and the survivors are representative of founders of the following generation, then quantification of differences will tell us something about rates on a generation-to-generation timescale. An analysis of Bumpus's measurements is outlined in Table 8.1. Measurements were first transformed to natural logarithms to normalize the variation, and then differences between all means and survivor means were expressed in units of pooled standard deviation for each particular trait.

Five of the trait means are larger in the surviving male sparrows than in the full sample, and four of the trait means are smaller in the survivors. Thus there is no clear effect of size on survival. Similarly, there is very little difference in the

Table 8.1 *Step rates h_0 calculated from Bumpus (1899) data for male sparrows surviving a winter sleet storm*

	Measurement								
	TL	AE	Wgt	BH	HL	FL	TTL	SW	SL
All adult males ($N = 59$)									
Mean	5.077	5.512	3.249	3.454	-0.310	-0.340	0.121	-0.506	-0.160
Standard deviation	0.019	0.015	0.053	0.020	0.030	0.031	0.033	0.022	0.042
Surviving adult males ($N = 35$)									
Mean	5.069	5.512	3.236	3.454	-0.304	-0.333	0.126	-0.507	-0.155
Standard deviation	0.018	0.016	0.050	0.020	0.027	0.031	0.032	0.023	0.043
Difference in means	-0.007	0.001	-0.013	-0.001	0.006	0.006	0.005	-0.001	0.006
Pooled standard deviation	0.018	0.015	0.051	0.020	0.029	0.031	0.033	0.023	0.043
Difference in std. dev. units	-0.410	0.033	-0.244	-0.028	0.198	0.195	0.169	-0.024	0.130
Rate (h_0, median = 0.169)	-0.410	0.033	-0.244	-0.028	0.198	0.195	0.169	-0.024	0.130
$\log_{10}$ lratel (median = -0.772)	-0.387	-1.483	-0.612	-1.559	-0.703	-0.709	-0.772	-1.624	-0.885

Thirty-five individuals survived out of a total of $N = 59$ males. Measurements are: total length (*TL*, mm); alar extent (*AE*, mm); body weight (*Wgt*, g); beak and head length (*BH*, mm); humerus length (*HL*, mm); femur length (*FL*, inches); tibiotarsus length (*TTL*, inches); skull width (*SW*, inches); and sternum length (*SL*, inches). All were transformed to natural logarithms for analysis. Step rates range in absolute value from $h_0 = 0.024$ to $h_0 = 0.410$ standard deviations per generation, with a median rate of $h_0 = 10^{-0.772} = 0.169$. Note that the rates here are selection differentials before reproduction.

161

standard deviations. Five of the trait standard deviations are larger in the surviving male sparrows than in the full sample, and four of the trait standard deviations are smaller in the survivors. This contradicts Bumpus's general claim that selection removed more-variable individuals.

Absolute values of the rates calculated in Table 8.1 range from -1.624 to -0.387 on a $\log_{10}$ scale, with a median of -0.772. These range in haldanes from $h_0 = 0.024$ to 0.410, with a median of 0.169 standard deviations per generation — rates that are comparable to the rates per generation on a one-generation timescale seen in the artificial selection experiments of the previous chapter.

Rates for the Bumpus sparrows are technically selection differentials and represent upper limits for per-generation change. These are not differences observed between successive generations but differences observed between one generation and the founders of the next. It is doubtful that the differences observed by Bumpus would be fully heritable; and thus change in the succeeding generation was almost certainly less than that expressed in Table 8.1.

8.1.2 House Sparrows Introduced into North America

House sparrows, *Passer domesticus*, were introduced into North America in 1851 or 1852, and their possible evolution after arrival has been investigated by several authors. David Lack (1940) studied house sparrows from the eastern United States, midwestern United States, and California, and concluded that sparrows in North America have been remarkably stable in wing and bill measurements. In response, John Calhoun (1947) carried out a more comprehensive investigation and published a table of wing-length means and standard deviations for large samples of male and female sparrows.

Calhoun's specimens came from different geographic regions and from four periods in time, centered on the years 1868, 1897, 1919, and 1938. He published an illustration of wing length changing through the four intervals and calculated this to be significant. The generation time of *P. domesticus* is 1.97 years (Jensen, et al., 2008), or, rounded, two years. The intervals between Calhoun's samples were 29, 22, and 19 years, representing, respectively, 14.5, 11, and 9.5 generations of evolutionary time.

Pooling samples within each temperature zone, Calhoun's statistics for wing length in males and females yield 16 base rates of change. The rates in haldanes range from $h_{0.978} = 0.002$ to $h_{0.978} = 0.150$ standard deviations per generation (median 0.034) on timescales of 9.5 to 14.5 generations. Ten of the changes are positive and six are negative, so the increasing size found by Calhoun was not universal.

8.1.3 African Monkeys Introduced on the Island of Saint Kitts

Saint Kitts is a Caribbean island colonized by Europeans in the 1620s. Green monkeys, *Cercopithecus aethiops sabaeus* or now *Chlorocebus sabaeus*, were introduced on St. Kitts from West Africa in the mid-seventeenth century. Eric H. Ashton and Solly Zuckerman (1950) made a thorough quantitative comparison of the dentition of green monkeys from St. Kitts with those of the West African parent population, based on collections made in the early twentieth century. By that time the two populations may have been separated by as many as 300 years. Ashton and Zuckerman (1950) estimated the generation time for *C. sabaeus* at three to four years and reasoned that 300 years would represent 75–100 generations. Recent estimates for the generation time are longer. Moorjani et al. (2016) listed the generation time for the green monkey at 11 years, and Pfeifer (2017) gave this as 8.5 years. A ten-year generation time implies 30 generations of separation between African and St. Kitts green monkeys.

Ashton and Zuckerman (1950) measured deciduous teeth and all of the upper and lower permanent teeth for male and female *C. sabaeus*. Sample sizes were small for the deciduous teeth, and relatively small for the females. The largest samples were for adult male specimens. Fifty-four measurements from Ashton and Zuckerman's tables 3 and 5 were chosen for analysis, which include three to five measurements for each tooth. Redundant measurements (maximum molar widths) and shape indices were not included.

The result is 54 base rates that range from $h_{1.477} = 0.0007$ to $h_{1.477} = 0.0349$ standard deviations per generation (median 0.0138) on a timescale of 30 generations.

8.1.4 House Mice Introduced on the Island of Skokholm

Skokholm is a small island some three kilometers off the coast of Wales. The island supports a feral population of the house mouse *Mus musculus*. In the 1960s, British geneticist R. J. Berry trapped and skeletonized large samples of the mice. After studying these and researching the history of Skokholm, Berry (1964) concluded that the mice were isolated on the island for about 70 years after being introduced from nearby mainland Pembrokeshire. Berry then compared the Skokholm mice and Pembrokeshire mainland mice, based on 35 skeletal variants (listed in his table 2) and on their average weights (listed in his table 6).

The weighted mean of the body weights Berry gave for mice from Skokholm and for mice from corn ricks in Pembrokeshire was 20.34 g. Inserting this weight in Equation 3.2 above points to an average generation time of 0.62 years. Dividing

70 years of isolation by a generation time of 0.62 years yields an estimate of about 112 generations for isolation of the Skokholm mice.

Berry (1964: table 2) recorded 35 skeletal variants for the Skokholm ($N = 332$) and Pembroke ($N = 231$) samples in terms of percentages, and provided the arcsine and other transformations necessary to normalize the means and calculate the expected binomial variances. The arcsine normalization of Berry (1964) is equivalent to that of Sokal and Rohlf (1981). Berry's 31 skeletal traits with nonzero rates have rates that range from $h_{2.050} = 0.000$ to $h_{2.050} = 0.313$ standard deviations per generation (median 0.030) on a timescale of 112 generations. These rates do not fit the pattern of temporal scaling of other rates observed in this chapter, presumably because the expected binomial standard deviation is a poor estimate of the (unknown) actual standard deviation (see discussion accompanying Equation 7.2). Rates for Berry's 31 skeletal traits are not included in the rate compilation analyzed later in this chapter.

Berry (1964: table 6) recorded mean body weights and empirical standard deviations for mature male and female Skokholm mice (20.8 g and 22.8 g, respectively, with $N = 135$ and 108 and $s = 3.5$ and 4.8) and for mature male and female Pembroke mice (17.9 g and 20.2 g, respectively, with $N = 113$ and 135 and $s = 2.8$ and 4.8). Rates for male and female body weights are $h_{2.050} = 0.008$ and 0.005 standard deviations per generation (median 0.007) on a timescale of 112 generations. These rates are included in the rate compilation analyzed later in the chapter.

8.1.5 Three-Spined Sticklebacks in a Lake in British Columbia

The three-spined stickleback, *Gasterosteus aculeatus*, is a small fish with a broad distribution on all three northern continents. Most are anadromous, live in the sea, and swim up into freshwater streams to spawn. Some *G. aculeatus* live permanently in streams, and many have become isolated in ponds and lakes where their evolution is often rapid. Donald McPhail investigated benthic and limnetic populations of three-spined sticklebacks living together in Enos Lake on Vancouver Island in the British Columbia province of western Canada. He published statistics for each population, describing seven meristic and eight measured traits in the years 1974, 1977, 1980, and 1981 (McPhail, 1984). The generation time for *G. aculeatus* in British Columbia is assumed to be two years like the modal age of freshwater females breeding in Alaska (Baker et al., 2008).

Thirty step rates for the *G. aculeatus* traits range from $h_0 = 0.100$ to $h_0 = 0.498$ standard deviations per generation (median 0.099) on a timescale of one generation. Step, base, and net rates in the entire set of 180 rates range from $h_{0.544} = 0.002$ to $h_{0.176} = 1.088$ standard deviations per generation (median 0.065) on timescales ranging from 1 to 3.5 generations.

8.1.6 Intertidal Periwinkles in the Gulf of Maine

The flat periwinkle, *Littorina obtusata*, is a low-spired marine gastropod that is widely distributed from the Gulf of Maine in the northwestern Atlantic Ocean to the northeastern Atlantic and Mediterranean Sea. As might be expected from the name *Littorina,* its species inhabit the nearshore or littoral zone of coastal waters. The European green crab *Carcinus maenas* is an invasive predator that reached the east coast of North America sometime in the eighteenth or early nineteenth centuries. Before 1900 *C. maenas* was known only from New Jersey to southern Massachusetts, but by 1905 it was reported in the Gulf of Maine (Vermeij, 1982). *C. maenas* is an active predator on *L. obtusata*.

Robin Hadlock Seeley studied *L. obtusata* shells in museum collections from three sites in northern Massachusetts and Maine. These older historical collections were made before *C. maenas* reached the Gulf of Maine. She then compared the museum shells with modern shells from the same localities collected just before and after 1983 (Seeley, 1986). As analyzed here, comparing older to younger samples, five of five younger samples had lower-spired shells and five of five younger samples had thicker shells. The generation time of *L. obtusata* is approximately two years (Hooks, 2013).

Birds and mammals are advantageous for study of evolutionary rates because they have determinate growth, and study of adults limits any confounding variation due to ontogenetic development. *Littorina obtusata*, which grows continuously through ontogeny, illustrates how rates can be studied in animals with indeterminate growth. The bivariate plot in Figure 8.1, for *L. obtusata* at one of Seeley's study sites, Nahant, shows how spire height changed during growth. Within each sample of shells, as for example the 1898 collection (open circles), spire height increased as shell width increased. This growth allometry is illustrated by the line of regression of ln spire height on ln shell width. Similar lines are shown for the 1915 and 1983 samples. There is some overlap of specimens representing different samples, but the regression lines for each sample are clearly separated.

The rate change of spire height for the interval from 1898 to 1915 is the first rate of interest here. This is an interval of 17 years and an interval of 8.5 generations. The regression lines for 1898 and 1915 are not perfectly parallel so it matters where the distance between them is calculated. The most representative shell width is arguably the mean width for the two pooled samples, and the vertical distance between the regression lines is appropriately calculated for this mean width. The length of the rate-1 arrow in Figure 8.1 shows the change of interest. The arrow corresponds to a change or difference of -0.622 mm in spire height. The pooled standard deviation for the 1898 and 1915 samples is 0.177 mm, calculated as the square root of the residual-mean-square deviation in an analysis of

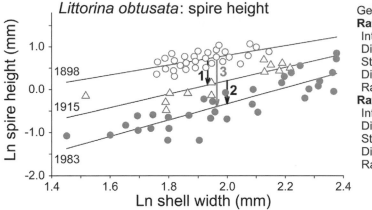

Figure 8.1 Spire height in time-successive samples of *Littorina obtusata* from Nahant at the southern margin of the Gulf of Maine (data from Seeley, 1986). Samples are known from 1898 (open circles), 1915 (open triangles), and 1983 (closed circles). Spire height decreased through time. Rates are calculated by comparing average spire heights at mean shell width for pairs of samples (arrows). Rate 1 quantifies change from 1898 to 1915, and rate 2 quantifies change from 1915 to 1983. Both are multigeneration base rates. Details are given in the right-hand column of text, where differences represent change in mm and in standard deviation units. Rates are in standard deviations per generation. Rate 3 (gray arrow) is a net rate on a longer timescale but dependent in part on rates 1 and 2. Pooled standard deviations are calculated as the square root of residual mean square deviations derived from analyses of covariance for each pair of samples.

covariance for the two samples. Thus the rate-1 arrow represents a difference of -3.514 standard deviations, and the absolute value of the corresponding base rate per generation is $h_{0.929} = 0.413$ standard deviations per generation on a timescale of 8.5 generations.

A second rate of interest is how spire height changed in the 34-generation interval from 1915 to 1983. The rate-2 arrow in Figure 8.1 shows the distance of interest. Following calculations like those just presented, the difference in spire height at the pooled mean shell width is -1.989 standard deviations, and the absolute value of the corresponding base rate per generation is $h_{1.532} = 0.059$ standard deviations per generation on a timescale of 34 generations.

A third rate of interest might be how spire height changed in the overlapping 42.5-generation interval between 1898 and 1983 (rate-3 arrow in Figure. 8.1). This is a net rate because of overlap with intervals of the two independent base rates, and it works out to $h_{1.628} = 0.114$ standard deviations per generation on a timescale of 42.5 generations.

Rates can also be calculated from Seeley's (1986) measurements for shell thickness at Nahant, and for spire height and shell thickness at Appledore Island

and Isle au Haut. All of the eight base rates for *L. obtusata* taken together range from $h_{1.532} = 0.005$ to $h_{0.929} = 0.413$ standard deviations per generation (median 0.062) on timescales ranging from 8.5 to 56 generations.

8.1.7 Laysan Finches on a Leeward Atoll of the Hawaiian Islands

Laysan is one of the small leeward islands in the Hawaiian Island chain. It is located 1,800 km northwest of the big island of Hawaii, and is home to an endemic Laysan duck and to the endemic Laysan finch. The latter, *Telespyza canans*, is actually a finch-billed Hawaiian honeycreeper. Males are larger than females.

In 1967 the U.S. Fish and Wildlife Service introduced Laysan finches to South-east Island in the Pearl and Hermes Atoll some 500 km farther northwest in the leeward islands. In 1973 *T. canans* reached North Island in the Pearl and Hermes Atoll. Later, in 1984 and 1985, Sheila Conant studied Laysan finches on Laysan itself and on Southeast Island and North Island in the Pearl and Hermes Atoll (Conant, 1988). She found significant differences between the islands in bill length and width measurements but not in tarsus measurements. Laysan finches on Southeast Island and North Island had been separated from the larger source population on Laysan for 18 years at the time they were studied. The IUCN Red List gives the generation time for *T. canans* as 4.9 years, meaning that the two atoll populations had been separated for some 3.7 generations.

Based on information in Conant (1988), 12 rates can be calculated for male and female tarsus length, bill length, and upper bill width for the Southeast Island and North Island *T. canans* populations compared to those on Laysan Island. The rates range from $h_{0.568} = 0.001$ to $h_{0.568} = 0.086$ standard deviations per generation (median 0.004) on a timescale of 3.7 generations.

8.1.8 Mysomela *Honeyeaters on Islands near New Guinea*

Long Island is a volcanic caldera in the Bismarck volcanic arc lying off the northeast coast of New Guinea. An explosive volcanic eruption in the mid-seventeenth century covered the island in meters of ash and debris. Long Island was then recolonized by plants and animals from New Guinea and neighboring islands. Jared Diamond and colleagues studied the avifauna 300 years later, in the 1970s, when they documented an interesting case of rapid character divergence involving the larger Bismarck black myzomela (*Myzomela pammelaena*) and the smaller Sclater's myzomela (*Myzomela sclateri*). Both are meliphagid honeyeaters. Where they are found together, on Long Island and surrounding islets, *M. pammelaena* is larger than conspecifics on islands lacking *M. sclateri*, and where they are found together *M. sclateri* is smaller than conspecifics on islands lacking *M. pammelaena* (Diamond et al. 1989).

Diamond et al. (1989) measured lengths of the wing, tail, bill, exposed culmen, and tarsus on museum specimens, and copied weights from field labels whenever these were available. They provided a full set of statistics for all measurements of males and females of *M. pammelaena* and *M. sclateri* on Long Island, and for each species where it occurs on islands that are plausible source populations for the recolonization that occurred on Long Island. Diamond et al. (1989, p. 693) assumed a generation time for *Myzomela* of three years, and 300 years (100 generations) of Long Island evolution from ancestral values. They then calculated rates in the form of selection differentials and selection responses per generation. The responses they calculated range from 0.01 to 0.03 standard deviations per generation for *M. pammelaena*, and from 0.005 to 0.01 standard deviations per generation for *M. sclateri*. Both sets of rates are on a 100-generation timescale.

We can make similar calculations of rates by analyzing the statistics of Diamond et al. (1989, table 3) comparing Long Island *M. pammelaena* to averages for its putative ancestors on other islands. We can then do the same for Long Island *M. sclateri* and averages for its putative ancestors on other islands. The 12 *M. pammelaena* rates calculated here range from $h_{2.000} = 0.018$ to $h_{2.000} = 0.041$ standard deviations per generation (median 0.029) on a timescale of 100 generations. This is close to the range inferred by Diamond et al. The 12 *M. sclateri* rates range from $h_{2.000} = 0.001$ to $h_{2.000} = 0.015$ standard deviations per generation (median 0.006) on a timescale of 100 generations. This is also close to the range inferred by Diamond et al. Both sets of rates support the Diamond et al. (1989) conclusion that size differences and character displacement observed in Long Island populations of *Myzomela* evolved in the short span of three centuries following explosive eruption of the Long Island volcano.

8.1.9 Eurasian Tree Sparrow Introduced in North America

The Eurasian tree sparrow, *Passer montanus*, is similar to its more familiar congener *P. domesticus* but differs in that it nests in trees and prefers wooded countryside. Its introduction to North America from Germany took place in the city of St. Louis in the state of Missouri in 1870. The species spread from there to neighboring states. In 1983, 1984, and 1985, Vincent St. Louis and Jon Barlow collected *P. montanus* specimens in Germany and in the U. S. state of Illinois. They measured 16 traits on skeletons of adult males and females, and showed, among other things, that after some 114 years of separation Illinois *P. montanus* are smaller than the German ancestral stock (St. Louis and Barlow, 1991). The generation time of *P. montanus* is reasonably considered to be similar to that of *P. domesticus* — which is very close to two years (Jensen et al., 2008).

Thirty-two rates can be calculated from measurements published by St. Louis and Barlow (1991, table 2). These range from $h_{1.756} = 0.0002$ to $h_{1.756} = 0.0258$ standard deviations per generation (median 0.0113) on a timescale of 57 generations.

8.1.10 Soapberry Bugs in Florida

The soapberry bug, *Jadera haematoloma*, is an insect that feeds on seeds of the soapberry plant in Florida and elsewhere. It does so by inserting its needle-like mouthparts or "beak" into fruit surrounding the seeds. Beak length is related to host-plant fruit size. Females are larger than males and have longer beaks. Scott P. Carroll and Christin Boyd studied the evolutionary response of *J. haematoloma* transitioning from native soapberry plants to an introduced Asian ornamental with smaller fruits. The Asian ornamental was introduced into Florida in 1926 and planted widely in the 1950s.

Carroll and Boyd (1992, figure 6) provided beak length measurements for samples of four or more female soapberry bugs collected in Florida in 1899, 1910, 1919, 1954–1958, 1971, and 1979. According to Carroll and Boyd, soap-berry bugs have two or three generations per year, making the average generation time about 0.4 years. When logged, the measurements yield 15 rates on timescales that range from 20 to 200 generations. The five base rates range from $h_{1.966} = 0.008$ to $h_{1.301} = 0.032$ (median 0.014) on timescales ranging from 20 to 92.5 generations.

In a follow-up study Carroll et al. (1997) confirmed that adaptation to the introduced host plants has a genetic basis and made comparisons yielding additional rates on an estimated 100-generation timescale. Eighteen rates for body length, thorax width, and beak length (Carroll et al., 1997, figure 2) range from $h_{2.000} = 0.0001$ to $h_{2.000} = 0.0196$ standard deviations per generation (median 0.0046) on a timescale of 100 generations. In the laboratory it is possible to compare growth in *J. haematoloma* feeding on native host plants with those feeding on introduced hosts (Carroll et al., 1997, figure 3), and the comparisons show similar differences and rates of change. Twelve rates for female thorax width, beak length, and development time range from $h_{2.000} = 0.0001$ to $h_{2.000} = 0.0247$ standard deviations per generation (median 0.0099) on a timescale of 100 generations.

8.1.11 Three-Spined Stickleback Isolated in Western Norway

The three-spined stickleback, *Gasterosteus aculeatus*, as noted above, is a small fish with a broad distribution on all three northern continents. Most are anadromous, live in the sea, and swim up into freshwater streams to spawn. Some *G. aculeatus* live permanently in streams, and many have become isolated in ponds and lakes where their evolution is often rapid. Tom Klepaker studied a population of marine sticklebacks isolated in a pond in the city of Bergen in western Norway. The pond was originally part of a narrow marine strait, but as the city developed it became a small lagoon and finally, in 1960, a freshwater pond.

Klepaker (1993) collected and measured 30-fish samples of *G. aculeatus* representing three stages of their evolution: (1) the marine source population in 1991, which probably evolved little since 1960, (2) the pond population in 1987 after 27 years of isolation, and (3) the pond population in 1991 after 31 years of isolation. Two of the traits studied were counts: One was a measurement of standard body length, and 21 were size-adjusted linear measurements of other parts of the body. The generation time for *G. aculeatus* in the Bergen pond is not known, but it is assumed to be two years like the modal age of freshwater females breeding in Alaska (Baker et al., 2008).

Rates of change for the marine source population compared to the 1987 sample range from $h_{1.130} = 0.004$ to $h_{1.130} = 0.172$ standard deviations per generation (median 0.030) on a timescale of 13.5 generations. Rates of change for the marine source population compared to the 1991 sample were similar at $h_{1.190} = 0.002$ to $h_{1.190} = 0.151$ standard deviations per generation (median 0.043) on a timescale of 15.5 generations. Finally, rates of change were higher for comparison of the 1987 and 1991 samples, ranging from $h_{0.301} = 0.029$ to $h_{0.301} = 0.536$ standard deviations per generation (median 0.270) on a timescale of two generations.

8.1.12 *Size of Snow Geese at La Pérouse Bay, Manitoba*

The snow goose, *Anser caerulescens*, is a North American species that nests in the Canadian Arctic and winters in the southern United States and Mexico. Snow geese were studied intensively by Fred Cooke and colleagues at La Pérouse Bay, an inlet on Hudson Bay, in northern Manitoba (Cooke et al., 1995). In an earlier report from this project, Cooch et al. (1991) documented a significant decrease in adult female body weight over time, and a parallel decrease in gosling body weight, tarsus length, and bill length (the Cooch et al. figures are reprinted in Cooke et al., 1995). In the earlier report Cooch et al. attributed the decreases in size to overexploitation of food resources at La Pérouse Bay, but the decline could also be an evolutionary response to global warming or some other environmental change.

The 17-year decline in the adult female body weight of snow geese shown by Cooke et al. (1995) is illustrated in Figure 8.2a. The generation time for

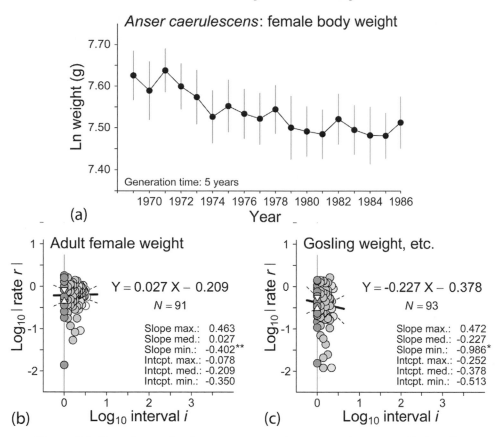

Figure 8.2 Size change in the snow goose *Anser caerulescens* studied at La Pérouse Bay, Manitoba, from 1969 through 1986 (Cooke et al., 1995). (a) Change in female body weight is shown through time. Solid circles are means; vertical lines span ±1 phenotypic standard deviation. Generation time is 5 years. (b) LRI log-rate-interval plot for 91 rates of change in female body weight spanning a full generation or longer. Median slope of 0.027 is consistent with directional change. A bootstrapped 95% confidence interval for the slope excludes both −0.500 expected for random change and −1.000 expected for stasis. (c) LRI plot for 93 rates of change in gosling weight, tarsus length, and beak length spanning a full generation or longer. Median slope of −0.227 lies between expectations for directional and random change. The 95% confidence interval excludes stasis.

A. caerulescens is about five years (Niel and Lebreton 2005, table 2). Thus the 17-year interval represents a total of about 3.4 generations. The Figure 8.2a time series yields a total of 91 rates on timescales in the range of a full generation to 3.4 generations. These have an LRI slope of 0.027 (Figure 8.2b), with a 95% confidence interval excluding the slopes of −1.000 and −0.500 expected for stasis and random change, respectively, affirming Cooke et al.'s interpretation of the decline as both directional and significant. The time series in Figure 8.2a yields 17 step rates that range from $h_0 = 0.015$ to $h_0 = 0.772$ standard deviations per generation (median 0.436) on a timescale of one generation.

The three Cooke et al. (1995) time series for gosling weight, gosling tarsus length, and gosling beak length are shorter, spanning 12, 10, and 12 years, respectively, equivalent to 2.4, 2.0, and 2.4 generations. Note that the sample ranges for gosling tarsus length and gosling beak length given by Cooke et al. (1995, figures 1b and 1c) span ±1 standard deviation and not ±1 standard error. Taken together the gosling weight, tarsus length, and beak length time series yield 93 rates in the range of a full generation to 3.4 generations. The LRI slope for all together is -0.227, lying between expectations for random and directional time series. The 95% confidence interval excludes only stasis. Step rates for gosling weight, tarsus length, and beak length range from $h_0 = 0.003$ to $h_0 = 1.663$ standard deviations per generation (median 0.398) on a timescale of one generation.

Change in the weights of snow goose eggs is also of interest (Cooke et al., 1995, p. 223). The time series is not long enough to constrain any interpretation of directional change, random change, or stasis, but the egg weights yield 13 step rates that range from $h_0 = 0.008$ to $h_0 = 0.493$ standard deviations per generation (median 0.222) on a timescale of one generation. Median rates for egg weight are lower than medians for adult female body weight and for gosling weight, tarsus length, and bill length.

8.1.13 *Nectivorous Honeycreepers on the Island of Hawaii*

The i'iwi, *Drepanis coccinea*, is a brightly colored nectivorous honeycreeper endemic to the Hawaiian Islands. It is distinctive in having a relatively long and downwardly curved bill. Thomas Smith and colleagues compared "old" i'iwi specimens in museum collections to "recent" specimens netted and released between 1988 and 1991, The former were older than 1902, and the average difference between old and recent birds was considered to be about 100 years (Smith et al., 1995). Five characteristics were measured: wing length, tarsus length, upper mandible length, lower mandible length, and lower mandible width. Smith et al. concluded that upper mandible length decreased significantly during the study interval. The IUCN Red List indicates a generation time of 4.9 years that was rounded to five years.

Smith et al. (1995, table 1) provide statistics for "old" and "recent" samples of *D. coccinea*. Tarsus length was unchanged during the study interval. Rates for the remaining characteristics range from $h_{1.301} = 0.007$ to $h_{1.301} = 0.033$ standard deviations per generation (median 0.0124), on a timescale of 20 generations.

8.1.14 Cliff Swallows in Southwestern Nebraska

Cliff swallows, *Petrochelidon pyrrhonota*, are migratory birds that breed in western North America and winter in southern South America. They feed exclusively on aerial insects and forage on the wing. Swallows generally arrive in southwestern Nebraska in early to mid-May. However, a six-day interval of cold and wet weather there in late May of 1996 killed thousands of *P. pyrrhonota*. During the three days following the event Charles R. Brown and Mary Bomberger Brown visited study sites in the area and collected some 1,800 dead swallows for comparison with more than 1,000 swallows that survived this natural selection event.

Brown and Brown (1998a, table 7) found statistical differences in tarsus length, beak length, beak width, wing length, and tail length distinguishing birds that survived the 1996 event from those in the population before the event. The five rates quantifying this change range from $h_0 = 0.262$ to $h_0 = 1.163$ standard deviations per generation (median 0.908) on a timescale of one generation. The survivors in this comparison were mature offspring of 1996 survivors, measured in 1997, so the differences reflect heritability and represent change on a generational timescale.

In a follow-up investigation Brown and Brown (2011) studied change in Nebraska *P. pyrrhonota* by measuring yearling birds annually from 1996 through 2006. Relevant statistics for male and female tarsus length, beak length, beak width, wing length, and tail length were published in figures (Brown and Brown, 2011: figures 2 and 3), and Mary Bomberger Brown supplied the relevant means and standard errors. The breeding life span for Nebraska *P. pyrrhonota* females, weighted by lifetime reproductive success, is 3.64 years (Brown and Brown, 1998b, table 10). A generation time cannot be longer, so three years is a reasonable estimate for generation time.

The time series of wing lengths and tarsus lengths for cliff swallows reported by Brown and Brown (2011) are shown in Figure 8.3a. These and time series for beak length, beak width, and tail length yield a total of 534 rates of change on timescales of one or more generations. A log-rate-interval LRI plot of rates for wing length that span a full generation or longer (Figure. 8.3b) shows that wing length scales with the slope expected for a directional time series, possibly with a random component. The 95% confidence interval for rate scaling of wing length excludes stasis. An LRI plot for tarsus length (Figure 8.3c) shows the opposite. It has the slope expected for a stationary time series, again possibly with a random component. The 95% confidence interval for rate scaling of tarsus length excludes directional change.

The five time series of Brown and Brown (2011) together yield 95 step rates that range from $h_0 = 0.023$ to $h_0 = 1.921$ standard deviations per generation (median $h_0 = 0.250$) on a timescale of one generation.

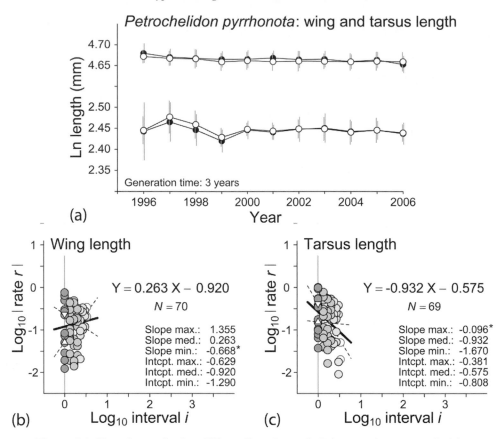

Figure 8.3 Size change in the cliff swallow *Petrochelidon pyrrhonota* studied in southwestern Nebraska from 1996 through 2006 (Brown and Brown, 2011). (a) Change in male and female wing length (above) and tarsus length (below) are shown through time. Solid circles are means for males and open circles are means for females; vertical lines span ± 1 phenotypic standard deviation. Generation time is three years. Note the composite scale on the ordinate. (b) LRI log-rate-interval plot for 70 rates of change in male and female wing length spanning a full generation or longer. Median slope of 0.263 is consistent with directional change. A bootstrapped 95% confidence interval for the slope includes -0.500 expected for random change but excludes -1.000 expected for stasis. (c) LRI plot for 69 rates of change in tarsus length spanning a full generation or longer. Median slope of -0.932 lies between expectation for random change and stasis. The 95% confidence interval excludes directional change.

8.1.15 Barnacle Geese at Laus Islet Reserve in the Baltic Sea

Laus Islet or Laus Holmar is a small ca. 5 km^2 Swedish nature reserve in the Baltic Sea off the east coast of Gotland. Barnacle geese, *Branta leucopsis*, nesting on Laus Islet were studied by Kjell Larsson and colleagues from 1984 through 1996. Head length and tarsus length were measured every year for an average of 679 birds

(Larsson et al., 1998, figure 4) and standard errors were reported for smaller samples (table 1). Larsson et al. (1998) reported that yearly mean measurements of head length and tarsus length declined significantly in males and females, by approximately 0.7 and 0.5 standard deviations, respectively, over the 12-year study interval. The generation time for *B. leucopsis* is about seven years (Niel and Lebreton, 2005, table 2). Thus the full 12-year interval represents a total of about 1.7 generations.

The change of 0.7 standard deviations for head length reported by Larsson et al. (1998) for a 1.7-generation interval yields a rate of $h_{0.234} = 0.408$ standard deviations per generation for this interval. The change of 0.5 standard deviations for tarsus length reported by Larsson et al. (1998) for a 1.7-generation interval yields a rate of $h_{0.234} = 0.292$ for this interval. Analysis here yields similar rates of $h_{0.234} = 0.416$ and 0.375 for male and female head length, and $h_{0.234} = 0.277$ and 0.302 for male and female tarsus length on timescales of 1.7 generations.

There are 21 male and 21 female rates for head length spanning timescales from a full generation to 1.7 generations. These have an average LRI slope of -0.266 representing directional or random change (stasis is excluded). There are similarly 21 male and 21 female rates for tarsus length spanning timescales from a full generation to 1.7 generations. These have an average LRI slope of 0.297, representing directional or random change (again stasis is excluded). Taken together these support the idea expressed by Larsson et al. (1998) of a directional change in yearly mean values.

Twenty-four male and female step rates for head length range from a minimum of $h_0 = 0.004$ to a maximum of $h_0 = 0.193$ standard deviations per generation (median 0.067) on a timescale of one generation. Twenty-four male and female step rates for tarsus length range from a minimum of $h_0 = 0.003$ to a maximum of $h_0 = 0.333$ standard deviations per generation (median 0.043) on a timescale of one generation.

8.1.16 Woodrats and Kangaroo Rats at Sevilleta in New Mexico

Sevilleta is a 900 km^2 U. S. National Wildlife Refuge spanning the Rio Grande River in the state of New Mexico. The white-throated woodrat, *Neotoma albigula*, is a relatively small woodrat common in arid environments. Felisa Smith and colleagues studied *N. albigula* body weight at Sevilleta in relation to time and environmental temperature through the seven-year interval from 1989 through 1996 (Smith et al., 1998). They found that body weight decreased over the interval as temperature increased. The generation time of *N. albigula* is not known but based on body size must be close to one year.

The body weight time series for *N. albigula* of Smith et al. (1998) yields 21 rates, including five step rates. The step rates range from $h_0 = 0.165$ to $h_0 = 1.963$ standard deviations per generation (median 0.466) on a timescale of one generation.

In a follow-up study Terri Koontz and colleagues studied the body weight of the smaller kangaroo rat *Dipodomys merriami* at Sevilleta through the same 1989–1996 interval. They had larger samples and were able to study change in females and males separately at three sites within the Sevilleta refuge (Koontz et al., 2001). Slow growth and long survival (Zeng and Brown, 1987) suggest that one year is a reasonable estimate for the generation time of *D. merriami*.

The six time series for body weight in male and female *Dipodomys merriami* at the three localities of Koontz et al. (2001) yield 162 rates on timescales ranging from 1 to 7 generations. On a log-rate-interval plot these scale with an LRI slope of -1.108 that is close to the slope of -1.000 expected for stasis. A 95% confidence interval excludes the slopes expected for directional or random change. The 39 step rates range from $h_0 = 0.119$ to $h_0 = 0.689$ standard deviations per generation (median 0.289) on a timescale of 1 generation.

8.1.17 Soay Sheep on the Island of Hirta, St Kilda

St Kilda is a small archipelago of North Atlantic islands, the outermost islands of the Outer Hebrides off the west coast of Scotland. One of the smaller St Kilda islands, Soay, 1 km^2 in area, is inhabited by a small feral sheep, *Ovis aries*, that is short tailed and hardy, with a naturally moulting fleece. Soay sheep are possibly remnants of the earliest sheep of Neolithic Europe. Today, Soay sheep are found on the largest of the St Kilda islands, Hirta, which is 6.7 km^2 in area, and this is where the life history and evolution of Soay sheep have been most thoroughly studied (Clutton-Brock and Pemberton, 2004).

Jos Milner and colleagues published statistics for annual measurements of body weight, hind leg length, and incisive arcade breadth for Soay sheep on Hirta for the 12-year interval from 1985 through 1996 (Milner et al., 1999). The measurement statistics are analyzed here. Coulson and Crawley (2004) estimated the generation time of high- and low-density female cohorts of Soay sheep at 2.95 and 4.92 years, respectively, and four years seems a reasonable estimate for the generation time.

The three time series for body weight, leg length, and incisive breadth are all too short to constrain any interpretation of temporal scaling. The time series for body weights of male and female *O. aries* yield a total of 22 step rates that range from $h_0 = 0.038$ to $h_0 = 1.469$ standard deviations per generation (median 0.252). The time series for leg lengths of male and female *O. aries* are shorter and yield 16 step rates that range from $h_0 = 0.001$ to $h_0 = 1.139$ standard deviations per generation (median 0.140). The time series for incisive breadths of male and female *O. aries* are even shorter and yield 13 step rates that range from $h_0 = 0.004$ to $h_0 = 0.786$ (median 0.281). All are on timescales of a single generation.

8.1.18 Deer Mice on the California Channel Islands

The Channel Islands lie in the eastern Pacific Ocean 30–100 km off the coast of southern California. They range in size from Santa Barbara and Anacapa at 3 km^2 to Santa Cruz at 250 km^2. Oliver Pergams and Mary Ashley measured museum specimens of endemic deer mice, *Peromyscus maniculatus*, from three of the islands, Santa Barbara, Anacapa, and Santa Cruz (Pergams and Ashley, 1999). These were collected at different times in the nineteenth and twentieth centuries, which Pergams and Ashley lumped into early (1897–1941) and late (1955–1988) composites. Early and late collections on the three islands differed by an average of 44, 38, and 38 years, respectively. The body size of *P. maniculatus* and its typical one-year lifespan point to a generation time of about 0.5 years, so rates here were calculated for intervals of 88, 76, and 76 generations.

Pergams and Ashley (1999) reported trait measurements for 22 early and late samples from Santa Barbara, Anacapa, and Santa Cruz islands. Base rates in haldanes for the 22 comparisons range from $h_{1.881} = 0.006$ to $h_{1.881} = 0.027$ standard deviations per generation (median 0.012) on timescales of 76 and 88 generations.

8.1.19 Indian Mongoose Introduced on Caribbean and Pacific Islands

The small Indian mongoose, *Herpestes auropunctatus* or sometimes *H. javanicus*, is a widely distributed invasive species that was introduced on islands around the world in an attempt to control rats and snakes, leading to the decline and extirpation of many native species. Daniel Simberloff and colleagues measured *H. auropunctatus* from St. Croix in the Caribbean Sea and from Hawaii, Oahu, Viti Levu, and Okinawa in the Pacific Ocean. This enabled a quantitative comparison with *H. auropunctatus* of Bengal in India, which was the source population for the island introductions (Simberloff et al., 2000). Most of the introductions took place in the 1880s, but *H. auropunctatus* was introduced earlier, in 1872, on St. Croix, and later, in 1910, on Okinawa. Specimens measured were collected after 1965, meaning that the time available to accrue differences from the source population was about 100 years for Hawaii, Oahu, and Viti Levu, 110 years for St. Croix, and 70 years for Okinawa. Little is known about the life history of *H. auropunctatus*, but a generation time of two years is consistent with its body size (Figure 3.3) and its longevity.

Measurements are available for the length of the cranium of *H. auropunctatus* and for the length of its upper canine tooth. Comparison of measurements for males and females on the islands with those for Bengal (Asia VI in Simberloff et al., 2000, table 1) yields 20 rates that range from $h_{1.699} = 0.023$ to $h_{1.544} = 0.112$ standard deviations per generation (median 0.053) on timescales ranging from 35 to 55 generations.

8.1.20 Mediterranean Fruit Flies in Western North America

The Mediterranean fruit fly, *Drosophila subobscura*, appeared in the southwestern part of South America in 1978 and in the northwestern part of North America in 1982. Geographic clines in chromosomal polymorphisms were established rapidly (Ayala et al. 1989). Glòria Pegueroles and colleagues investigated the clines in Chile and the western United States. They published measurements for female and male *D. subobscura* from sites collected in 1986 and 1988 (Pegueroles et al., 1995). George Gilchrist and colleagues followed this up by making new collections of *D. subobscura* from six of the same or neighboring North American sites in 1997 (Gilchrist et al., 2001), and new collections from five of the same South American sites in 1999 (Gilchrist et al., 2004). *D. subobscura* is assumed to cycle through about five generations in a year, for an average generation time of 0.2 years (Gilchrist et al., 2001).

Measurements are available for three traits in male and female *D. subobscura*, for specimens collected in South America in 1986 (Pegueroles et al., 1995) and then again in 1999 (Gilchrist et al., 2004). The specimens came from five sites in Chile, on a geographic cline from Santiago in the north to Coyhaique in the south. Taken together these provide 30 base rates that range from $h_{1.813} = 0.016$ to $h_{1.813} = 0.092$ standard deviations per generation (median 0.044) on a timescale of 65 generations.

Measurements are available for three traits in male and female *D. subobscura*, for specimens collected in North America in 1986–1988 and then again in 1997. The specimens came from six sites in the United States, on a geographic cline from Davis, California, in the south to Arlington or Bellingham, Washington, in the north. Taken together these provide 36 base rates that range from $h_{1.740} = 0.011$ to $h_{1.653} = 0.093$ standard deviations per generation (median 0.049) on timescales of 45 and 55 generations.

8.1.21 Collared Flycatchers on the Island of Gotland

Gotland is a large 3,000 km^2 island province of Sweden in the Baltic Sea. It is part of the summer nesting range of collared flycatchers, *Ficedula albicollis*, which were intensively studied by Loeske Kruuk and colleagues from 1981 through 1998. Kruuk et al. (2001) measured the tarsus length of more than 18,000 young flycatchers 10–12 days old, just before fledging, by which time the tarsus is adult in size. Means and standard errors can be digitized from figure 2 in Kruuk et al., and sample sizes are given in their table 3. The generation time for *F. albicollis* is relatively short, with the mean estimated to be 2.1 years (Przybylo et al., 2000). This was rounded down to two years in the analyses carried out here.

Log-rate-interval scaling of rates yields a slope of -1.042 with a 95% confidence interval that includes stasis but excludes both random and directional change. The 17 samples of tarsus length in *F. albicollis* yield 135 rates of change on timescales ranging from one to 8.5 generations. Fifteen of these are step rates, which, in haldanes, range from $h_0 = 0.021$ to $h_0 = 0.924$ standard deviations per generation (median 0.137) on a timescale of one generation.

8.1.22 Three-spined Sticklebacks in a Dammed Fjord in Iceland

The three-spined stickleback, *Gasterosteus aculeatus*, as noted above, is a small fish with a broad distribution on the three northern continents. Most are anadromous, live in the sea, and swim up into freshwater streams to spawn. Some *G. aculeatus* live permanently in streams, and many have become isolated in lakes where their evolution is often rapid. A small fjord in northwestern Iceland, Hraunsfjördur or "lava fjord," is partially dammed by a lava flow. In 1987 damming of the narrow part of the fjord was completed to create a $1–2 \text{ km}^2$ freshwater lake for recreational fishing. Addition of the man-made dam separated *G. aculeatus* in the lake from its much larger ancestral population in the fjord. Twelve years later, in 1999, Bjarni Kristjánsson and colleagues sampled and compared sticklebacks in the freshwater lake to those in the marine part of the fjord. The generation time for *G. aculeatus* in the Hraunsfjördur lake is not known, but it is assumed to be two years like the modal age of freshwater females breeding in Alaska (Baker et al., 2008).

Kristjánsson et al. (2002, table 1) provided statistics for male, female, and combined-sex samples of *G. aculeatus* from three sites that permit a longitudinal comparison of the marine source population and two derived lacustrine representatives, one living on a lava substrate and one living on a muddy substrate. Thirty fish were collected at each of the three sites sampled, but sample sizes are not given for the sexes separately, so we can only compare the combined samples. Kristjánsson et al. standardized 18 of the morphological traits measured on the sticklebacks to a common body length, and these were not ln-transformed before calculating rates. The remaining eight measurements given were treated in the normal way (Box 8.1).

Rates of adaptation to the lava bottom and muddy bottom differ little and the differences are probably not significant. When combined, 50 of the 52 rates are nonzero. These range from $h_{0.778} = 0.009$ to $h_{0.778} = 0.198$ standard deviations per generation (median 0.058) on a timescale of six generations.

8.1.23 Red Deer on the Isle of Rum in Western Scotland

The Isle of Rum is a mountainous 100 km^2 island in the Inner Hebrides off the west coast of Scotland. Red deer, *Cervus elaphus*, have been studied there for many years (Clutton-Brock et al., 1982). Male *C. elaphus* have antlers, and in 2002 Loeske Kruuk and colleagues reported that the weight of red deer antlers decreased by 202.38 g over a 30-year period, giving this as an average decrease of 6.746 grams per year (Kruuk et al., 2002, p. 1690). The standard deviation of antler weight is 162.55 g, so a decrease of 202.38 g is a decrease of 1.245 standard deviations. The generation time for *C. elaphus* is 8.33 years (Kruuk et al., 2002), making the rate of decrease $h_{0.556} = 0.346$ standard deviations per generation on a timescale of 3.6 generations.

8.1.24 Red Foxes in Poland

The red fox, *Vulpes vulpes*, is a common carnivorous mammal that is widely distributed in Poland. Elwira Szuma studied some 1,450 museum specimens collected in the interval from 1927 through 1996. She grouped these into five collection periods with modal years of 1931, 1963, 1967, 1971, and 1995 (Szuma, 2003). The first period was omitted here because it spans many years and includes relatively few specimens. The remaining periods are narrowly defined and include many specimens. The generation time for *V. vulpes* is four years (Kerk et al., 2013).

Rates can be calculated for the eight dental measurements summarized in Szuma (2003, table 2). Each yields a step rate from 1963 to 1967 and from 1967 to 1971, and a base rate from 1971 to 1995. The 15 nonzero step rates range from $h_0 = 0.029$ to $h_0 = 0.311$ standard deviations per generation (median 0.105) on a timescale of one generation. The seven nonzero base rates range from $h_{0.778} = 0.021$ to $h_{0.778} = 0.073$ standard deviations per generation (median 0.061) on a timescale of six generations.

8.1.25 Three-spined Sticklebacks Colonizing Loberg Lake in Alaska

The three-spined stickleback, *Gasterosteus aculeatus*, as noted above, is a small fish with a broad distribution on all three northern continents. Most are anadromous, live in the sea, and swim up into freshwater streams to spawn. Some *G. aculeatus* live permanently in streams, and many have become isolated in lakes where their evolution is often rapid. Loberg Lake in southcentral Alaska provides a well-studied example. The lake was poisoned in 1982 to remove unwanted fish before it was restocked with trout and salmon. Some years later, in 1990, "complete-morph" sticklebacks were found in the lake. These are fish that are fully armored with bony lateral plates. Finding complete-morph fish clearly indicates that Loberg Lake was invaded by anadromous sticklebacks.

Michael Bell and Windsor Aguirre studied a sample of the now-extinct sticklebacks that lived in Loberg Lake in 1982, and have sampled the lake continuously since 1990. They co-authored a series of reports on sticklebacks in Loberg Lake (Bell et al., 2004, Aguirre et al., 2004, Aguirre and Bell, 2012, Bell and Aguirre, 2013), and generously provided soon-to-be published annual counts of gill rakers and lateral plates for *G. aculeatus* in the system through 2017. The modal age of breeding females and generation time for freshwater sticklebacks in Alaska is two years (Baker et al., 2008).

The 28-year time series for gill raker counts in Loberg Lake yields a total of 26 nonzero step rates that range from $h_0 = 0.015$ to $h_0 = 1.095$ standard deviations per generation (median 0.150) on a timescale of one generation. The 27-year time series for low-morph lateral plate counts also yields 26 step rates, and these range from $h_0 = 0.011$ to $h_0 = 0.859$ standard deviations per generation (median 0.171) on a time scale of one generation.

8.1.26 Blue Tits on Corsica and Mainland France

The blue tit, *Parus caeruleus* or *Cyanistes caeruleus*, is a distinctively colored passerine with a deep blue crown, wings, and tail, and lighter underparts. It lives year-round in Europe and feeds preferentially on insects. Anne Charmantier and colleagues studied the evolution of blue tits at three sites, one in deciduous forest (Muro or site M) and one in evergreen forest (Pirio or site P) on the island of Corsica, and the third in deciduous forest near Montpellier in southern France proper (La Rouvière or site R). They monitored nests from 1989 through 2002, and measured tarsus length and body weight in large samples of fledgling birds. Resulting time series ranged from 8 to 12 years at the three sites. Generation times are 1.9 and 2.0 years (here rounded to 2) for the Corsican sites, and 2.6 (rounded to 3) for the Montpellier site (Charmantier et al., 2004).

The three time series of Charmantier et al. (2004) yield a total of 359 nonzero rates. The slope calculated for log-rate-interval scaling of fledgling tarsus length, combining the three time series, is -1.169. The slope calculated for LRI scaling of fledgling body weight is -1.243. Both slopes are characteristic of stationary time series, and a 95% confidence interval excludes directional change and random change for both.

Thirty-one step rates for fledgling tarsus length ranged from $h_0 = 0.004$ to $h_0 = 1.020$ standard deviations per generation (median 0.220). Similarly, 31 step rates for fledgling body weight ranged from $h_0 = 0.045$ to $h_0 = 1.330$ standard deviations per generation (median 0.317). All are on a timescale of one generation.

8.1.27 Barn Swallows at Kraghede in Northern Denmark

The barn swallow, *Hirundo rustica*, is the most cosmopolitan of swallows. Those that nest in Denmark spend the European winter in southern Africa. Males and females are similar, but males differ in having longer streaming tail feathers. Anders Pape Møller and Tibor Szép carried out a long-term study of barn swallows from 1984 through 2002, at Kraghede in northern Denmarck. They published tail lengths for males and females (Møller and Szép, 2005, figure 3a), and enough information about numbers of birds studied and sample standard deviations to enable calculation of rates. Møller and Szép estimated the generation time for *H. rustica* at 1.5 years, which is here conservatively rounded up to two years.

Time series for tail length in female swallows and male swallows, measured as outermost feather length, are shown in Figure 8.4a. Møller and Szép (2005) found that tail length in females increased by 0.5 standard deviations from 1984 through 2002, an interval they interpreted as representing 13.1 generations. Tail length in males increased by 1.3 standard deviations during the same interval. Møller and Szép's numbers yield a rate for females of $h_{1.119} = 0.038$ and a rate for males of $h_{1.119} = 0.099$. Both represent change on a timescale of 13.1 generations. The rates calculated here on the longest timescale are comparable: $h_{0.954} = 0.034$ for females and $h_{0.954} = 0.131$ for males. Both represent change on a timescale of nine generations.

The LRI temporal scaling slope for females in Figure 8.4b is -0.964, which is close to -1.000 expected for a stationary time series, and the confidence interval for females excludes both directional and random change. The temporal scaling slope for males in Figure 8.4c is -0.421, which is close to -0.500 expected for random change, and the confidence interval for males excludes both stasis and directional change. Tail length in male *H. rustica* changed over the course of the study interval while that in females did not.

Eighteen step rates for tail length in females range from $h_0 = 0.006$ to $h_0 = 0.882$ standard deviations per generation (median 0.096). Eighteen step rates in males range from $h_0 = 0.005$ to $h_0 = 0.527$ standard deviations per generation (median 0.162). Both represent change on a timescale of one generation.

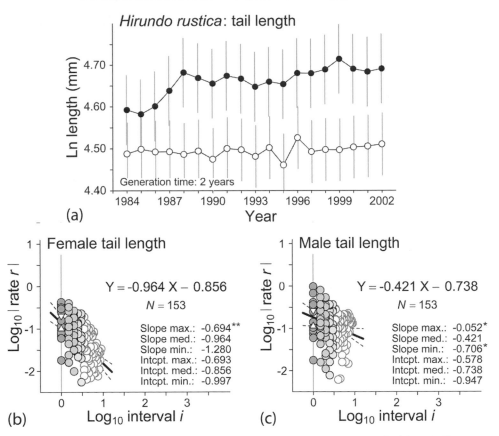

(a)

(b)

(c)

Figure 8.4 Size change in the barn swallow *Hirundo rustica* studied by Møller and Szép (2005) in northern Denmark. (a) Change in male and female tail feather lengths are shown through time. Solid circles are means for males and open circles are means for females; vertical lines span ±1 phenotypic standard deviation. Generation time is two years. (b) LRI log-rate-interval plot for 153 rates of change in female tail length spanning a full generation or longer. Median slope of −0.964 is consistent with stasis. A bootstrapped 95% confidence interval for the slope excludes both 0.000 expected for directional change and −0.500 expected for random change. (c) LRI log-rate-interval plot for 153 rates of change in male tail length. Median slope of −0.421 is near the slope of −0.500 expected for random change. A 95% confidence interval excludes both directional change and stasis.

8.1.28 Whirligig Beetles from Louisiana and Surrounding States

Gyretes sinuatus is an aquatic "whirligig" beetle of the family Gyrinidae that inhabits streams in the southern United States including Louisiana and surrounding states. Jennifer Babin-Fenske and colleagues surveyed museum collections and analyzed the size and shape of 79 *G. sinuatus* beetles collected from 1928 through

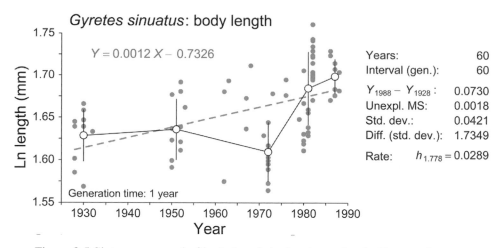

Figure 8.5 Sixty-year record of body length in the stream beetle *Gyretes sinuatus* based on measurements of Babin-Fenske et al. (2008; gray symbols). The dashed line represents regression of ln length on year. Calculations on the right side of the figure yielding $h_{1.778} = 0.0289$ are explained in the text. As an alternative, the open circles with black connecting lines are fit to five successive groups of samples. Rates for the fit to groups are described in the text.

1988, an interval of 60 years. *Gyretes* was interpreted as univoltine, meaning that it has one generation per year. Babin-Fenske et al. (2008) found that the body size of *G. sinuatus* increased by 8% in 60 years based on a linear regression of measurements on time. "Fineness," the ratio of length to width, increased by 6%. Dr. Babin-Fenske sent the measurements used in this study, which enables illustration of two complementary approaches to quantification of rates.

Conversion of the percent change reported by Babin-Fenske et al. to rates in haldanes, standard deviations per generation, is illustrated in Figure 8.5. The figure shows ln body length for *G. sinuatus* plotted against the year of collection, and a regression of the former on the latter. Regression values for body length, *Y* in the figure, are 1.6118 for 1928 and 1.6848 for 1988, indicating a change of 0.0730 ln units. The pooled standard deviation of ln body length is the square root of the unexplained mean square variance for the regression, 0.0421, which is the standard deviation of the residuals. Dividing 0.0730 by 0.0421 yields a change of 1.7349 standard deviation units. Finally, dividing the change by the number of generations, 60, the overall haldane rate for body length is $h_{1.778} = 0.0289$ standard deviations per generation on a timescale of 60 generations. Similar calculations for body width and fineness yield rates of $h_{1.778} = 0.0076$ and $h_{1.778} = 0.0346$.

It is sometimes advantageous to confine analysis to groups with substantial sample sizes, and five groups of *G. sinuatus* specimens stand out among the measurements plotted in Figure 8.5. These were collected in or near 1930, 1951, 1972, 1981, and 1987, and their means are represented by open circles in the figure. Grouping yields rates on timescales ranging from 6 to 57 generations. The resulting ten rates for body length range from $h_{1.322} = 0.011$ to $h_{1.176} = 0.210$ standard deviations per generation, on timescales of 21 and 15 generations respectively. The median rate for length of 0.044 combines rates on timescales of 30 and 57 generations. Ten rates for body width range from $h_{1.322} = 0.005$ to $h_{0.954} = 0.805$ standard deviations per generation, on timescales of 21 and 9 generations respectively. The median rate for width of 0.010 combines rates on timescales of 30 and 51 generations. Ten rates for fineness range from $h_{1.322} = 0.002$ to $h_{0.954} = 0.304$ standard deviations per generation, on timescales of 21 and 9 generations respectively. The median rate for fineness of 0.060 combines rates on timescales of 30 and 51 generations. Median rates for length, width, and fineness described here are systematically higher than comparable rates calculated on longer timescales in the previous paragraph for regression fits to all of the observations.

A fit to groups requires somewhat arbitrary formation of groups and ignores information in any intervening samples. However, on the positive side, a fit to groups yields more rates on a range of shorter timescales. Shorter timescales mean that change in the underlying time series is captured in greater detail. The fit to groups in Figure 8.5 suggests that much of the observed change happened between 1972 and 1987.

8.1.29 House Sparrows on Islands of Northern Norway

House sparrows, *Passer domesticus*, nest in dairy barns on islands off the coast of northern Norway. In a ten-year five-generation study, Henrik Jensen and colleagues investigated the possible role of sexual selection affecting the size of the black badge on the chest of male sparrows. Total badge size did not change, but visible badge size increased slightly. Five other characteristics of males changed, as did five characteristics of females (Jensen et al., 2008, table 5). The 11 nonzero rates of change ranged from $h_{0.699} = 0.001$ to $h_{0.699} = 0.015$ standard deviations per generation (median 0.003) on a timescale of five generations.

8.1.30 Ground Squirrels in the Sierra Nevada Mountains of California

Ground squirrels are widely distributed in the Sierra Nevada mountains of California. Two species, *Urocitellus beldingi* and *Callospermophilus lateralis*, are found at higher elevations in the Sierra Nevada. A third, *Otospermophilus beecheyi*, is known from lower elevations. All three species were studied early in the nineteenth century, and again early in the twentieth century. Lindsey Eastman and colleagues compared historical samples collected from 1902 through 1950 (mean 1926) to modern samples collected from 2000 through 2008 (mean 2004; Eastman et al., 2012)). They quantified differences in skull length, maxillary toothrow length, body length, and body weight. Generation times of 1.5, 2.5, and 3.0 years were rounded from Millar and Zammuto (1983) or, in the case of *O. beecheyi*, inferred from body size.

The three species and four traits yield 12 rates that range from $h_{1.414} = 0.0001$ to $h_{1.724} = 0.0168$ standard deviations per generation (median 0.0018) on timescales ranging from 26 to 53 generations.

8.1.31 Darwin's Finches on the Island of Daphne Major, Galápagos

Daphne Major, hereafter Daphne, is a small volcanic island in the Galápagos Archipelago of Ecuador. It is a conical crater less than a square kilometer in area rising 120 meters above the eastern Pacific Ocean off the coast of South America. Daphne lies just a half degree of latitude south of the equator. Charles Darwin visited the larger Galápagos Islands in September and October, 1835, during the voyage of the *Beagle*. He found the birds he called "finches" to be the "most singular" of birds on the archipelago, and noted that a "nearly perfect gradation of structure ... can be traced in the form of the beak, from one exceeding in dimensions that of the largest gros-beak to another differing but little from that of a warbler" (Darwin. 1839, pp. 461–462). Galápagos finches are now classified as tanagers, but they are nevertheless finch-like in morphology.

The ornithologist and evolutionary biologist David Lack linked Darwin, Galápagos, and finches forever when he published *Darwin's Finches* (Lack. 1947). This was based on a season of field work in the Galápagos Islands and a survey of specimens in museum collections. Lack's study was cross-sectional with respect to time. Later Peter Grant, seemingly frustrated by cross-sectional studies and sensing opportunity in the "perfect gradation of structure" in Darwin's finches, initiated a long-term longitudinal field study of finch ecology and evolution in the Galápagos. This became a 10-year project (Grant, 1986), a 20-year project (Grant and Grant, 1993), a 30-year project (Grant and Grant, 2002), and eventually a 40-year project (Grant and Grant, 2014).

Early in the research Boag and Grant (1981) documented the evolutionary response of the Daphne medium ground finch *Geospiza fortis* to a 1977 drought. Seeds are a staple of the dry season diet. In 1977 seed abundance declined, the birds did not breed, and the population declined by 85%. Measurements of Daphne *G. fortis* before and after the drought showed that survivors were distinctly larger on average than conspecifics before the drought. Nine rates of change for Boag and Grant's measurements of body size, bill size, and male and female principal component I, range from $h_0 = 0.464$ to $h_0 = 0.670$ standard deviations per generation (median 0.544) on a timescale of one generation.

Later Grant and Grant (2014) illustrated a 40-year time series for measurements of beak length, beak depth, and beak width for two long-term residents of Daphne, *Geospiza fortis* and the cactus finch *Geospiza scandens*. Beak measurements were derived from some 12,700 individual birds for the *G. fortis* series and from some 5,500 individuals for the *G. scandens* series. Birds of both species are similar in size, averaging about 16–17 g in weight, and both have beaks of similar depth and width. *G. fortis* is distinctive in having a shorter beak relative to its depth and width (making it deeper and wider relative to length). *G. scandens* is distinctive in having a longer beak relative to depth and width (making it shallower and narrower relative to length). *G. scandens* is also distinctive in exhibiting greater sexual dimorphism: males are heavier on average than those of *G. fortis*, but females are lighter than those of *G. fortis*.

Grant and Grant (2014, with accompanying online data) provided sample sizes, means, and standard errors for each sample in the *G. fortis* time series and for each in the *G. scandens* time series. These enable calculation of ln mean as an estimate of the mean of ln-transformed measurements for each sample, and calculation of the coefficient of variation as an estimate of the standard deviation of ln-transformed measurements. Natural-log transformed means for beak length, beak depth, and beak width are plotted against the year of measurement for each species in Figures 8.6 and 8.7.

Grant (1986, p. 170) estimated the generation times for *G. fortis* and *G. scandens*, the average interval between birth of a mother and birth of her offspring, as "rather long for small passerine birds, on the order of 3–5 years." Later Grant and Grant (2010, p. 20157) described the average generation length of *G. fortis* as 4.5 years and the average generation length of *G. scandens* as 5.5 years. Here generation times of four years for *G. fortis* and five years for *G. scandents* were used to calculate rates of change in haldanes.

Drought intervals of unusually low rainfall are represented by vertical bars of gray shading in Figures 8.6 and 8.7. Examination of beak measurements during and following drought show that sometimes survivors have distinctly larger beaks, as Boag and Grant (1981) found for the 1977 drought, but sometimes, as in 2004, survivors have smaller beaks. At other times there is no clear response to drought. A simple variable like rainfall has multiple effects, direct and indirect, that shape any measurable response (Grant and Grant, 2014).

The LRI log-rate-interval temporal scaling slope for *G. fortis* beak length (Figure 8.6b) is -0.449, which is close to -0.500 expected for a random time series. A bootstrapped 95% confidence interval excludes 0.000 expected for directional change and -1.000 expected for stasis. Beak depth has a temporal

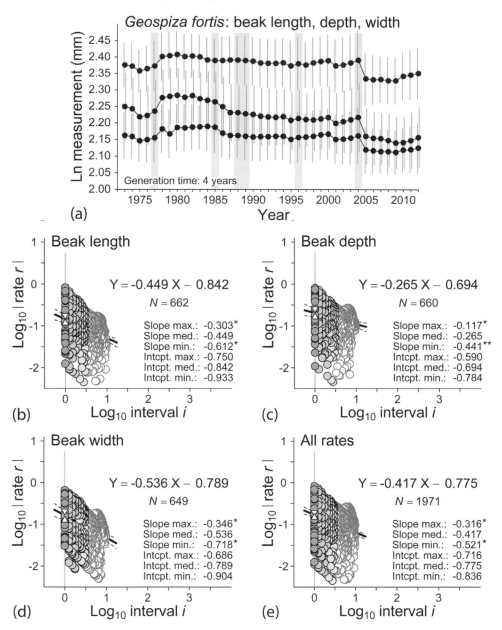

(a)

(b)

(c)

(d)

(e)

Figure 8.6 Size change in the Galápagos finch *Geospiza fortis* reported by Grant and Grant (2014). (a) Changes in beak length (top), beak depth (middle), and beak width (bottom) are shown through time. Solid circles are means; vertical lines span ±1 phenotypic standard deviation. Gray shading in the background marks years of drought (from Grant and Grant, 2014, figure 4.12). Generation time is four years. (b) LRI log-rate-interval plot for beak length showing 662 rates of change spanning a full generation or longer. A median slope of −0.449 is consistent with random change. Directional change and stasis are

scaling slope of -0.265 (Figure 8.6c), which lies between expectations for directional (0.000) and random (-0.500) change. It is significantly different from both, and even more different from stasis. Beak width has a temporal scaling slope of -0.536 (Figure 8.6d), which is again close to -0.500 expected for a random time series. A 95% confidence interval excludes 0.000 expected for directional change and -1.000 expected for stasis. Finally, all of the rates considered together have a temporal scaling slope of -0.417 (Figure 8.6e), which is close to -0.500 expected for a random time series. A 95% confidence interval excludes directional change and stasis. Changes in *G. fortis* beak length and width are indistinguishable from random, but over the same interval of time beak depth decreased with a clear directional component.

Thirty-five step rates for each of the successive nonequal beak lengths in the *G. fortis* time series range from $h_0 = 0.013$ to $h_0 = 0.687$ standard deviations per generation (median 0.056) on a timescale of one generation. Step rates for *G. fortis* beak depth and beak width are similar. All of the 108 step rates for beak length, depth, and width in *G. fortis* taken together range from $h_0 = 0.012$ to $h_0 = 0.687$ standard deviations per generation (median 0.059) on a timescale of one generation.

The LRI log-rate-interval temporal scaling slope for *G. scandens* beak length (Figure 8.7b) is 0.059, which is very close to 0.000 expected for a directional time series. A bootstrapped 95% confidence interval excludes both -0.500 expected for random change and -1.000 expected for stasis. Beak depth has a temporal scaling slope of -0.778 (Figure 8.7c), which lies between expectations for random change (-0.500) and stasis (-1.000). It is significantly different from both, and even more different from directional change. Beak width has a temporal scaling slope of -0.912 (Figure 8.7d), which is again close to -1.000 expected for stasis. A 95% confidence interval excludes 0.000 expected for directional change and -0.500 expected for random change. Finally, all of the rates considered together have a temporal scaling slope of -0.496 (Figure 8.7e), which is close to -0.500 expected for a random time series. A 95% confidence interval excludes directional change

---

Figure 8.6 (*cont.*) excluded. (c) LRI plot for beak depth showing 660 rates of change spanning a generation or longer. A median slope of -0.265 indicates directional change with a random component. Purely directional change, purely random change, and stasis are all excluded. (d) LRI plot for beak width showing 649 rates of change spanning a generation or longer. A median slope of -0.536 is consistent with random change. Directional change and stasis are excluded. (e) LRI plot for all rates of change spanning a generation or longer. An overall median slope of -0.417 is consistent with a general pattern of random change. Purely directional change and pure stasis are excluded.

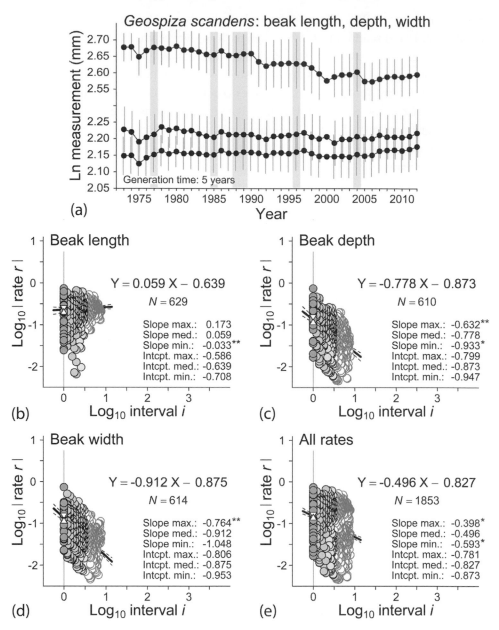

(a)

(b)

(c)

(d)

(e)

Figure 8.7 Size change in the Galápagos finch *Geospiza scandens* reported by Grant and Grant (2014). (a) Changes in beak length (top), beak depth (middle), and beak width (bottom) are shown through time. Solid circles are means; vertical lines span ±1 phenotypic standard deviation. Gray shading in the background marks years of drought (from Grant and Grant, 2014, figure 4.12). Generation time is five years. Note the composite scale on the ordinate. (b) LRI log-rate-interval plot for beak length showing 629 rates of change spanning a full generation or longer. A median slope of 0.059 is consistent with directional

and stasis. Changes in *G. scandens* beak depth and width are indistinguishable from stasis, but over the same interval of time beak length decreased directionally.

Thirty-eight step rates for each of the successive nonequal beak lengths in the *G. scandens* time series range from $h_0 = 0.009$ to $h_0 = 0.586$ standard deviations per generation (median 0.107) on a timescale of one generation. Step rates for *G. scandens* beak depth and beak width are similar. All of the 107 step rates for beak length, depth, and width in *G. scandens* taken together range from $h_0 = 0.009$ to $h_0 = 0.586$ standard deviations per generation (median 0.115) on a timescale of one generation.

8.1.32 Yellow Dung Fly in Switzerland

The yellow dung fly, *Scathophaga stercoraria*, is a widely distributed cool-climate predator on other insects. The species is commensal with cows and other large mammals in the sense that females lay their eggs in fresh mammal dung. Wolf Blanckenhorn collected and measured large numbers of *S. stercoraria* on farms in Switzerland from 1994 through 2009 (Blanckenhorn, 2015a). Measurements of hind limb tibia length were published for 3,961 female flies and 11,753 male flies (Blanckenhorn, 2015b). Female *S. stercoraria* lay eggs more or less weekly during the spring and again in the autumn, and adult flies require about three weeks to develop. The flies produce about four generations per year (Blanckenhorn, 2015a), and thus the generation time is about 0.25 years.

Measurements of flies were ln-transformed and combined into annual samples. Male and female *S. stercoraria* are dimorphic, so rates for the sexes were calculated separately. LRI log-rate-interval scaling slopes, with the sexes separated or combined (slope −0.791), indicate stationary to random time series. In each of the three cases the 95% confidence interval excludes the zero slope expected for directional change.

The 45 rates for females and 66 rates for males have similar distributions. When combined the 111 rates range from $h_{1.716} = 0.001$ to $h_{0.602} = 0.298$ standard deviations per generation (median 0.029) on timescales of four to 60 generations. The 20 base

Figure 8.7 *(cont.)* change. Random change and stasis are excluded. (c) LRI plot for beak depth showing 610 rates of change spanning a generation or longer. A median slope of −0.778 indicates stasis with a random component. Purely directional and random change are excluded, as is stasis. (d) LRI plot for beak width showing 614 rates of change spanning a generation or longer. A median slope of −0.912 is consistent with stasis. Directional and random change are excluded. (e) LRI plot for all rates of change spanning a generation or longer. An overall median slope of −0.496 is consistent with a general pattern of random change. Purely directional change and pure stasis are excluded.

rates for *S. stercoraria* range from $h_{0.602} = 0.006$ to $h_{0.602} = 0.298$ standard deviations per generation (median 0.083) on timescales of 4 to 16 generations.

8.1.33 Woodland Salamanders in the Southeastern United States

Plethodontids are distinctive among salamanders in lacking lungs, respiring through their skin, and lacking an aquatic larval stage. They feed on springtails and other invertebrate foragers in leaf litter. Woodland or mountain salamanders in the genus *Plethodon* are species-rich and best known from the Appalachian Mountains of the southeastern United States. Amphibians generally are ectotherms and moisture-dependent, making them vulnerable to environmental change. With this in mind, Nicholas Caruso and colleagues studied body size in 15 species of *Plethodon* for decades from the 1950s to the present. Reduction in size was significant in six of the species and increase in size was significant in one (Caruso et al., 2014, table 1). These changes can also be compared in terms of rate. Nicholas Caruso provided the sample variances necessary to do this.

Generation times for salamanders are poorly known. Here they are predicted from regression of ln generation time G in years on snout-vent length L in mm, where $G = \exp{(0.77 \cdot \ln L - 1.14)}$. Generation times for nine plethodontids in the regression were taken from Hairston (1987, pp. 54–55) and Bruce (1988, p. 24).

Rates can be calculated for 14 of the 15 species of *Plethodon* listed in Caruso et al. (2014, table 1), the omission being *P. aureoles* with only one sizeable sample. Size declined in 87 of the 124 rate comparisons (70%). The 14 species yield 51 base rates that range from $h_{0.173} = 0.003$ to $h_{0.110} = 1.023$ standard deviations per generation (median 0.104) on timescales ranging from 1.24 to 7.18 generations.

8.1.34 Vlei Rats in the Northern Provinces of South Africa

Vlei rats of the genus *Otomys* are small but large-eared murine rodents that are widely distributed across sub-Saharan Africa. Aluwani Nengovhela and colleagues studied two species, *O. auretus* and *O. angoniensis*, in the four northern provinces of South Africa (formerly grouped as Transvaal). They studied museum specimens collected as long ago as 1906, and new specimens collected in recent years. Regression of skull length on the year of collection showed that both species decreased in size over the 106 year intervals of study (Nengovhela et al., 2015). Peter Taylor generously provided original measurements permitting ln-transformation before analysis of the rates involved. The small size of vlei rats means that their generation time was probably about one year (Figure 3.3).

Nengovhela et al. (2015) calculated an adjusted skull length (residual) based on multiple regression of measured skull length on tooth-wear stage and the latitude and longitude of the source locality. Here the same was done for ln-transformed skull length. Regression of this residual on year of collection yielded a decrease of 0.85 standard deviations and a net rate of $h_{2.025} = 0.008$ standard deviations per generation for *O. auretus*. A similar analysis yielded a decrease of 0.69 standard deviations and a net rate of $h_{2.025} = 0.007$ standard deviations per generation for *O. angoniensis*. Both are on timescales of 106 generations.

The specimens studied by Nengovhela et al. (2015) can be divided into five groups by year of collection. The *O. auretus* groups have median collection years of 1923, 1958, 1974, 1991, and 2013, and the *O. angoniensis* groups have median collection years of 1923, 1955, 1973, 1989, and 2011. Comparisons between different medians yielded an additional four base rates and six net rates for each species. Collectively the eight base rates ranged from $h_{1.505} = 0.001$ to $h_{1.342} = 0.049$ standard deviations per generation (median 0.011) on timescales ranging from 16 to 35 generations.

8.1.35 Black Rats on Anacapa in the California Channel Islands

Black rats, *Rattus rattus*, are thought to have been introduced on Anacapa Island off the coast of southern California in the mid-1800s. These were studied by Oliver Pergams and colleagues, who measured crania of 59 museum specimens collected between 1940 and 2000 (Pergams et al., 2015). The collections fall into four time intervals: 1940, 1975–1979, 1986, and 2000.

Erickson and Halvorson (1990) showed that pregnancy and lactation on Anacapa and neighboring San Miguel islands peak in April and again in August. They weighed 284 male and female rats from the two islands that averaged 154 g. This body weight points to a one-year generation time (Equation 3.2). Thus the collection intervals of Pergams et al. are separated by 37, 9, and 14 generations, or some combination of these.

Measurements of cranial length yield six rates that range from $h_{1.362} = 0.011$ to $h_{0.954} = 0.084$ standard deviations per generation (median 0.022) on timescales of 23 and 9 generations, respectively. Measurements of cranial width yield an additional six rates that range from $h_{1.146} = 0.020$ to $h_{0.954} = 0.032$ standard deviations per generation (median 0.023) on timescales of 14 and 9 generations, respectively.

8.1.36 Deer Mice in the Sierra Nevada Mountains of California

Deer mice, *Peromyscus maniculatus*, are widely distributed in the Sierra Nevada mountains of California, where they were studied early in the nineteenth century, and again early in the twentieth century. Michael Holmes and colleagues compared the early and late samples in relation to climate change in national parks of the northern, central, and southern Sierra Nevada (at Lassen, where the collection spanned 81 years; Yosemite, spanning 89 years; and Sequoia–Kings Canyon, spanning 95 years). The body size of *Peromyscus* and its typical lifespan of one year make a generation time of 0.5 year more likely than that of a full year.

Multivariate comparisons of cranial shapes yielded Mahalanobis distances between early and late samples in the three areas of 2.76, 2.55, and 2.18, respectively (Holmes et al., 2016). The corresponding rates of change are $h_{2.210} = 0.017$, $h_{2.250} = 0.014$, and $h_{2.279} = 0.012$ standard deviations per generation on timescales of 162, 178, and 190 generations.

8.1.37 Geckos in the Brazilian Cerrado of Goiás

In 1996 water began filling the reservoir associated with the Serra da Mesa hydroelectric plant in the Brazilian state of Goiás. Rising water turned hilltops into islands, isolating samples of the endemic fauna. Fifteen years later Mariana Eloy de Amorim and colleagues collected samples of the gecko *Gymnodactylus amarali* from five of the new islands for comparison with five mainland sites. They found *G. amarali* on islands ate larger termites and had longer heads relative to body length than *G. amarali* on the surrounding mainland (Eloy de Amorim et al., 2017). The differences in morphology necessarily evolved in parallel, independently, on each island. Here a generation time of one year is assumed.

Eloy de Amorim et al. (2017) published their measurements as supplemental information, which enables calculation of rates of change of head size. Four of the islands and three of the mainland sites have samples of four or more female *G. amarali*. Three of the islands and three of the mainland sites have samples of four or more males. Island and mainland samples, female and male, are compared in terms of residuals relative to a mainland baseline regression of ln head length on ln body length.

Twenty-one rates can be calculated comparing each island sample to each mainland sample. All of the rates are positive, confirming that head size increased in parallel on each of the islands sampled. All of the rates together range from $h_{1.176} = 0.003$ to $h_{1.176} = 0.392$ standard deviations per generation (median 0.056) on a timescale of 15 generations.

8.1.38 Deer Mice and White-Footed Mice in Southern Quebec

Two species of field- and forest-dwelling mice occur together in parts of southern Quebec Province in eastern Canada. These are the deer mouse, *Peromyscus maniculatus*, and the white-footed mouse, *Peromyscus leucopus*. Both were studied by Virginie Millien and colleagues in an investigation of character displacement and response to climate change. Millien et al. (2017, p. 855) reported rates of evolution for the two species of 47 and 128 haldanes, standard deviations per generation, on a timescale of 50 generations. Something is amiss, because a change of 47 or 128 standard deviations per generation is unthinkable on any scale of time.

Millien et al. (2017) published their measurements as supplemental information, making it possible to recalculate rates. Mont Saint-Hilaire is a site with good historical and modern samples of *P. maniculatus* and *P. leucopus*. The historical samples were made just before and after 1963, and the modern samples were made just before and after 2008. This is a difference of 45 years. The body size of *Peromyscus* and its typical lifespan of one year make a generation time of 0.5 year more likely than that of a full year, so rates were calculated for an interval of 90 generations.

Calculation of rates for the two species (*P. maniculatus* and *P. leucopus*) at Mont Saint-Hilaire, two sexes (female and male), and four morphological characteristics (body weight, body length, shape axis 1 and shape axis 2) yield 16 rates of change. These range from $h_{1.954} = 0.0006$ to $h_{1.954} = 0.017$ standard deviations per generation (median 0.005) on a timescale of 90 generations.

8.1.39 House Mice in a Barn in Northern Switzerland

In 2002 the Institute for Evolutionary Biology and Environmental Studies at the University of Zurich established a free-living population of wild house mice, *Mus musculus*, in a barn in the nearby countryside. The population has been closely monitored ever since, with virtually all mice born there weighed and measured as 13-day-old pups. Madeleine Geiger and colleagues analyzed these data for mouse pups born in the 10-year interval from 2007 through 2016 (Geiger et al., 2018). Dr. Geiger generously provided the weights and head lengths for 2,633 mouse pups analyzed here.

The entire *M. musculus* time series spans 3,432 days. The generation time for the barn population is variously given as nine months (ca. 270 days; Manser et al., 2011) or as 263 days (Geiger et al., 2018). A generation time of 264 days is used here because it divides evenly into the total span of 3,432 days. With this slight modification, the mouse pup time series of Geiger et al. spans 13 generations. Weights and head lengths were ln-transformed, and sample sizes, means, and standard deviations then computed for each generation.

Log-rate-interval scaling for body weight yields a slope of −0.250, and a 95% confidence interval excluding −1.000 expected for a stationary time series. Twelve step rates for body weight range from $h_0 = 0.011$ to $h_0 = 0.672$ standard deviations per generation (median 0.099) on a timescale of one generation. LRI scaling for head length yields a slope of −0.430, and a 95% confidence interval excluding both 0.000 expected for purely directional change and −1.000 expected for stasis.

Twelve step rates for head length range from $h_0 = 0.089$ to $h_0 = 0.598$ standard deviations per generation (median 0.183) on a timescale of one generation.

8.1.40 Carabid Beetles in Southern British Columbia

Ground beetles of the family Carabidae are relatively large terrestrial beetles that are specialized to feed on a variety of smaller invertebrates including slugs and snails that are often considered garden and agricultural pests. Michelle Tseng and colleagues measured elytron length in large and small samples of eight carabid species from field sites in southern British Columbia, western Canada. The species are *Pterostichus algidus*, *Scaphinotus angusticollis*, *Pterostichus melanarius*, *Carabus nemoralis*, *Harpalus fraternus*, *Euryderus grossus*, *Cymindis planipennis*, and *Amara quenseli*. Tseng et al. (2018a) measured specimens in museum collections spanning the past 100 years. Their study was designed to investigate how carabid body size is linked to environmental temperature, but the resulting measurements (Tseng et al., 2018b) are equally interesting for analysis of rates.

Carabid beetles undergo complete metamorphosis in the course of a year, developing from an egg through larval instars to a pupa, finally emerging as an adult. The entire cycle takes a year. Some carabids overwinter as adults and breed in the spring. Others overwinter as larvae. In either case the generation time is reasonably considered to be one year.

For analysis of rates the Tseng et al. measurements were grouped by species and by year. A time series was assembled for each species from the successive samples of 10 or more individual specimens. Eletron length was ln-transformed and matrices of sample sizes, mean values, and standard deviations for each species were converted to rates. Analyzed this way, time series for the species listed above include 10, 4, 9, 4, 11, 16, 13, and 13 samples, respectively, separated by intervals of time ranging from 1 to 53 years.

There are 45 step rates in the eight time series combined. The step rates range from $h_0 = 0.023$ to $h_0 = 1.756$ standard deviations per generation (median 0.293) on a timescale of one generation. There are an additional 27 base rates in the eight time series. The base rates range from $h_{1.114} = 0.001$ to $h_{0.301} = 0.774$ standard deviations per generation (median 0.107) on timescales ranging from 2 to 53 generations.

8.2 Observational Studies: Cross-Sectional with Respect to Time

8.2.1 House Sparrows Introduced into North America

A longitudinal study of North American house sparrows by Calhoun (1947) was mentioned above. In 1962 Richard F. Johnston and Robert K. Selander initiated a more comprehensive study of *Passer domesticus*. Their first report provided metrical comparisons of wing length, bill length, and body weight for samples of adult male sparrows that are inferred to have diverged beginning in 1851, some 111 years before sampling started in 1962 (Johnston and Selander, 1964). The generation time of *P. domesticus* is 1.97 years (Jensen et al., 2008), which, rounded, is two years. Sparrow divergences in North America are cross-sectional with respect to time, evolving in parallel, so the time of separation totals about 222 years or, for our purposes, 111 generations.

The three wing length measurements yield rates in haldanes of $h_{2.045} = 0.0083$, 0.0108, and 0.0243 standard deviations per generation on a timescale of 111 generations. The three bill length measurements (leaving out Honolulu) yield rates in haldanes of $h_{2.045} = 0.0005$, 0.0090, and 0.0111 standard deviations per generation on a timescale of 111 generations. The 14 body weights (leaving out Oaxaca City, Progreso, and Mexico City) yield 91 rates, which range from $h_{2.045} = 0.0001$ to $h_{2.045} = 0.0206$ standard deviations per generation (median 0.0066) on a timescale of 111 generations.

In a longer report, Johnston and Selander (1971) provided lists of means of sternum length, skull length, and femur length for widely scattered populations of male and female sparrows. Analysis here is confined to sites in Canada and the continental United States. Coefficients of variation for males and females are listed in table 8 of Selander and Johnston (1967). Weighted means for the coefficients associated with male linear measurements and female linear measurements indicate a phenotypic standard deviation of 0.032 for the logged measurements of both. The generation time is again two years, and samples are again separated by 111 generations.

When male and female sparrows are grouped together, 678 rates for sternum length range from $h_{2.045} = 0.0001$ to $h_{2.045} = 0.0149$ standard deviations per generation (median 0.0043). Similarly, 645 rates for skull length range from $h_{2.045} = 0.0001$ to $h_{2.045} = 0.0135$ standard deviations per generation (median 0.0026), and 647 rates for femur length range from $h_{2.045} = 0.0002$ to $h_{2.045} = 0.0280$ standard deviations per generation (median 0.0034). All are on timescales of 111 generations.

8.2.2 Rabbits Introduced to Australia

The European rabbit, *Oryctolagus cuniculus*, was introduced to Australia in 1859 on an estate near the port city of Geelong, near Melbourne, on the southern coast of Victoria. European settlers thought they would be fun to hunt. Within a decade rabbits occupied much of the surrounding countryside and they continued dispersing, decade by decade, until by the 1920s they reached the west coast. Now rabbits range across virtually all of central and southern Australia (Myers, 1971). McCluskey et al. (1974) made an initial attempt to quantify rates, comparing a sample from Wanneroo in the west and a sample from Tidbinbilla in the east to a hypothetical intermediate as a possible ancestor to both. McCluskey et al. calculated rates in darwin units with no consideration of sample variation. They concluded that substantial morphological differences between rabbit populations had developed rapidly in Australia.

A more comprehensive morphometric study of Australian rabbits was carried out simultaneously by Jane Taylor, who recorded 21 dimensions of the rabbit cranium and five dimensions of the mandible. These were measured on representative samples from six sites across southern Australia. Statistics for the raw measurements were recorded in a Bachelor's thesis (Taylor, 1974). Dr. Taylor did not quantify rates but generously allowed me to reanalyze the statistics in her thesis. The dispersal history of rabbits in Australia is well documented (Myers, 1971, figure 1 map), and ages for the nodes separating pairs of samples can be calibrated using this history.

Rabbits from Werribee near Geelong have evolved separately from those at the other five sites since about 1865, yielding a two-way divergence interval of 210 years. Rabbits from Pine Plains evolved separately from those at the remaining four sites since about 1875 and the divergence interval is 190 years. Rabbits from Tero Creek evolved separately from those at the remaining three sites since about 1885 and the divergence interval is 170 years. Rabbits from Canberra in the east evolved separately from those at the two sites in the far west since about 1885 and the divergence interval is again 170 years. Finally rabbits from Chidlow evolved separately from those at Cape Naturaliste since about 1915 and the divergence interval is 110 years.

Temporal calibration introduces a longitudinal component to the evolution of differences, but the approach taken here is still fundamentally cross-sectional. Generation times for *O. cuniculus* in Australia range from 1.17 to 1.91 years, with a mean of 1.55 and a median of 1.46 (Myers, 1970). A generation time of 1.5 years is representative.

Comparison of 26 measured traits for the 15 paired combinations of six sites outlined above yields an expectation of 390 rates. Three comparisons showed no difference, leaving a total of 387 nonzero rates. These ranged from $h_{2.146} = 0.0001$ to $h_{2.054} = 0.0335$ standard deviations per generation (median 0.0058) on timescales ranging from 73 to 140 generations.

The final publication of these measurements compared a slightly different set of sites, with the Canberra site separated into three (Taylor et al., 1977). Differences between these eight sites were summarized as Mahalanobis D values in units of multivariate standard deviation. D values can be converted to rates by again calibrating nodes separating pairs of samples. The D values of Taylor et al. (1977) yield rates for Australian rabbits ranging from $h_{2.146} = 0.017$ to $h_{2.054} = 0.045$ standard deviations per generation (median 0.029) on timescales ranging from 73 to 140 generations. The range of rates for Mahalanobis D values is, as expected, narrower than the range for all of the individual comparisons. The median for the D values is about five times greater than the median for all of the individual comparisons.

8.2.3 Rats Introduced to the Galápagos Islands

Black rats, *Rattus rattus*, were introduced to the Galápagos Islands in four phases (Patton et al., 1975): (1) early colonization of Isla Santiago by English buccaneers about 1684; (2) colonization of Islas Floreana, San Cristóbal, and possibly Pinzón about 1832; (3) dispersal to Isla Isabela about 1893; and finally (4) colonization of Islas Baltra and Santa Cruz about 1942. Sources of the rats are not known with certainty, but those in the first phase may have come from England; rats in the second and third phases presumably came from coastal South America; and rats in the fourth phase probably came from North America.

Galápagos *Rattus rattus* average 120 g in body weight and Equation 3.2 yields a generation time of one year. This is consistent with their rapid development of sexual maturity, their short lifespan, and field observations that Galápagos rats breed from January through July (Clark, 1980). Patton et al. (1975) finished their study of Galápagos rats in 1974, by which time the rats had lived on Santiago for about 290 years and diverged for about 580 generations; lived on Floreana, San Cristóbal, and possibly Pinzón for about 142 years and diverged for about 284 generations; lived on Isabela for about 81 years and diverged for about 162 generations, and lived on Baltra and Santa Cruz for about 32 years and diverged for about 64 generations. Colonization of groups of islands from different sources means that divergences between groups happened earlier at unknown times, precluding calculation of rates between the groups.

Patton et al. (1975) recorded frequencies of occurrence for five nonmetrical characteristics of cranial foramina. The frequencies were normalized using the arcsine transformation of Sokal and Rohlf (1981), with differences expressed in units of standard deviation. Rates can be calculated by dividing each difference by the appropriate divergence in generations. This yields rates that range from a low of $h_{2.435} = 0.0004$ standard deviations per generation on a timescale of 284 generations to a high of $h_{1.806} = 0.0982$ standard deviations per generation on a timescale of 64 generations. The median is 0.0062 on an intermediate timescale. The rates fit the pattern of temporal scaling of other rates, but they are not included in the rate compilation analyzed later in the chapter because of uncertainty about the binomial representation of the standard deviation (see discussion accompanying Equation 7.2).

Patton et al. (1975) also recorded a series of 15 length measurements, mostly cranial, for all of the Galápagos *Rattus rattus* specimens studied. Measured differences between specimens were then summarized in a matrix of Mahalanobis *D* distances (Patton et al., 1975: table 6). Mahalanobis *D* distances are multivariate differences expressed in terms of multivariate standard deviations. Rates in haldanes can be calculated by dividing each value of *D* by the appropriate divergence time. Rates for the *D* values range from a low of $h_{2.763} = 0.0035$ standard deviations per generation on a timescale of 580 generations to a high of $h_{1.806} = 0.0575$ standard deviations per generation on a timescale of 64 generations. The median is 0.0151 on an intermediate timescale.

8.2.4 Mosquitofish Introduced to the Hawaiian Islands

The mosquitofish, *Gambusia affinis*, is a small freshwater cyprinodontiform or carp-tooth fish that is closely related to the guppy. As its name suggests, it preys on mosquitos. The mosquitofish is endemic to the Mississippi River basin of North America, but it was widely introduced in many parts of the world in an attempt to control mosquitos. *G. affinis* was introducd to the Hawaiian Islands in 1905. Stephen Stearns studied the life history of six samples of Hawaiian *G. affinis*, two from named reservoirs on Hawaii itself, and four from numbered reservoirs on Maui.

Stearns (1983) measured a number of traits in the six samples and then quantified the differences between each trait in a sample and the mean of the trait for all samples. Divergence took place approximately 70 years or about 140 generations before the samples were collected. Stearns calculated divergences over the 140-generation interval and found these to average 0.06% (0.0006) to 0.49% (0.0049) as a proportion of the assumed initial value. Direct comparison as proportions ignores differences in the variability of the samples.

The mosquitofish study is a classic cross-sectional study, and it makes sense to compare the measured samples to each other rather than comparing each to a hypothetical mean. Structured this way the total divergence time is the sum of the time in each diverging lineage, or about 280 generations. Rates were quantified for five of Stearns (1983) traits: female age at maturity, female length at maturity, female offspring weight, male age at maturity, and male length at maturity. Comparisons between all samples yield 15 rates for each trait, for a total of 75 rates. These range from $h_{2.447} = 0.00003$ to $h_{2.447} = 0.00644$ standard deviations per generation (median 0.00184) on a timescale of 280 generations.

8.2.5 Chaffinches Introduced to New Zealand

The common chaffinch, *Fringilla coelebs*, is an attractive sparrow-like bird that is widely distributed in Europe. Males are conspicuous in having a rust-colored breast and face, with a blueish cap. Through much of the year chaffinches forage for seeds on the ground, often in large flocks. In around 1865 chaffinches were introduced to the North Island and South Island of New Zealand, and later, around 1900, they reached the Chatham Islands. Allan Baker and colleagues made a large collection of New Zealand chaffinches in 1985. These came from eight widely distributed localities. Twelve traits were measured on skeletons of 15 to 30 males from each site. These represent some 120 years of diversification on the two main islands of New Zealand, and some 90 years of separation on Chatham (Baker et al., 1990). Chaffinches have a two-year generation time (Baker and Marshall, 1999), so the comparisons span either 120 or 90 generations.

Comparison of the Baker et al. (1990) chaffinch measurements between Chatham and other localities yields 82 nonzero differences and rates. The rates range from $h_{1.954} = 0.0004$ to $h_{1.954} = 0.0084$ standard deviations per generation (median 0.0030) on a timescale of 90 generations. Comparison of chaffinch measurements between the other sites, within and between the North and South islands, yields 247 nonzero differences and rates. The rates range from $h_{2.079} = 0.0001$ to $h_{2.079} = 0.0100$ standard deviations per generation (median 0.0024) on a timescale of 120 generations.

8.2.6 Sockeye Salmon Introduced in Lakes near Seattle

Sockeye salmon, *Oncorhyncus nerka*, are native to the west coast of North America, where they range from coastal California to Alaska. They are also commonly stocked in freshwater lakes. In 1992 and 1993, Andrew Hendry and Thomas Quinn studied one set of *O. nerka* introductions from fish hatcheries at Baker Lake near Seattle, Washington, in the western United States. These were introduced at three study sites in the same region: Pleasure Point in Lake Washington, Cedar River, and Issaquah Creek. According to Hendry and Quinn (1997, p. 77) the Baker-derived populations have been isolated since 1945. The three populations sampled in 1992 and 1993 were isolated for 47 and 48 years, which, with a four-year generation time, are equivalent to two-way isolation intervals of 23.5 and 24 generations.

Hendry and Quinn (1997) provided tables with sample sizes, means, and standard errors for each sample studied. These yield 30 base rates, ranging from $h_{1.371} = 0.001$ to $h_{1.380} = 0.115$ standard deviations per generation (median 0.032) on timescales of 23.5 and 24 generations.

8.2.7 Grayling Introduced in Streams of Northern Norway

The European grayling, *Thymallus thymallus*, is a trout-like fish that is native to streams in central and northern Europe. It was introduced progressively to a set of three Norwegian mountain lakes in the Atlantic Ocean drainage: to the first in about 1880, the second in 1910, and the third in 1920. Thrond Haugen and Leif Vøllestad collected and studied grayling from the lakes in 1998. At the time of collection, grayling in the second and third lakes, Hårrtjønn and Aursjøen, had been separated from those in the first lake, Lesjaskogsvatn, for 88 years, and grayling in Hårrtjønn and Aursjøen had been separated from each other for 78 years (Haugen and Vøllestad 2000; 2001). The fish in each of the lakes were introduced and each population evolved independently from the time of separation, so comparisons between lakes are cross-sectional with respect to time. The generation time for grayling is estimated to be six years (Haugen and Vøllestad, 2000), so the two-way separations of the populations are about 29, 29, and 26 generations.

Haugen and Vøllestad (2000, table 1) published statistics for three measurements of grayling from each of the three lakes: body length, body weight, and dry egg weight (calculations here assume that unstated sample sizes were equal). Body lengths and weights increased progressively from Lesjaskogsvatn to Hårrtjønn to Aursjøen, but dry egg weights decreased from Lesjaskogsvatn to the other two lakes. Nine base rates can be calculated from these comparisons, and the rates range from $h_{1.415} = 0.007$ to $h_{1.462} = 0.104$ standard deviations per generation (median 0.052) on timescales of 26 and 29 generations.

8.3 Experimental Field Studies

8.3.1 Transplantation of California River Guppies

Age at maturation and body size at maturation are two variable life history traits subject to natural selection, and predation has an influence on each. River guppies, *Poecilia reticulata*, from high-predation communities attain maturity at an earlier age and smaller size than their counterparts from low-predation communities. David Reznick and co-authors designed experiments to test this in the field by transplanting natural populations of guppies in tributaries of two California rivers. Guppies living in high-predation streams were compared to guppies transplanted to headwaters above rapids or waterfalls that excluded predators. The experiments described by Reznick et al. (1997) were cross-sectional with respect to time, but they also included a longitudinal component. Cross-sectional comparisons included control and experimental guppies below and above barriers of the Aripo River after a separation of 11 years (18.1 generations), and control and experimental guppies below and above barriers of the El Cedro River separated for

4 years (6.9 generations) and then 7.5 years (12.7 generations). In addition, a longitudinal comparison is possible in the El Cedro River, where control guppies and experimental guppies were sampled twice, at two times separated by 3.5 years (5.8 generations).

Reznick et al. (1997) reported rates in darwins but provided all of the information necessary to calculate rates in haldanes. Rates for male age, male size, female age, and female size can be calculated for the three cross-sectional comparisons and for the two longitudinal comparisons, for a total of 20 rates on four scales of time. These range from a minimum of $h_{0.839} = 0.014$ to a maximum of $h_{0.763} = 0.407$ standard deviations per generation (median 0.091) on timescales ranging from 5.8 to 18.1 generations. The highest rates are those on the shortest scales of time.

8.3.2 Transplantation of Three-Spined Sticklebacks in Iceland

The three-spined stickleback, *Gasterosteus aculeatus*, as noted above, is a small fish with a broad distribution on all three northern continents. Most are anadromous, live in the sea, and swim up into freshwater streams to spawn. Some *G. aculeatus* live permanently in streams, and many have become isolated in lakes where their evolution is often rapid.

Bjarni Kristjánsson followed his 2002 field study, described above, with a simple experiment. In June 2003 he collected *G. aculeatus* from marine tide pools in southwestern Iceland and, following acclimatization to fresh water, released these into two recently created freshwater ponds in northern Iceland. The ponds lacked fish except for arctic char in the larger pond introduced for recreational fishing. One year later, in June of 2004, Kristjánsson collected first-generation *G. aculeatus* from both ponds and compared these to the marine sample (Kristjánsson, 2005).

Rates of change for body length in the experiment are $h_0 = 1.844$ and 5.412 standard deviations per generation in the larger and smaller ponds, respectively. Rates of change for armor plates are $h_0 = 1.201$ and 0.586 standard deviations per generation in the larger and smaller ponds. All are on timescales of one generation.

These rates confirm that phenotypic change can be fast, but the one for body length in the smaller pond is so extreme that something more than natural selection must be involved. Kristjánsson (2005) attributed the observed rapid change to a combination of phenotypic plasticity (reaction of a genotype to a different environment) and natural selection.

8.3.3 Mustard Flowering Times Altered by California Drought

Climate change alters the life history characteristics of plants and animals. This is illustrated by an experimental field study involving an annual plant, the field mustard *Brassica rapa*. Steven Franks and colleagues compared the flowering times of *B. rapa* grown from seed collected before and after a 2000–2004 drought in California. Seed collected in 1997, before the drought, yielded longer flowering times, and seed collected seven generations later, in 2004 after surviving the drought, yielded shorter flowering times. Hybrid crosses of 1997 and 2004 plants had intermediate flowering times, demonstrating an additive genetic basis for the change (Franks et al., 2007).

Franks et al. (2007) compared mustard grown from seed collected at two sites, one dryer and one wetter. Seed collected from the dry site in 1997, raised with a long water treatment time to minimize mortality, had a mean flowering time of 42.8 days with a standard deviation of 6.6 days. Logged, this translates to a mean of 3.757 and a standard deviation of 0.154. Seed from the dry site collected in 2004 had a mean flowering time of 41.0 days (ln 3.714). The resulting rate in haldanes is $h_{0.845} = 0.040$ standard deviations per generation on a seven- generation timescale.

Seed collected from the wet site in 1997, again with a long water treatment time, had a mean flowering time of 58.8 days with a standard deviation of 12.3 days. Logged, this translates to a mean of 4.074 and a standard deviation of 0.209. Seed from the wet site collected in 2004 had a mean flowering time of 50.1 days (ln 3.914). The resulting rate in haldanes is $h_{0.845} = 0.109$ standard deviations per generation, again on a seven-generation timescale.

8.3.4 *Carolina Anoles on Florida Dredge-Spoil Islands*

The dredge-spoil islands of interest here are artificial islands in Mosquito Lagoon in eastern Florida. The islands were created by the United States Army Corps of Engineers at various times in the twentieth century while excavating and maintaining a shipping route called the Atlantic Intracoastal Waterway. Soon after creation the islands were colonized by Carolina anoles, *Anolis carolinensis*, an arboreal lizard native to coastal Florida. In 1995 an experiment was initiated involving introduction of a competitive congener, the invasive brown anole *Anolis sagrei*, to selected dredge-spoil islands. The two species partitioned arboreal space almost immediately, with *A. carolinensis* moving to higher perches and *A. sagrei* occupying lower perches.

Then in 2010 Yoel Stuart and colleagues returned to the Mosquito Lagoon islands to capture and study *A. carolinensis* on islands with and without the invasive *A. sagrei* (Stuart et al., 2014). The expectation was that Carolina anoles on islands with the invasive competitor would have (1) a larger toepad area relative to body size; and (2) more lamellae on toepads. Larger toepads with more lamellae enhance the ability of *A. carolinensis* to grasp the smaller branches of higher perches. Both expectations were confirmed by measurement. Stuart et al. (2014) converted their measurements to rates, assuming a 20-generation divergence time. They found mean rates of change of $h_{1.301} = 0.091$ for toepads and $h_{1.301} = 0.077$ for lamellae on a timescale of 20 generations. These rates seem reasonable for the timescale involved and they have not been recalculated.

8.4 Discussion

The 57 field studies reviewed in this chapter yielded 814 independent step rates of change on a timescale of one generation, 3,627 independent base rates of change on timescales longer than a generation, and 8,011 net rates compounded from these on timescales of a generation or longer. The total number of field-study rates recorded here is thus 12,452. Rates from field studies are important in themselves but also important in comparison to the rates of change in experimental lineages reviewed in Chapter 7. Finally, at the end of the discussion, it is interesting to compare rates observed in field studies with rates of change in what anthropologists call the "secular trend" of increasing human stature.

8.4.1 Rates of Evolution in Field Studies

The temporal scaling of differences and intervals on a log-difference-interval LDI plot and the temporal scaling of rates and intervals on a log-rate-interval LRI plot are shown side by side in Figure 8.8. As was the case for experimental rates, histograms for the untransformed step differences and rates recorded in field studies (not shown) are strongly skewed in a positive direction with a long right tail. Logarithmic transformation does not remove the skew but reduces it and reverses it to negative skew, as shown in Figures 8.8a and 8.8b.

The histogram of differences in Figure 8.8a shows the distribution of $\log_{10}$ values for step differences in the field studies reviewed here. There are 814 independent one-generation step differences, which yield rates having a median $\log_{10}$ value of -0.814. This corresponds to a median difference of 0.15 standard deviations per generation on a timescale of one generation. In Figure 8.8c, this is the distribution of differences for interval lengths of one generation, and the distribution of shaded circles that have a zero coordinate on the $\log_{10}$ interval axis.

The histogram of rates in Figure 8.8b shows the distribution of $\log_{10}$ values for step rates in the field studies reviewed here. It is the distribution of rates with interval lengths of one generation from Figure 8.8d. It is again the distribution of shaded circles with a $\log_{10}$ values of zero on the interval axis. Differences on a one-generation timescale are equivalent to rates on a one-generation timescale, or,

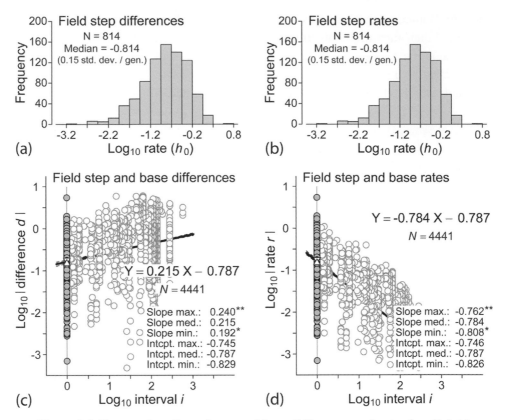

Figure 8.8 Temporal scaling of step and base differences and rates for all field studies in this chapter. (a) Histogram of step differences (column of shaded circles in panel c below). (b) Histogram of step rates (column of shaded circles in panel d below). (c) LDI log-difference-interval plot for step and base differences. (d) LRI log-rate-interval plot for step and base rates. Differences d are in standard deviation units, rates r are in standard deviation units per generation, and interval lengths i are in generations. Confidence intervals were bootstrapped with 1,000 replications. Note that the median slopes of 0.215 in panel c and -0.784 in panel d are complementary in differing by one unit, while median intercepts of -0.787 are the same in both panels. The corresponding step rate is $h_0 = 10^{-0.787} = 0.163$ standard deviations per generation.

inverting this statement, rates on a one-generation timescale are equivalent to differences on a one-generation timescale.

The LDI temporal scaling of step and base differences shown in Figure 8.8c is complementary to the LRI temporal scaling of step and base rates shown in Figure 8.8d. In the LDI plot the differences are plotted against the corresponding number of generations over which each difference was measured. In the LRI plot the rates are the differences divided by their corresponding intervals, with these quotients plotted against the corresponding number of generations over which each difference was measured and each rate calculated.

The LDI distribution in Figure 8.8c gives us the same information about the underlying set of multiple time series as the LRI distribution in Figure 8.8d. The median slope of 0.588 for the LDI distribution and the median slope of -0.412 for the LRI distribution differ from each other by one unit, reflecting their complementarity. The two slopes lie between the expectation for a stationary time series and the expectation for a random time series (between expectations of 0.000 and 0.500, respectively, in the LDI case, and between expectations of -1.000 and -0.500, respectively, in the LRI case). Both slopes are significantly different from expectations for a purely stationary time series and for purely random change, and both slopes are even farther from expectation for a directional time series (1.000 for LDI and 0.000 for LRI).

Temporal scaling slopes that are significantly different from expectation for a purely stationary time series and from expectation for purely random change tell us two things: (1) Time series in a field setting are predominantly stationary and (2) there is in addition a substantial component of random change, which must be due to some combination of stasis, chance, and directional change.

The median LDI and LRI intercepts in Figures 8.8c and 8.8d provide the same information because both medians have the same value (-0.787). The 95% confidence intervals differ slightly because each was bootstrapped independently. A median intercept of -0.787 for an LDI distribution of high- and low-line differences points to a median rate of $h_0 = 0.16$ standard deviations per generation on a timescale of one generation. This is a little higher but close to the median step rate of $h_0 = 0.15$.

What constitutes rapid change in a field study? This is a comparative question of how a given rate relates to expectation. And it is an empirical question because the rate expected depends on what has been observed before. The LRI distribution of haldane rates and intervals in Figure 8.8d represents a set of rates observed to date, indexed by the corresponding interval length. A rate for a given interval length that is higher than the line fit to the points in Figure 8.8d is a rate that is higher than expectation. Similarly, a rate for a given interval length that is lower than the line fit to the points is a rate that is lower than expectation.

8.4.2 Comparison of Rates in Experimental and Field Studies

Figure 8.9 provides a comparison of rates for laboratory selection experiments from Chapter 7 with rates for field studies reviewed in this chapter. Control experiments are not included in the laboratory selection rates of Figure 8.9 to avoid mixing rates and temporal scaling related to directional selection with those related to stabilizing selection.

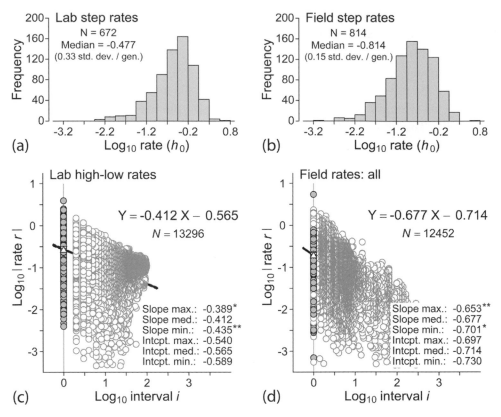

Figure 8.9 Comparison of step and base rates for laboratory and field studies analyzed here. (a) Histogram of step rates for laboratory studies reviewed in Chapter 7 (column of shaded circles in panel c below). (b) Histogram of step rates for field studies reviewed here in Chapter 8 (column of shaded circles in panel d below). (c) LRI log-rate-interval plot for all rates from laboratory high- and low-selection lines. (d) LRI log-rate-interval plot for all field rates. Rates r are in standard deviation units per generation, and interval lengths i are in generations. Confidence intervals were bootstrapped with 1,000 replications. Note that the median slope of -0.412 in panel c is shallower than expected for a random time series (-0.500), and the median slope of -0.784 in panel d is steeper than expected for a random time series.

The step rates resulting from directional selection in an experimental laboratory setting (Figure. 8.9a) occupy the same range of values as the step rates observed in field studies (Figure. 8.9b). Both lie between -3.2 to 0.8 on a $\log_{10}$ scale. Step rates for laboratory selection have a median value of $h_0 = 0.33$ that is slightly more than twice the median value of $h_0 = 0.15$ observed in field studies. We know that the laboratory step rates result from directional selection, meaning that directional selection can explain the lower step rates found in field studies. A higher median value for rates in an experimental setting may result from higher selection differentials, but it is also plausible that field studies have a lower median rate because they include intervals of random change and stabilizing selection.

Step rates for the field studies reviewed here corroborate the experimental laboratory step rates found in Chapter 7, showing that there is nothing artificial about the laboratory selection step rates. Similarly, the experimental laboratory step rates in Chapter 7 corroborate the step rates found in the field studies reviewed here, showing that the field rates lie within a range to be expected for natural selection.

The LRI temporal scaling of rates derived from experimental laboratory selection is different from the temporal scaling of rates observed in a field setting. First, the slopes are different. The median temporal scaling slope of -0.412 for laboratory selection (Figure 8.9c) is shallower and hence more directional than -0.500 expected for random time series. In contrast, the median temporal scaling slope of -0.677 for field studies (Figure. 8.9d) is steeper and hence more stationary than -0.500 expected for random time series. This again can be explained by inclusion of intervals of stabilizing selection with random and directional change in the field studies.

In addition, the LRI median intercept for experimental laboratory selection is different from that observed in a field setting. Laboratory selection rates have an intercept of -0.565 (Figure. 8.9c), which translates to a median $h_0 = 0.27$. Field rates have an intercept of -0.714 (Figure 8.9d), which translates to a median $h_0 = 0.19$. These differences parallel the differences observed in step rates, and the explanation is probably the same.

8.4.3 Proportion of Zero Rates in Field Studies

Distributions of rates, like those plotted on $\log_{10}$ scales in Figures 8.8d and 8.9d, have an upper limit dictated by the standard-deviation width of a normal curve, and rates greater than 2.0 standard deviations per generation (0.30 on a $\log_{10}$ scale) are extremely rare. One out of the 4,441 step and base rates in Figure 8.8d is greater than 2.0 (a proportion of 0.02%), and three out of all 12,452 rates in Figure 8.9d are greater than 2.0 (again 0.02%).

However, distributions of rates, like those plotted on $\log_{10}$ scales in Figures 8.8 and 8.9, have no real lower limit. A rate of change reflects a difference between observations in different generations, and some rates are zero because no difference can be detected. This occurs when there is no change and hence no difference, but it also occurs when any change is less than our precision of measurement. Sample means have more power than individual measurements to detect a difference, but this power is lost when means are rounded too conservatively.

We know that the expected number of rates for a time series is given by $0.5 \cdot n \cdot (n - 1)$, where n is the number of samples in a time series (Equation 5.3): two samples yield one rate, three samples yield three rates, etc. This expectation can be lower if standard deviations are missing. The observed number of rates will also be less than the total expected if some rates are zero. Missing rates are necessarily ignored in LRI temporal scaling, and if missing rates were a substantial proportion of rates being analyzed, they would have an effect on the median rates calculated in Figures 8.9a and 8.9b.

The numbers of rates expected, observed, and zero for each of the 57 empirical studies analyzed here are listed in Table 8.2. One study has a high proportion missing (20%), but this is a study with only five expected. Another study has a large number missing (99), but this is a small proportion of the total. Overall, the number of missing rates is 1.34% of the total expected. If this proportion holds for the 814 step rates in Figure 8.9b, then 11 zero rates are missing. Adding these would move the median of the $\log_{10}$ distribution from -0.814 to -0.828, and move the corresponding median rate from $h_0 = 0.153$ to $h_0 = 0.149$ standard deviations per generation on a timescale of one generation. Addition of a small number of zero rates has a negligible effect on our estimation of rates that are representative in nature.

8.4.4 Step Rates on an Arithmetic Scale

Rate analyzes here are carried out on a logarithmic or geometric scale because differences in the values of rates and their intervals are distributed proportionally. Once analyzed, we can back calculate how rate distributions appear on an arithmetic scale. The distributions of Figures 8.9a and 8.9b are recalculated in Figures 8.10a and 8.10b, first by exponentiating the $\log_{10}$ rates in Figures 8.9a and 8.9b to recover the positive rates represented, and then by adding a negative reflection of each to recover the full spectrum of rates represented. The result is a symmetrical distribution in each instance, centered on zero.

The experimental laboratory step-rate distribution (Figure. 8.10a) has a mean of $h_0 = 0$, and a distribution of step rates about this mean with a standard deviation of $s = 0.60$. The field step-rate distribution (Figure 8.10b) has a mean $h_0 = 0$, and a narrower distribution of step rates about this mean with a standard deviation of $s = 0.38$. Both resemble the unimodal distribution of rates derived from Darwin's diagram in the *Origin of Species* (Figure 1.3).

8.4.5 Human "Secular Trend"

One way to appreciate what the rates in Figures 8.9a and 8.9b mean is to relate these to ourselves. In 1932, Gordon Townsend Bowles, a research student at Harvard University, wrote a classic monograph on "Old Americans" and what

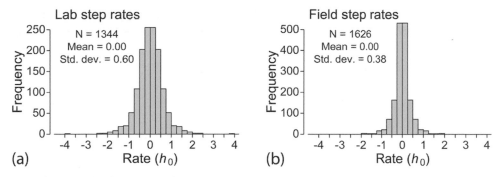

Figure 8.10 Histograms of step rates on arithmetic scales for (a) laboratory selection (high and low selection lines), and (b) field rates. These are reconstructed, including reflection, from the corresponding histograms of step rates on logarithmic scales shown in Figure 8.9. Note that the mean (and median) rates are 0 in both, representing stasis. The standard deviation of $s = 0.60$ rate units for selection experiments is greater than $s = 0.38$ rate units for field studies, reflecting greater response, positive and negative, to directional selection in a laboratory setting.

anthropologists and human biologists call "the secular trend." Bowles (1932) compared two anthropomorphic measurements, stature and body weight, for a large set of more than 1,000 Harvard College father–son pairs, matched for age, and compared 29 anthropomorphic measurements for a smaller set of ca. 400 Harvard father–son pairs. Bowles also examined two decade-by-decade time series for the stature and weight of Harvard male students in the decades from 1836 through 1915.

Bowles completed his study by comparing six anthropomorphic measurements for a set of ca. 200 Wellesley College mother–daughter pairs and by comparing 13 measurements for a set of ca. 200 Vassar College mother–daughter pairs. Bowles also examined two decade-by-decade time series for the stature and weight of Wellesley, Vassar, Smith, and Mount Holyoke College female students for the decades from 1856 to 1915.

We can calculate per-generation rates for all of Bowles' (1932) one-generation comparisons. The result is a set of 74 secular changes that range in rate from $h_0 = 0.018$ to $h_0 = 1.403$ standard deviations per generation (median $h_0 = 0.467$) on a timescale of one generation. Log_{10} values for Bowles' 74 rates fit nicely within the histogram of rates for laboratory studies in Figure 8.9a and within the histogram of rates for field studies in Figure 8.9b. The median rate of $h_0 = 0.467$ for humans is about 1.5 times the median lab rate ($h_0 = 0.33$) and about three times the median field rate ($h_0 = 0.15$), which themselves differ by a factor of just over two. In unlogged form, Bowles' rates fit nicely within the histograms in Figures 8.10a and 8.10b.

Table 8.2 *Number of rates expected and observed for time series in the field studies analyzed here.*

Study	Author	Year	Text section	Taxon	Trait	Expected	Observed	Zero	Zero %
1	Calhoun	1947	8.1.02	*Passer domesticus*	Wing length	16	16	0	0.0
2	Ashton and Zuckerman	1950	8.1.03	*Chlorocebus sabaeus*	Male tooth meas.	54	54	0	0.0
3	Berry	1964	8.1.04	*Mus musculus*	Weight	2	2	0	0.0
4	Johnston and Selander	1964	8.2.01	*Passer domesticus*	Various	97	97	0	0.0
5	Johnston and Selander	1971	8.2.01	*Passer domesticus*	Various	1990	1970	20	1.0
6	Taylor et al.	1974	8.2.02	*Oryctolagus cuniculus*	Various	390	387	3	0.8
7	Patton et al.	1975	8.2.03	*Rattus rattus*	Mahalanobis D	12	12	0	0.0
8	Taylor et al.	1977	8.2.02	*Oryctolagus cuniculus*	Mahalanobis D	26	26	0	0.0
9	Boag and Grant	1981	8.1.31	*Geospiza fortis*	Various	9	9	0	0.0
10	Stearns	1983	8.2.04	*Gambusia affinis*	Various	75	75	0	0.0
11	McPhail	1984	8.1.05	*Gasterosteus aculeatus*	Various	180	180	0	0.0
12	Seeley	1986	8.1.06	*Littorina obtusata*	Various	10	10	0	0.0
13	Conant	1988	8.1.07	*Telespyza cantans*	Various	12	12	0	0.0
14	Diamond et al.	1989	8.1.08	*Myzomela* spp.	Various	24	24	0	0.0
15	Baker et al.	1990	8.2.05	*Fringilla coelebs*	Various	336	329	7	2.1
16	St. Louis and Barlow	1991	8.1.09	*Passer montanus*	Various	32	32	0	0.0
17	Carroll and Boyd	1992	8.1.10	*Jadera haematoloma*	Beak length	15	15	0	0.0
18	Klepaker	1993	8.1.11	*Gasterosteus aculeatus*	Various	72	70	2	2.8
19	Cooke et al.	1995	8.1.12	*Anser caerulescens*	Various	455	455	0	0.0
20	Smith et al.	1995	8.1.13	*Drepanis coccinea*	Various	5	4	1	20.0
21	Carroll et al.	1997	8.1.10	*Jadera haematoloma*	Various	30	30	0	0.0
22	Hendry and Quinn	1997	8.2.06	*Oncorhyncus nerka*	Various	30	30	0	0.0
23	Reznick et al.	1997	8.3.01	*Poecilia reticulata*	Various	20	20	0	0.0
24	Brown and Brown	1998	8.1.14	*Petrochelidon pyrrhonota*	Various	5	5	0	0.0
25	Larsson et al.	1998	8.1.15	*Branta leucopsis*	Various	312	310	2	0.6
26	Smith et al.	1998	8.1.16	*Neotoma albigula*	Body weight	21	21	0	0.0
27	Milner et al.	1999	8.1.17	*Ovis aries*	Various	253	253	0	0.0
28	Pergams and Ashley	1999	8.1.18	*Peromyscus maniculatus*	Various	22	22	0	0.0
29	Haugen and Vøllestad	2000	8.2.07	*Thymallus thymallus*	Life history	9	9	0	0.0
30	Simberloff et al.	2000	8.1.19	*Herpestes auropunctatus*	Various	20	20	0	0.0

#	Author	Year	Code	Species	Measure				
31	Gilchrist et al.	2001	8.1.20	*Drosophila subobscura*	Various	36	36	0	0.0
32	Koontz et al.	2001	8.1.16	*Dipodomys merriami*	Body weight	168	162	6	3.6
33	Kruuk et al.	2001	8.1.21	*Ficedula albicollis*	Tarsus length	135	135	0	0.0
34	Kristjánsson et al.	2002	8.1.22	*Gasterosteus aculeatus*	Various	52	50	2	3.8
35	Kruuk et al.	2002	8.1.23	*Cervus elaphus*	Antler weight	1	1	0	0.0
36	Szuma	2003	8.1.24	*Vulpes vulpes*	Various	48	46	2	4.2
37	Bell et al.	2004	8.1.25	*Gasterosteus aculeatus*	Various	729	722	7	1.0
38	Charmantier et al.	2004	8.1.26	*Parus caeruleus*	Various	360	359	1	0.3
39	Gilchrist et al.	2004	8.1.20	*Drosophila subobscura*	Various	30	30	0	0.0
40	Kristjánsson	2005	8.3.02	*Gasterosteus aculeatus*	Various	4	4	0	0.0
41	Møller and Szép	2005	8.1.27	*Hirundo rustica*	Tail length	344	344	0	0.0
42	Franks et al.	2007	8.3.03	*Brassica rapa*	Flowering time	2	2	0	0.0
43	Babin-Fenske et al.	2008	8.1.28	*Gyretes sinuatus*	Various	33	33	0	0.0
44	Jensen et al.	2008	8.1.29	*Passer domesticus*	Various	12	11	1	8.3
45	Brown and Brown	2011	8.1.14	*Petrochelidon pyrrhonota*	Various	550	534	16	2.9
46	Eastman et al.	2012	8.1.30	*Urocitellus beldingi*	Various	12	12	0	0.0
47	Grant and Grant	2014	8.1.31	*Geospiza* spp.	Various	4680	4581	99	2.1
48	Stuart et al.	2014	8.3.04	*Anolis carolinensis*	Various	2	2	0	0.0
49	Blanckenhorn	2015	8.1.32	*Scathophaga stercoraria*	Hind tibia length	111	111	0	0.0
50	Caruso et al.	2015	8.1.33	*Plethodon* spp.	Snout-vent length	124	124	0	0.0
51	Nengovhela et al.	2015	8.1.34	*Otomys* spp.	Skull length	22	22	0	0.0
52	Pergams et al.	2015	8.1.35	*Rattus rattus*	Various	12	12	0	0.0
53	Holmes	2016	8.1.36	*Peromyscus maniculatus*	Mahalanobis D	3	3	0	0.0
54	Eloy de Amorim et al.	2017	8.1.37	*Gymnodactylus amarali*	Head/body length	21	21	0	0.0
55	Millien et al.	2017	8.1.38	*Peromyscus* spp.	Various	16	16	0	0.0
56	Geiger et al.	2018	8.1.39	*Mus musculus*	Various	156	156	0	0.0
57	Tseng et al.	2018	8.1.40	Carabidae (8 spp.)	Elytron length	424	424	0	0.0
					Totals	12621	12452	169	1.34

Sixteen of the rates calculated from the Bowles' (1932) study quantify the change observed in human stature from one generation to the next. These range from $h_0 = 0.254$ to $h_0 = 0.657$ standard deviations per generation, with a median of $h_0 = 0.547$). This median is a little higher than the rate for secular changes in general. All of the differences are positive, meaning stature increased through time in every comparison. To put this in perspective, the median increase in stature for male students was 3.20 cm per generation, or about 1 cm per decade. The median increase in stature for female students was 3.21 cm per generation, or again about 1 cm per decade.

What caused a sustained increase in human stature in both men and women? Bowles (1932, p. 142) listed the possibilities as improved medicine, "cultural modernization and a general speeding up process," better food in more abundance, more exercise, possible assortative mating, occupational change, and finally a "non-ascertainable element of climatological and meteorological effect."

The late Stanley M. Garn was an esteemed human biologist. When asked whether the cause of the human secular trend might be natural selection, he retorted, "No, it's much too fast!" We don't know the cause for sustained increase in human stature, but we can't rule out natural selection on the theory that the increase is too fast. Adding the rates from Bowles' study to Figures 8.9a and 8.9b shows that rates for the increase in human stature lie within the range we should expect for other evolutionary change.

8.5 Summary

1. Analysis of 57 field studies yields a total of 814 independent step rates quantifying change from one generation to the next, 3,627 independent base rates of change on timescales longer than a generation, and 8,011 net rates compounded from these. The median step rate here is $h_0 = 0.15$ standard deviations per generation on a timescale of one generation.

2. Log-difference-interval LDI temporal scaling provides the same information about a time series as log-rate-interval LRI temporal scaling. Step differences on a one-generation timescale are equivalent to step rates on a one-generation timescale.

3. Step rates following directional selection in a laboratory setting have the same range of values as step rates observed in field studies. This consistency shows: (*a*) experimental selection is not artificial but represents what we see in field studies; and (*b*) the change we see in field studies is what we expect for experimental and hence natural selection.

4. The median step rate of $h_0 = 0.33$ standard deviations per generation for laboratory selection is slightly more than twice the median step rate found in field studies. This is not surprising considering that selection differentials are high in experimental settings.

5. LDI and LRI slopes for field studies lie between expectation for a stationary time series and expectation for a random time series. LDI and LRI slopes for experimental selection in the laboratory lie between expectation for a directional time series and expectation for a random time series. Again, selection differentials are high in experimental settings, and they are more consistent in sign.

6. Rapid change in a field study is change at a rate lying above a line fit to an empirical LRI distribution of step rates and base rates: change lying above $Y = -0.784 \cdot X - 0.787$, where Y is $\log_{10}$ of the rate r and X is $\log_{10}$ of the corresponding interval i (Figure 8.8d).

7. Zero rates occur when there is no difference in traits measured at different times. These are necessarily ignored when rates are transformed to logarithms. Empirically, the proportion of zero rates in field studies is 1.34%, mostly due to rounding, and zero rates have negligible effect on the rate statistics presented here.

8. Step rates in selection experiments and field studies reflected as positive and negative rates on an arithmetical scale, have unimodal distributions centered on zero (stasis), with standard deviations of 0.60 and 0.38 rate units, respectively. These resemble the unimodal distribution of rates derived from Darwin's (1859) partially scaled diagram of evolution in the *Origin of Species* (Chapter 1).

9. The anthropologist's "secular trend" of increasing human stature may be due to nutrition or other factors, but it has rates in the range found here for evolution by natural selection.

9

Phenotypic Change in the Fossil Record

Quantification of phenotypic change in the fossil record started with Alexei Petrovich Pavlov (reported in Sewertzoff, 1931) and George Gaylord Simpson (1944). Both attempted to quantify change in a long-standing evolutionary classic: evolution of the horse. Simpson's study, in his book *Tempo and Mode in Evolution*, attracted the attention of J. B. S. Haldane (1949). Haldane proposed a rate unit, now named for him, that quantified change in units of phenotypic standard deviation per generation. A rate unit is meaningless on any timescale other than the one that anchors it. Hence the *haldane* as a rate unit is appropriately subscripted with $\log_{10}$ of the timescale involved: h_0 for step rates on a one-generation timescale, $h_{0.301}$ for rates on a two-generation timesale, and so forth.

Birds, fish, and insects are common extant organisms in field investigations like those in the previous chapter because they are both abundant and tractable for study. However, all require special and unusual conditions for preservation in the fossil record. In this chapter most case studies involve mammals. Mammals have highly mineralized bones and teeth that preserve well in many environments. Most have determinate skeletal growth. Mammalian teeth are especially useful because they are often complex, readily identifiable to genus and species, and most develop and erupt fully formed in the jaws. Teeth that do not grow simplify quantification of sample means, sample variation, and change through time.

The paleontological studies included here are many and diverse, but they do not begin to exhaust the material available for analysis. Some studies are omitted because they lack good time control, some are omitted because they lack sufficient sample statistics (sample sizes, sample means, and sample variances, standard deviations, or standard errors). Other studies are omitted for no reason: it was simply impractical to include all. Readers are encouraged to calculate rates for any study omitted here as a way to compare and possibly extend generalizations at the end of the chapter. All rates calculated in this chapter can be found at www.cambridge.org/evolution.

9.1 Geological Timescale

Planet Earth has a long history, and animal life fills a substantial portion of the history. Many animals were preserved as fossils when they died and were buried in successive layers of sediment. Each layer or stratum accumulated on older strata. The resulting vertical sequence provides a reliable framework for determining the relative ages for the strata and any fossils they contain. Fossils in stratigraphic context provide direct and irrefutable evidence that life on Earth has changed through time.

The timescale is built up from local sequences of fossils and strata correlated from place to place using changes in the fossils, reversals of the Earth's magnetic field, changes in sea level, isotopic signatures in carbon and other elements, and orbital tuning of astronomically controlled sedimentary cycles. Numerical calibration of eras, periods, epochs, and ages in the sequence of fossils and strata comes from radiometric dating of interbedded lavas and ashes. The calibrations are often in "megannum" or million-year units, abbreviated "Ma." The result is now a dense network of numerical ages in addition to the basic superposition of fossils and strata.

The Earth is some 4.5 billion years old, and evidence of abundant life fills the last 541 Ma (million years of Earth history), an eon called the Phanerozoic. The name itself means literally the time of "visible" or "evident animals." The change in fossils through time is so clear that the Phanerozoic is divided into three eras, the Paleozoic, Mesozoic, and Cenozoic. These names mean "ancient animals," "middle animals," and "recent animals," but they are also called, informally, the age of trilobites, the age of dinosaurs, and the age of mammals. Major extinction and origination events mark the end of the Paleozoic and beginning of the Mesozoic at 252 Ma, and the end of the Mesozoic and beginning of the Cenozoic at 66 Ma. An outline of the geological timescale is provided in Figure 9.1, with calibration ages from Gradstein et al. (2012).

9.2 Longitudinal Studies in the Fossil Record

The presentation of case studies and subjects here is chronological, paralleling the organization of Chapter 8. One difference from the previous chapter is that all of the studies included here are longitudinal in the sense that they compare traits in an evolving population at successive times in the geological past. Ages in the fossil record are rarely known with precision, but their accuracy is often sufficient to provide reasonable estimates of the intervals of time between successive samples. Conversion of intervals expressed in years, thousands of years, or even millions of

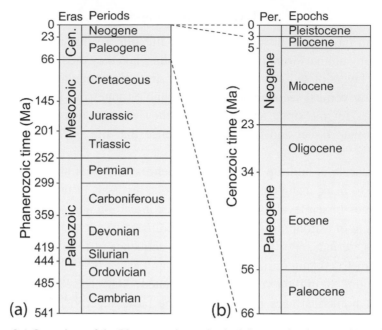

Figure 9.1 Overview of the Phanerozoic geological time scale, the past 541 million years comprising the Paleozoic, Mesozoic, and Cenozoic eras when the fossil record shows multicellular life on earth to have been diverse and abundant. (a) Whole Phanerozoic. (b) Enlarged Cenozoic. Numerical calibration of era, period, and epoch boundaries follows Gradstein et al. (2012).

years to intervals expressed in generations often depends on the relationship of generation time to body weight in living models as described in Chapter 3.

9.2.1 Pavlov, Sewertzoff, and Cenozoic Horses

Alexei Petrovich Pavlov's early estimate for the rate of phenotypic evolution in horses (reported in Sewertzoff, 1931) was described and illustrated in Chapter 1. Pavlov considered early Eocene horses to have stood about a half-meter high at the shoulder, in contrast to recent horses that stand about 1.5 m high. The difference in time between Pavlov's early horses (evidently based on "*Eohippus resartus*" from the middle part of the early Eocene) and modern horses today is about 53 million years (m.y.).

Ln values for the shoulder heights are −0.69 and 0.41, respectively, which differ by 1.10 natural-log units. Shoulder height is a linear measurement, with an expected standard deviation on a natural-log scale of 0.05, so the difference in ln shoulder heights corresponds to a difference of 22.0 standard deviations. A horse standing 0.5 m high at the shoulder should weigh about 11 kg (Richard-Hanson et al., 1999), and have a generation time of about three years (Equation 3.2). Modern horses in the wild have a generation time of about eight years (Berger, 1986, figure 5.2). Thus the interval involved in Pavlov's comparison is on the order of 9.4 million generations. Change of 22 standard deviations through this interval yields an average rate of $h_{6.971} = 2.35 \times 10^{-6}$ standard deviations per generation on a timescale of 9.4 million generations.

The credibility of estimates like these, we shall see, depends on their consistency with other estimates. Note that the timescale here is profoundly different from the timescale of laboratory and field studies.

9.2.2 Simpson and Cenozoic Horses

One of George Gaylord Simpson's most influential books was *Tempo and Mode in Evolution* published in 1944. In the book Simpson analyzed rates of evolution for dental characteristics in horses. His rates were compared, as described and illustrated in Chapter 1, by plotting the logarithms of tooth measurements against their age in geological time. It is possible to recalculate these, based on measurements in Simpson (1944, table 1), in units of standard deviation per generation.

Five equid species are involved in two evolutionary lineages: (1) Eocene *Hyracotherium* (or *Eohippus*) *borealis*; Oligocene *Mesohippus bairdi*, Miocene *Merychippus paniensis*, and Miocene *Neohipparion* or *Cormohipparion occidentale* in one lineage; and (2) early Eocene *Hyracotherium* (or *Eohippus*) *borealis*; Oligocene *Mesohippus bairdi*, and Miocene *Hypohippus osborni* in the second lineage. The lineages include species that MacFadden (1986) numbered 1, 8, 22, and 28 in the first lineage, and 1, 8, and 17 in the second lineage. Body weights for Simpson's species were taken from MacFadden (1986, table 4) and substituted in Equation 3.2 to estimate generation times. Numerical ages for each on a modern geological timescale were taken from MacFadden (1986, figure 4).

Simpson (1944, table 1) provided statistics for two measurements of upper third molars and one ratio of these for each of the five equid species. The three traits analyzed are: (1) paracone height as a measure of molar crown height, (2) ectoloph length as a measure of molar crown size, and (3) 100 · (paracone height / ectoloph length) as an index of relative crown height or hypsodonty. This information can be analyzed in the normal way (Box 8.1) to yield four base rates for each of the three traits in successive pairs of species in the two lineages. The transition from Eocene *H. borealis* to Oligocene *M. bairdi* is found in both lineages, but the corresponding rates were recorded once. There are an additional four compound net rates for each trait calculated by stepping over one or more species in a lineage.

The 12 base rates for Simpson's statistics for horse teeth range from $h_{6.726} = 9.92 \times 10^{-7}$ to $h_{5.871} = 1.25 \times 10^{-5}$ standard deviations per generation (median $h_{6.510} = 2.95 \times 10^{-6}$) on timescales ranging from hundreds of thousands to millions of generations.

9.2.3 Mesozoic Dinosaurs

Edwin H. Colbert published a review of the horned or ceratopsian dinosaurs in 1948. There he compared the rate of body size increase in ceratopsians to rates of size increase in other large-bodied dinosaur groups. Colbert did this graphically, by comparing slopes after plotting estimates of body length against geological time calibrated in millions of years (Colbert, 1948, figures 4–5). He did not log the length axis, nor did he make any comparison to rates (slopes) for horses calculated by Simpson (1944). This is a little surprising because Colbert was a younger colleague of Simpson's at the American Museum of Natural History in New York, and he was surely familiar with Simpson's recently published *Tempo and Mode in Evolution*.

We can quantify Colbert's comparisons in haldanes by transforming his size axis to natural logarithms and, again, as in Section 9.1.1 above, assuming a standard deviation of 0.05 for the transformed measurements. Generation times for dinosaurs are uncertain and questionable. Erickson et al. (2006, figure 1) suggested reproductive ages for the theropod *Albertosaurus sarcophagus* in the range of 15–25 years. Lee and Werning (2008) calculated reproductive ages for the ornithopod *Tenontosaurus tilletti* and theropod *Tyrannosaurus rex* in the range of 10–20 and 20–30 years, respectively. Erickson et al. (2009, figure 5) suggested reproductive ages for the small ceratopsian *Psittacosaurus lujiatunensis* in the range of 8–12 years. Sander et al. (2011, p. 134) indicated that sexual maturity was achieved in large sauropods in the second or third decade of life. Twenty years is the generation time used here to quantify rates of evolution for all of Colbert's dinosaurs.

Seven base rates can be calculated from the information given by Colbert (1948). These range from $h_{6.301} = 1.38 \times 10^{-6}$ to $h_{5.477} = 2.88 \times 10^{-5}$ standard deviations per generation (median 8.84×10^{-6}) on timescales ranging from hundreds of thousands to millions of generations.

Colbert (1948, p. 152) found that "evolutionary rates in the ceratopsian dinosaurs were rapid," a claim born out here, in a relative sense, where the two base rates in his lineage of ceratopsians are $h_{5.903} = 2.57 \times 10^{-5}$ and $h_{5.477} = 2.88 \times 10^{-5}$, respectively. Colbert's ceratopsian rates were highest among the dinosaurs he studied, but they were also his rates on the shortest timescales. However, given what we have seen in laboratory and field studies, it is hard to imagine 0.0000288 standard deviations per generation being fast on any timescale.

9.2.4 Haldane and the Pleistocene Evolution of Humans

Simpson (1944) and Colbert (1948) provided most of the examples that J. B. S. Haldane analyzed in developing methods for quantification of evolutionary rates. Haldane calculated rates in *darwin* units (factors of *e* per year), and these are recalculated here in *haldane* units (standard deviations per generation).

Haldane (1949, pp. 54–55) went a step farther by adding an example from the fossil record of human evolution, based on measurements of *Sinanthropus pekinensis* (now included in *Homo erectus*) published by Weidenreich (1943). Haldane calculated that the length-opisthion height index evolved from a value of 0.541 in *S. pekinensis* to 0.736 in modern humans. He assumed that this required 0.5 million years, for a rate in darwins of $[\ln (0.736) - \ln (0.541)] / 500,000 = 6.16 \times 10^{-7}$.

In a second calculation Haldane assumed the geological age of *S. pekinensis* to be 0.5 million years, a generation time of 20 years, and the coefficient of variation for the length-opisthion height index to be 0.04 (without giving a source). From these numbers he calculated the rate for the length-opisthion height index, from *S. pekinensis* to modern humans, to be 1/3500 or 2.86×10^{-4} standard deviations per generation. Calculation here for the same geological interval and generation time, assuming 0.541 and 0.736 are already proportions and the standard deviation for the length-opisthion height index is 0.014 (from Weidenreich, 1943, p. 110), yields an interval of 25,000 generations, a difference in standard deviation units of 13.93 generations, and a base rate of $h_{4.398} = 5.57 \times 10^{-4}$ standard deviations per generation. This rate is much higher than the rates above for horses and for dinosaurs, but of course the interval involved is much shorter.

9.2.5 Miocene Merycoidodontidae

Merycoidodontidae are oreodonts, extinct ruminating pig-like artiodactyls that are now extinct. Oreodonts were a major constituent of mammalian faunas in North America during the Oligocene and Miocene epochs. In 1954 Robert Smith Bader completed a doctoral dissertation on variation and evolutionary rates in Miocene oreodonts. In the following publication Bader (1955) focused on two common forms: (1) smaller and more lightly built Merychyinae including the species of the genus *Merychyus*; and (2) larger and more heavily built Merycochoerinae including species of the genera *Merycochoerus* and *Brachycrus*.

Bader (1955) published statistics for 23 measurements of skulls for five species of Merychyinae and five species of Merycochoerinae. Statistics for one of the

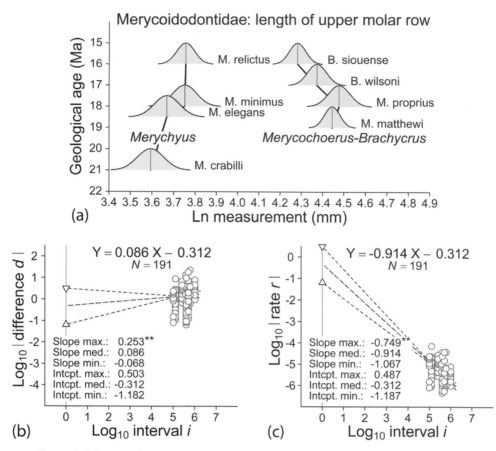

Figure 9.2 Rates of evolution in two Miocene merycoidodont mammal lineages spanning five million years of geological time. (a) Time-calibrated stratophenetic graph of species in the lineages analyzed by Bader (1955). Independent variable, time, is plotted on the vertical or *y* axis. The base of each normal curve spans ±3 standard deviations. (b) Log-difference-interval LDI plot for 23 measurements showing that differences between species scale with a slope (0.086) close to that expected for a stationary time series (0.000). The slope is significantly different from expectation for both random and directional change. (c) Log-rate-interval LRI plot shows that rates of change scale with a slope (−0.914) close to that expected for a stationary time series (−1.000). This slope too is significantly different from expectations for both random and directional change. Intercepts in (b) and (c) are the same; both have wide 95% confidence intervals approaching a span of two orders of magnitude (open triangles).

measurements, length of the upper molar tooth row, are shown graphically in Figure 9.2 to illustrate the merycoidodont lineages analyzed here. Each lineage has four species. Bader showed each lineage to have a fifth species branching from it, but the branching species are closely similar to a contemporaneous congener and are omitted here. The geological age for each species follows Hilgen et al. (2012, figure 29.9). Generation times of five years for *Merychyus* and eight years for *Merycochoerus* are estimated here (Equation 3.2) from body weights of 41–42 and 275–309 kg, respectively, taken from Janis (1990, table 13.4).

Bader's measurements yield a total of 191 rates, all nonzero. Ninety-seven of these are base rates that range from $h_{5.602} = 1.65 \times 10^{-7}$ to $h_{5.097} = 7.06 \times 10^{-5}$ standard deviations per generation (median 5.99×10^{-6}) on timescales of hundreds of thousands of generations.

9.2.6 Long-Term Evolution of Neogene Bears

Bears have a reasonably good fossil record in Europe and Björn Kurtén was for many years a leading authority on their evolution. In an early study, Kurtén (1955) traced a long lineage of six species starting in the early Miocene with *Ursavus elmensis* and ending with the Pleistocene cave bear *Ursus spelaeus*. He quantified change, following Haldane (1949), in terms of darwins and millidarwins. In the early stages of bear evolution Kurtén found rates comparable to Simpson's rates for horses but documented an abrupt change to higher rates in the Pleistocene. Kurtén attributed the increase in rates to rapidly changing Ice Age climate, but of course the time intervals being sampled were much shorter.

Kurtén (1955, table 6) provided statistics for moderate to large samples of first upper molar lengths as a measure of the size of the animal, and provided ages for the samples in millions of years. Kurtén's ages for the samples have changed slightly, from 22, 15, 10, 0.9, 0.5, and 0.1 Ma in 1955 to 18, 15, 6, 1.5, 0.5, and 0.06 Ma today, as calibration of the geological timescale has improved. Generation times can be estimated by first calculating body weights from mean molar length using comparative data for Urside in Van Valkenburgh (1990, table 10.1). The body weights can then be inserted in Equation 3.2 here to calculate generation times. We can calculate a new set of rates in haldanes from Kurtén's standard deviations for each species and our estimates of their generation times.

Kurtén's measurements yield a total of 15 rates, all nonzero. Five of these are base rates that range from $h_{6.187} = 2.46 \times 10^{-6}$ to $h_{4.784} = 5.03 \times 10^{-5}$ standard deviations per generation (median 7.84×10^{-6}) on timescales of tens of thousands to millions of generations.

9.2.7 Short-Term Evolution in Pleistocene Bears

After completion of the 1955 study, Björn Kurtén continued to explore rates of evolution in bears. He focused on bear evolution in the Pleistocene, studying a lineage including *Ursus etruscus* to *U. arctos* (Kurtén, 1959). In the Pleistocene study Kurtén found rates substantially higher than those in the earlier study.

Kurtén (1959, table 2) provided means for samples of second lower molar lengths as a measure of the size of the animal but no sample sizes, standard deviations, nor coefficients of variation. The latter can be taken as 0.05, which is the weighted mean from the 1959 study. Here again, Kurtén's ages for the samples have changed slightly as temporal calibration improved, from the 800, 400, 375, 350, 300, 200, 150, 80, 30, 8, and 0 thousand years he reported in 1959 to ages of 2300, 530, 480, 450, 400, 180, 120, 100, 30, 10, and 0 kyr today. Virtually all of the taxa involved in the 1959 study are subspecies of *Ursus arctos*, so it makes sense to use a 13.5 year generation time published for this species (COSEWIC, 2012).

Measurements in Kurtén's 1959 study of bears yield a total of 55 rates. Ten of these are base rates that range from $h_{5.118} = 2.32 \times 10^{-5}$ to $h_{2.970} = 2.85 \times 10^{-3}$ standard deviations per generation (median 3.61×10^{-4}) on timescales of thousands to hundreds of thousands of generations.

9.2.8 Change in Postglacial-to-Recent Mammals

Björn Kurtén was intrigued by differences in the rates he found himself for long-term and short-term evolution in bears when compared to the rates for "Tertiary" Eocene through Pliocene mammals published by Simpson (1944, 1953) and Bader (1955). This prompted him to investigate change in late- and post-Pleistocene "postglacial" mammals in the second part of his 1959 study.

Kurtén (1959, table 4) lists a series of proportional differences in size for postglacial mammals in relation to the age of each sample in years before present. The differences are based on linear measurements, normalized to the modern sample, so change in standard deviation units can be calculated as pair-wise differences in natural-log values divided by 0.05. Generation times can be taken from Millar and Zammuto (1983) or calculated from body weight using Equation 3.2. Generation-unit differences between samples can then be calculated from the ages and the generation times. Finally, rates in standard deviations per generation can be calculated for all pair-wise differences in each lineage.

There are 33 rates embedded in Kurtén's table 4. The 21 base rates range from $h_{3.168} = 1.36 \times 10^{-6}$ to $h_{2.569} = 5.39 \times 10^{-3}$ standard deviations per generation (median 5.54×10^{-4}) on timescales of hundreds to thousands of generations.

9.2.9 Silurian Brachiopod Eocoelia

Eocoelia is a Silurian representative of the articulate brachiopod order Rhynchonellida. These are small mollusk-like brachiopods with a short hinge line and biconvex ribbed shells. *Eocoelia* is an important component of early Silurian nearshore faunas and it has been used effectively in biostratigraphic correlation.

Alfred Ziegler published an exemplary biometrical study of a lineage comprising four species. He did not attempt to quantify rates of change but provided information that invites this.

Ziegler (1966, tables 3, 4, and 6) provided statistics for rib counts, rib angles, and rib shapes in 13 samples in overlapping stratigraphic successions. The samples span all of Upper Llandovery epoch of Silurian time, ranging from about 438 to 434 Ma. If we assume that the samples are equally spaced, then each is separated from the next by about a third of a million years. Rickwood (1977) studied life history in the brachiopod *Waltonia*. She found that the life span is typically eight years, and reproduction starts in the fourth year. Thus it is reasonable to expect a generation time of about five years. Curry (1982) studied life history in the brachiopod *Terebratulina* and found a life span of seven years. Reproduction starts in year three, and it is reasonable to expect a generation time of about four years. Here a generation time of four years is assumed for *Eocoelia*.

Analysis yields a total of 234 rates, all nonzero. Thirty-six base rates range from $h_{4.929} = 3.40 \times 10^{-7}$ to $h_{4.917} = 1.93 \times 10^{-5}$ standard deviations per generation (median 6.69×10^{-6}) on timescales of tens of thousands to a million generations.

9.2.10 Pliocene-to-Pleistocene Elephantidae

Elephants are large mammals that are often found as fossils. Vincent Maglio studied their temporal and geographic distributions in the late Miocene through Recent of Africa, Europe, and Asia. He traced the evolution of *Loxodonta*, *Elephas*, and *Mammuthus* from a common ancestry in *Primelephas*.

Maglio (1973) was able to study reasonably large samples for 18 species or stages within a species, and his statistics for molar size and other traits were compared in the lineages he described to calculate the rates compiled here. Each pairwise comparison had at least one sample that included five or more individuals. A generation time of 17 years for *Loxodonta africana* from Moss (2001) was assumed to represent the family.

Maglio's measurements of dental traits in Elephantidae yield 270 rates, of which 267 are nonzero. The 166 base rates calculated here range from $h_{5.000} = 9.46 \times 10^{-7}$ to $h_{4.469} = 1.28 \times 10^{-4}$ standard deviations per generation (median 1.70×10^{-5}) on timescales of tens of thousands to hundreds of thousands of generations.

9.2.11 Pleistocene Radiolarian Pseudocubus vema

Radiolaria are protozoans that are widely distributed in the world's oceans. Many produce intricate and beautiful skeletons of silica that enhance their potential for preservation and identification as fossils. Davida Kellogg analyzed an evolutionary lineage of the Pleistocene radiolarian *Pseudocubus vema* (sometimes called *Helotholus vema*, or *H. praevema* and *H. vema*) from an Antarctic deep-sea core. Core samples of microfossils are advantageous for such studies because they usually contain many specimens from a single locality. The specimens are stratigraphically ordered in time, and in this case have an accompanying record of paleomagnetic reversals enabling a reliable determination of ages.

Kellogg (1975, figure 4) summarized change in the width of the thorax of *P. vema* in a series of 34 samples ranging from 1750 to 1082 meters deep in the core. Calibration from the ages of four paleomagnetic reversals indicate that the samples range from 4.843 to 2.366 Ma. Kellogg herself listed rates for six intervals in the sequence, ranging in length from 221 to 729 thousand years, and, with one generation per year, 221 to 729 thousand generations. Kellogg (1975, table 2) calculated rates in terms of the time required for change through one standard deviation. These are net rates. Inverting, Kellogg's rates range from $h_{5.696} = 2.21 \times 10^{-10}$ to $h_{5.122} = 1.33 \times 10^{-7}$. She noted three intervals of stasis separated and followed by three intervals of more rapid change.

The six rates of Kellogg lie within the distribution of 561 base and net rates calculated here for all combinations of her samples. Of these 33 are base rates, which range from $h_{4.851} = 2.36 \times 10^{-7}$ to $h_{4.785} = 1.70 \times 10^{-5}$ standard deviations per generation (median 3.83×10^{-6}) on timescales of tens of thousands to just over a hundred thousand generations.

9.2.12 Eocene Condylarth Hyopsodus Spp.

Hyopsodus is a small, generalized, herbivorous mammal known from thousands of specimens in lower and middle Eocene strata of western North America. Mammals such as *Hyopsodus* are advantageous for evolutionary study because they have molar teeth that grow to definitive size before eruption and do not continue to grow later in ontogeny. Gingerich (1974; 1976) attempted to trace diversification of early Eocene species of *Hyopsodus* through successive strata in the central Bighorn Basin of Wyoming. Temporal calibration has been improved by Clyde et al. (2007), enabling rates to be calculated. *Hyopsodus* weighed approximately 1–2 kg and had a generation time averaging about two years (Equation 3.2).

The three lineages in Figure 9.3a yield a total of 318 nonzero rates. Together these have a log-rate-interval LRI temporal scaling slope of -0.775 (Figure 9.3b), which differs from, and lies between, slopes expected for a stationary time series and a random time series (-1.000 and -0.500, respectively). The observed slope is closer to -1.000 than to -0.500. Hence change in the lineages combined is interpreted as neither stationary nor random but a combination of both. Note that the *H. miticulus–H. lysitensis* lineage by itself (Figure 9.3c) has rates scaling with a

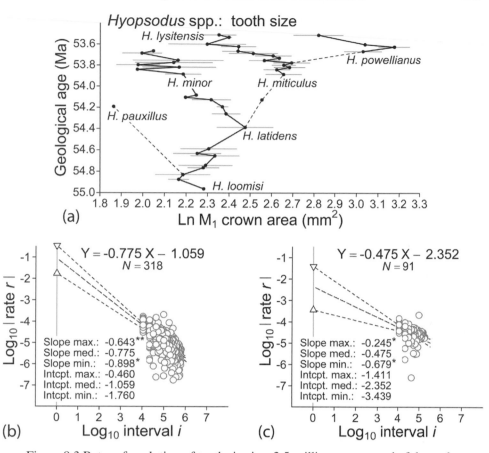

(a)

(b)

(c)

Figure 9.3 Rates of evolution of tooth size in a 2.5-million-year record of the early Eocene condylarthran mammal *Hyopsodus* (Gingerich 1974, 1976). (a) Time-calibrated stratophenetic graph shows the principal lineages *H. loomisi–H. minor*, *H. miticulus–H. lysitensis*, and *H. powellianus*. Independent variable time is on the vertical or *y* axis. Horizontal lines span ±1 standard deviation. (b) Log-rate-interval LRI plot shows that rates of change in all of the *Hyopsodus* together scale with a slope of −0.775. This lies between slopes expected for a stationary time series (−1.000) and a random walk (−0.500), significantly different from both, and a little closer to −1.000. The lineages are interpreted to combine stasis and random change. (c) LRI plot shows that rates of change in the *H. miticulus–H. lysitensis* lineage scale with a slope of −0.475. This lies between slopes expected for a stationary time series (−1.000) and directional change (0.000) but is not significantly different from the slope expected for random change (−0.500).

slope of −0.475, which is significantly different from stationary but not significantly different from random.

There are 36 base rates in the principal lineages, and these range from $h_{4.342} = 1.74 \times 10^{-6}$ to $h_{4.041} = 1.71 \times 10^{-4}$ standard deviations per generation (median 4.39×10^{-5}) on timescales of 11 thousand to 102 thousand generations.

9.2.13 *Miocene-to-Pliocene Foraminiferan* Globorotalia inflata

Globorotalia is a small, coiled, species-rich, marine planktonic foraminiferan widely used in Cenozoic biostratigraphy. It was first studied from an evolutionary point of view by James Kennett (1966) and then examined in greater detail by Björn Malmgren and Kennett (1981). Malmgren and Kennett described a lineage of four species of late Miocene and Pliocene *Globorotalia* (*G. conoidea, G. conomiozea, G. puncticulata,* and *G. inflata*) from a deep-sea core in the southwest Pacific Ocean.

Malmgren and Kennett (1981, appendix) listed means and coefficients of variation for seven characteristics based on samples of 40–60 individual specimens from each successive sample. Ages for individual samples can be interpolated from the ages of New Zealand stage boundaries (Hilgen et al., 2012). Age control is uncertain for the four Pleistocene samples from the top of the core, and these were omitted. Similarly, following Malmgren and Kennett, the outlying general-size sample from 176.4 m core depth was omitted. The remaining samples range in age from 8.02 to 2.59 Ma. Rates are calculated for two characteristics: chamber number and general size. Chamber number is represented by 68 successive samples, and general size is represented by 67 samples. Interval length in generations assumes an average generation time of one year.

Change in the number of chambers in Miocene–Pliocene *Globorotalia* yields 2,258 nonzero rates. The LRI temporal scaling slope is -0.457, which is close to -0.500 expected for a random walk and significantly different from both 0.000 expected for directional change and -1.000 expected for stasis. The base rates for chamber number range from $h_{5.663} = 1.43 \times 10^{-7}$ to $h_{4.000} = 5.41 \times 10^{-5}$ standard deviations per generation (median 5.74×10^{-6}) on timescales of tens of thousands to millions of generations.

Change in the general size of Miocene–Pliocene *Globorotalia* yields 2,141 nonzero rates. The LRI temporal scaling slope is -0.807, which lies between -0.500 expected for a random walk and -1.000 expected for stasis, and is significantly different from both. This scaling differs from that for chamber number, and indicates a combination of stasis and random change. The base rates for chamber number range from $h_{4.477} = 7.18 \times 10^{-8}$ to $h_{4.000} = 3.91 \times 10^{-5}$ standard deviations per generation (median 4.53×10^{-6}) on timescales of tens of thousands to millions of generations.

9.2.14 *Pleistocene-to-Recent* Bison *Species*

Bison are large mammals with distinctive horns that are conspicuous as fossils because of their size. The oldest well-dated species in North America are the steppe bison, *Bison priscus*, dated at about 130,000 years before present (130 kyr), and the slightly younger long-horned bison, *B. latifrons*, dated at about 120 kyr (Froese, 2017). Younger species include *B antiquus* ranging in age from about 18–11 kyr, *B. occidentalis* from about 11–5 kyr, and *B. bison* from about 5 kyr to the present.

Jerry McDonald (1981) made a thorough biometrical study of 36 cranial and postcranial characteristcs in North American male and female bison, fossil and living. Results are included in nine tables, from which we can calculate a series of rates. All five of the species just mentioned are represented. The samples span a range of known and inferred ages, and considering the species to be 130, 120, 14, 8, and 2.5 kyr in age, respectively, is an approximation. The generation time for extant bison is eight years (COSEWIC, 2013a).

The measurements of McDonald (1981) yield 339 nonzero rates. The 205 base rates in this sample range from $h_{4.161} = 1.46 \times 10^{-5}$ to $h_{3.097} = 7.63 \times 10^{-3}$ standard deviations per generation (median 1.80×10^{-3}) on timescales of hundreds to some thousands of generations.

9.2.15 *Jurassic Ammonite* Kosmoceras *Species*

David Raup and Rex Crick (1981) analyzed several time series of measurements of the Callovian-age late Middle Jurassic ammonite *Kosmoceras*. Roland Brinkmann studied the specimens in the field and published statistical summaries for samples from narrow stratigraphic intervals (Brinkmann, 1929). These came from the lower 14 meters (1,400 cm) of a 17-meter stratigraphic section at Peterborough in England. The section was later described in detail by Hudson and Martill (1994). Brinkmann recognized four lineages, which he considered subgenera of *Kosmoceras*. Most common were the microconch lineage *Anakosmoceras* and the parallel macroconch lineage *Zugokosmoceras*.

Patterns of change through time that Brinkmann saw in *Kosmoceras* at Peterborough are found at other middle and upper Callovian localities, and the section yielding Brinkmann's ammonites probably ranges in age from about 165.2 to 163.5 Ma on the geological time scale of Ogg et al. (2012). This 1.7-million-year range for a section 17 meters thick means we can expect a meter of stratified sediment to represent about 100,000 years of geological time, and each centimeter to represent, on average, about 1,000 years. Modern *Nautilus* has a generation time of about 15 years (Saunders, 1984), and we can expect the same for *Kosmoceras* and *Zugokosmoceras*.

Raup and Crick (1981) had access to Brinkmann's original measurements and grouped these into 14 equal 100-cm (100,000-year) bins. The measurements were deliberately binned to facilitate analysis using runs tests for reversals of evolutionary direction (implicitly assuming some underlying evolutionary process to operate on this 100,000-year timescale). Brinkmann himself published statistics for 42 successive samples in the *Anakosmoceras* lineage and 48 samples in the *Zugokosmoceras* lineage (Brinkman 1929, tables 39 and 63). Brinkmann died in 1995 and the measurements Raup and Crick analyzed in 1981 are now lost, so analysis depends on the statistics Brinkmann published.

The microconch "*Anakosmoceras*" and macroconch "*Zugokosmoceras*" lineages of Brinkmann (1929) are illustrated in Figure 9.4a. Diameters of their shells change in parallel, inviting interpretation as a single sexually dimorphic lineage. The log-rate-interval LRI plot in Figure 9.4b shows that rates of change in "*Anakosmoceras*" scale with a slope of −0.806, which lies between the slopes expected for a stationary time series (−1.000) and a random walk (−0.500). The LRI plot in Figure 9.4c shows that rates of change in "*Zugokosmoceras*" scale with a slope of −0.704, which is again between those expected for a stationary time series and a random walk. Taken as a whole, each time series is significantly different from stationary and random but is some combination of both. Here, contrary to Raup and Crick (1981, table 3), we can reject a simple random walk hypothesis. Change accumulates on longer time scales more rapidly than would be expected for a lineage in stasis but not as rapidly as expected for a random walk.

Forty-two samples in the "*Anakosmoceras*" lineage yield 858 nonzero rates. The 39 base rates calculated here range from $h_{2.824} = 2.13 \times 10^{-5}$ to $h_{2.523} = 1.12 \times 10^{-2}$ standard deviations per generation (median 2.31×10^{-4}) on timescales of hundreds to thousands of generations. Forty-two samples in the "*Zugokosmoceras*" lineage yield 1,126 nonzero rates. The 47 base rates range from $h_{3.315} = 3.51 \times 10^{-6}$ to $h_{2.802} = 6.18 \times 10^{-3}$ standard deviations per generation (median 4.90×10^{-4}) on timescales of hundreds to thousands of generations.

9.2.16 *Miocene-to-Pliocene Foraminiferan* Globorotalia tumida

In 1983 Björn Malmgren, William Berggren, and George Lohmann published an investigation of *Globoratalia plesiotumida* and *G. tumida* as a follow-up to the *Globoratalia inflata* study by Malmgren and Kennett (1981) described above. The second study had the advantage of two clearly defined species, one in the Miocene and the other in the Pliocene, with a rapid transition between them (Figure 9.5a). Malmgren et al. (1983) labeled the resulting evolutionary pattern "punctuated gradualism." Here we examine and quantify the pattern in terms of rate.

Malmgren et al (1983, appendix I) listed sample ages, sample sizes, means, and standard deviations for a measure of size (cross-sectional area of the test oriented in edge view) and a measure of shape (second eigenshape) in the *G. plesiotumida* through *G. tumida* lineage, including the transition between the two.

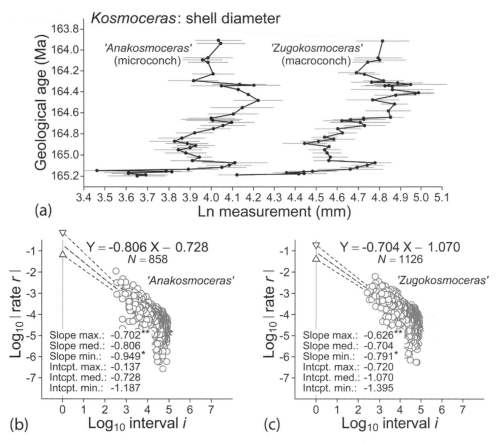

(a)

(b)

(c)

Figure 9.4 Rates of evolution for microconch and macroconch shell diameter in a 1.4-million-year record for the late Middle Jurassic ammonite *Kosmoceras* (Raup and Crick, 1981). (a) Time-calibrated stratophenetic graph shows lineages of the subgenera "*Anakosmoceras*" and "*Zugokosmoceras*" based on measurements published by Brinkmann (1929). Independent variable time is on the vertical or *y* axis. Horizontal lines span ±1 standard deviation. (b) Log-rate-interval LRI plot shows that rates of change in "*Anakosmoceras*" scale with a slope of −0.806. This lies between slopes expected for a stationary time series (−1.000) and a random walk (−0.500), and is significantly different from both. (c) LRI plot shows that rates of change in "*Zugokosmoceras*" scale with a slope of −0.704. This lies between slopes expected for a stationary time series (−1.000) and a random walk (−0.500), and is again significantly different from both. Each time series as a whole is interpreted as combining stasis and random change.

G. plesiotumida samples for which ages are available range from 6.500 to 5.650 Ma (nine samples), the transition ranges from 5.650 to 5.040 Ma (44 samples), and *G. tumida* ranges from 5.040 to 0.100 Ma (44 samples). Transition samples are spaced more closely in time than samples within species. Here again interval length in generations assumes an average generation time of one year.

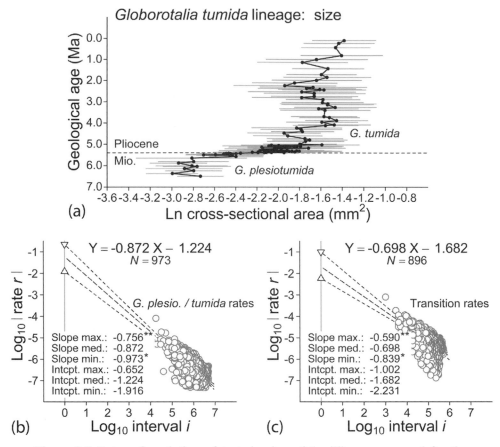

Figure 9.5 Rates of evolution of test size in a 6.5-million-year record for the *Globorotalia tumida* foraminiferal lineage crossing the Miocene–Pliocene boundary. Size here is the cross-sectional area of the foram test oriented in edge view (Malmgren et al., 1983). (a) Time-calibrated stratophenetic graph of size change in *G. plesiotumida* and *G. tumida*. Independent variable time is on the vertical or *y* axis. Horizontal lines span ± 1 standard deviation. (b) Log-rate-interval LRI plot shows that rates of change within *G. plesiotumida* and *G. tumida* scale with a slope of -0.872. This lies between slopes expected for a stationary time series (-1.000) and a random walk (-0.500), significantly different from both, and change within species is a combination of stasis and random change. (c) LRI plot shows that rates of change in the transition from *G. plesiotumida* to *G. tumida* scale with a slope of -0.698. This lies between slopes expected for a stationary time series (-1.000) and a random walk (-0.500), significantly different from both, and the transition is a combination of random change and stasis.

Change in the size of Miocene–Pliocene *G. plesiotumida* and *G. tumida* yields 1,869 nonzero rates. The LRI temporal scaling slope for rates within the two species is -0.872 (Fiigure 9.5b), which is significantly different from both -0.500 expected for a random walk and -1.000 expected for stasis but closer to the latter. Base rates for size within the two species range from $h_{5.477} = 8.83 \times 10^{-8}$ to $h_{4.255} = 8.05 \times 10^{-5}$ standard deviations per generation (median 2.70×10^{-6}) on timescales of tens to hundreds of thousands of generations.

The LRI temporal scaling slope for rates in the transition between the two species (Figure. 9.5c) is -0.698, which is significantly different from both -0.500 expected for a random walk and -1.000 expected for stasis but closer to the former. Base rates for size in the transition between the two species range from $h_{3.954} = 3.53 \times 10^{-6}$ to $h_{3.000} = 7.94 \times 10^{-4}$ standard deviations per generation (median 3.52×10^{-5}) on timescales of a thousand to tens of thousands of generations.

From the temporal scaling slopes we see that changes in size within *G. plesiotumida* and *G. tumida* are more stationary, and changes between the species are more random. To compare rates directly they must be compared on the same timescale. The temporal scaling intercepts for rates within *G. plesiotumida* and *G. tumida* lie within the range of -1.9 to -0.6 on a $\log_{10}$ scale, and the temporal scaling intercepts for rates between *G. plesiotumida* and *G. tumida* lie within the range of -2.2 to -1.0 on a $\log_{10}$ scale. The median values differ, but the ranges overlap considerably and the difference in medians may not be significant (see MacLeod, 1991 for a review of related investigations).

LRI analysis of changes in shape within and between *G. plesiotumida* and *G. tumida* (not shown) yield results similar to those for changes in size. The "punctuated gradualism" pattern in Figure 9.5a is real, but it is doubtful that punctuated change across the Miocene–Pliocene boundary involved any process different from that producing background gradual change before and after the boundary.

9.2.17 *Pliocene-to-Recent* Cryptopecten vesiculosus

Cryptopecten vesiculosus is a medium-sized marine pectininid bivalve or scallop that today inhabits sublittoral to bathal waters of the western Pacific and smaller seas around Japan. It has a fossil record starting in the Pliocene. Itaru Hayami studied *C. vesiculosus* in detail, measuring thousands of fossil and living specimens.

Two sets of statistics recorded by Hayami (1984) are amenable to quantification of rates. Hyami measured shell size at the first (annual) growth line in Pleistocene and Recent *C. vesiculosus* that have growth lines. He also counted the number of ribs on both valves of all specimens. Hayami stated that *C. vesiculosus* individuals reach sexual maturity in their second summer. Some live for five years or more, so a generation time of three years seems conservative.

Thirty-six rates of change in shell size have an LRI temporal scaling slope of -0.728, with a confidence interval for the slope that excludes only directional change and includes both random change and stasis. Fifty-four rates of change in rib number have a temporal scaling slope of -1.053, indicating stasis, with a confidence interval that excludes both directional and random change.

Base rates of change for size and rib number range from $h_{5.523} = 7.07 \times 10^{-8}$ to $h_{2.957} = 7.28 \times 10^{-4}$ standard deviations per generation (median 7.68×10^{-6}) on timescales of about one thousand to hundreds of thousands of generations.

9.2.18 *Miocene Stickleback* Gasterosteus doryssus

Gasterosteus aculeatus is the three-spined stickleback fish familiar from studies reviewed in the last chapter, and the species analyzed here, *G. doryssus*, is a member of the *G. aculeatus* species complex. In 1982, Michael Bell and Thomas Haglund described temporal variation in Miocene *G. doryssus* from six sampling pits at different levels in a varved diatomaceous lake deposit of the Truckee Formation of western Nevada. The Bell and Haglund (1982) study led to a more intensive collecting effort, and a second publication by Bell et al. (1985).

The Bell et al. (1985) study provided statistics for six morphological traits of *G. doryssus* based on large samples of fish collected from 26 successive levels in a relatively short interval of about 108 thousand years of geological time. The traits included standard length of the fish, a score based on development of the pelvic girdle, and counts of the number of dorsal spines, predorsal pterygiophores, dorsal fin rays, and anal fin rays. All traits were ln-transformed before analysis except for the pelvic development score. Rates were calculated on the assumption that *G. doryssus* had a two-year generation time like that of *G. aculeatus* (COSEWIC, 2013b).

The Bell et al. (1985) study is interesting because there is a large difference in pelvic scores between the successive samples numbered 20 and 21, calibrated at approximately 92,006 and 93,445 years, respectively (Figure 9.6a). Dorsal spine number increased greatly too from sample 20 to 21. Bell et al. (1985, p. 264) called the change in pelvic score an "apparently instantaneous change," and interpreted the difference as indicating local extinction and recolonization by a different population of *G. doryssus*. However, the change was not literally instantaneous, and a more neutral characterization seems appropriate. In Figure 9.6a the "instantaneous" change of Bell et al. is simply labeled the "highest rate" of change.

The highest rate change in the pelvic score of *G. doryssus* (Figure. 9.6a) is a change of 6.46 standard deviation units in a time interval of 1,439 years. In the last chapter we found the median step rate for field studies to be $h_0 = 0.15$ standard deviations per generation, meaning that a change of 6.46 standard deviation units in *G. doryssus* could be accomplished in as few as 44 generations or 88 years of directional selection. Bell et al. (1985, p. 267) noted that concomitant reduction of

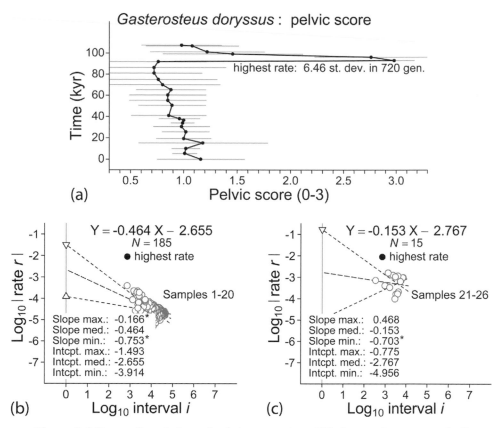

Figure 9.6 Rates of evolution of pelvic scores in a 120-thousand-year record of Miocene *Gasterosteus doryssus* stickleback fish from the Truckee Formation of western Nevada (Bell et al., 1985). (a) Time-calibrated stratophenetic graph of pelvic score change in *G. doryssus*. Independent variable time is on the vertical or *y* axis. Horizontal lines span ±1 standard deviation. (b) Log-rate-interval LRI plot shows that rates of change in the first 20 samples scale with a slope of −0.464. This is close to the slope expected for a random time series (−0.500) and significantly different from expectation for a stationary or directional time series. (c) LRI plot shows that rates of change in the last six samples scale with a slope of −0.153. This is close to the slope expected for a directional time series (0.000) and significantly different from the slope expected for stasis (−1.000). Note that the highest rate, from samples 20 to 21 (represented by closed circles in panels b and c) is substantially higher than rates for the preceding and following series of samples, but it is not too high to be explained by directional selection at step rates observed in the field studies compiled in Chapter 8.

the pelvis and dorsal spines in *G. doryssus* probably indicates selection for reduction of defensive structures, and it is easy to envision augmentation of both to indicate selection for enhancement of defensive structures. In a Truckee Formation inland lake setting it is just as parsimonious to invoke selection favoring defensive structures as it is to invoke replacement by a new population of *G. doryssus* of unknown origin.

Rates for the lineage segment comprising samples 1 through 20 are shown graphically on the LRI plot of Figure 9.6b, where the temporal scaling slope is -0.464, which is close to -0.500 expected for a random time series and significantly different from both stasis and directional change. Rates for the lineage segment comprising samples 21 through 26 are shown in Figure 9.6c, where the temporal scaling slope is lower, -0.153, lying close to 0.000 expected for a directional time series and again significantly different from stasis.

Twenty-three base rates of change for pelvic score range from $h_{3.682} = 1.87 \times 10^{-5}$ to the highest rate of $h_{2.857} = 8.98 \times 10^{-3}$ standard deviations per generation (median 8.69×10^{-5}) on timescales of hundreds to thousands of generations. The other five traits were studied similarly, in lineage segments comprising samples 1 through 20, 20 through 21, and 21 through 26. These yielded an additional 123 base rates that range from $h_{3.494} = 3.44 \times 10^{-6}$ to $h_{2.857} = 3.49 \times 10^{-3}$ standard deviations per generation (median 1.36×10^{-4}) on timescales of, again, hundreds to thousands of generations.

9.2.19 Neogene Bivalve Mollusca

Bivalve mollusks or clams have an excellent Neogene fossil record and many genera and species remain alive today. Steven Stanley and Xiangning Yang studied rates in a representative sample of bivalve lineages and concluded that "shape, as opposed to size, has been highly stable in bivalve evolution over millions of years and 10^6–10^7 generations" (Stanley and Yang, 1987, p. 113).

Stanley and Yang (1987) presented results for the evolution of size as Mahalanobis D multivariate standard deviation units, which can be scaled against interval length and compared as differences and rates in the ordinary way. Results for shape were presented as "nonoverlap percentages" for eigenshapes with no indication of the variability inherent in the samples used to calculate the overlap. Thus here we can only quantify changes in size, and we cannot test the idea that shape has been more stable than size. The Mahalanobis D values for size are not linked to individual lineages, so we cannot be certain which rates are base rates and which are net rates compounded from these. The idea that 1–17 million years of geological time is equivalent to 10^6 to 10^7 bivalve generations suggests an average generation time for bivalves of about five years, which is the number used in calculations here.

The log-difference-interval LDI plot in Figure 9.7a is redrawn from Stanley and Yang's figure 12 with values on both axes converted to logarithms. Log difference and interval values scale with a slope of 0.175, which is close to the slope expected

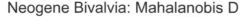

Neogene Bivalvia: Mahalanobis D

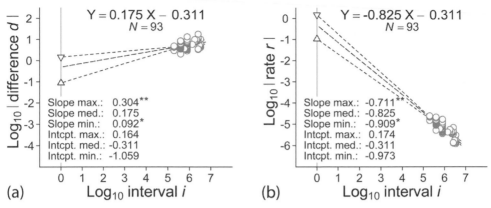

Figure 9.7 Rates of evolution in a 17-million-year record of shell size in Neogene bivalve mollusks (Stanley and Yang, 1987). Differences in size were quantified in Mahalanobis D multivariate standard deviation units. (a) Log-difference-interval LDI plot shows that differences scale with a slope of 0.175, which is close to the slope expected for a stationary time series (0.000) but significantly different from this and from expectation for a random or directional time series. (b) Log-rate-difference LRI plot shows that rates of change scale with a slope of −0.825, which is close to the slope expected for stationary time series (−1.000) but significantly different from this and from expectation for a random or directional time series. Temporal scaling indicates that evolution in the lineages studied by Stanley and Yang was neither purely stationary nor purely random but a combination of both.

for a stationary time series (0.000) but significantly different from this, and significantly different from the slopes expected for a random (0.500) or directional time series (1.000). The log-rate-interval LRI plot in Figure 9.7b shows that rates of change scale with a slope of −0.825, which is close to the slope expected for stationary time series (−1.000) but significantly different from this, and significantly different from the slopes expectated for both random and directional time series. Overall, the evolution of size in the bivalve lineages studied by Stanley and Yang (1987) was neither stationary nor random but a combination of both.

The 93 rates involved range from $h_{6.532} = 1.22 \times 10^{-6}$ to $h_{5.301} = 2.44 \times 10^{-5}$ standard deviations per generation (median 6.08×10^{-6}) on timescales of hundreds of thousands to millions of generations.

9.2.20 Eocene-to-Recent Equidae

We have already reviewed several studies of rates of evolution in mammalian horses of the family Equidae. Most comprehensive, in some ways, are the rate studies of Bruce MacFadden published in the 1980s and reviewed in his now-classic book on *Fossil Horses* (MacFadden,1992). Three studies stand out in particular: MacFadden (1985, 1986, and 1988). MacFadden calculated rates in darwins but provided information necessary to convert the rates to haldanes.

Body weights in MacFadden (1986, table 4) and MacFadden and Hulbert (1990, table 15.4) can be used to estimate generation times and to calculate rates of change in body weight. The standard deviation of ln-transformed body weight, 0.1395, is provided by MacFadden (1986, table 1). A good estimate for the standard deviation of ln-transformed tooth measurements is 0.05. Generation times were calculated from body weight using Equation 3.2.

The comparisons of MacFadden (1985, 1986; 1988) are between a total of 30 species pairs. The 30 species pairs, body weights, and four tooth measures yield a total of 143 base rates (a few entries are missing and there are two zero rates). Base rates for the species pairs range from $h_{6.022} = 9.66 \times 10^{-8}$ to $h_{5.725} = 3.83 \times 10^{-5}$ standard deviations per generation (median 5.13×10^{-6}) on timescales of hundreds of thousands to millions of generations.

9.2.21 Dwarfing in the Pleistocene Red Deer Cervus elaphus

Dwarfing is a common response for large mammals isolated on islands, but this can rarely be quantified in terms of rate. Adrian Lister developed an interesting case involving Pleistocene red deer, *Cervus elaphus*, on the island of Jersey, which lies in the English Channel off the coast of Normandy. During cooler phases of the Pleistocene, lowered sea level meant that Jersey was part of a broad continental plain in what is now northwestern France. During the warmest part of the Eemian interglacial, about 120 kyr before present, a high sea stand isolated Jersey as an island for an interval of about 5.8 thousand years. Small *C. elaphus* found as fossils in a raised beach deposit in Belle Hougue cave date from this interval.

Lister (1989) published measurements of 10 *C. elaphus* skeletal elements from Belle Hougue cave and compared these with corresponding elements from contemporary Ipswichian British specimens. Postcranial remains of the dwarfed deer of Jersey are about 60% the length of contemporary mainland specimens, indicating body weights about one sixth of those from the mainland. Generation times for the dwarfed deer were probably about 4.5 years (Equation 3.2), compared with a generation time of 8.33 years for British *C. elaphus* today (Kruuk et al., 2002). These average about 6.4 years, and it is reasonable to expect that dwarfing in 5.8 thousand years spanned about 900 generations.

Base rates for the 10 skeletal measurements range from $h_{3.097} = 4.33 \times 10^{-3}$ to 2.35×10^{-2} standard deviations per generation (median 9.63×10^{-3}), all on a timescale of 900 generations. All of the changes are negative as expected for a dwarfing lineage.

9.2.22 Pleistocene Horse Equus germanicus

Šandalja II is an important archaeological and paleontological site in Croatia (Jan-ković et al., 2012). The site is stratified, with eight layers of bone breccia and finer sediment that accumulated in a cave setting. Bone in the breccias is principally from auroch (*Bos*) and horse (*Equus*) hunted by Epipaleolithic humans on the surrounding plains (Miracle, 1995). Anne-Marie Forstén studied the horses, which she identified as *Equus germanicus* (or *Equus caballus germanicus*). Some intervals have a smaller European ass, *Equus hydruntinus*, in addition to *E. germanicus*.

Forstén (1990, tables 1 and 2) provided statistics for a suite of 34 measurements of premolar and molar teeth of Šandalja II *E. germanicus*. Most were measured at the bases of tooth crowns and on worn occlusal surfaces. Forstén also measured something she called "protoconal length" and she counted the number of enamel folds on worn occlusal surfaces. Horses are found in seven layers that are radio-carbon dated at, respectively, 27.800, 22.660, 16.940 (averaged from two ages), 13.070, 12.695 (averaged from underlying and overlying ages), 12.320, and 10.830 kyr before present (Miracle, 1995, Jancović et al., 2012, and references therein). Modern horses in the wild have a generation time of about eight years (Berger, 1986, figure 5.2).

Forstén's measurements of dental traits in *E. germanicus* yield 714 rates, of which 666 are nonzero. The 193 base rates calculated here range from $h_{2.808} = 8.01 \times 10^{-5}$ to $h_{1.671} = 3.07 \times 10^{-2}$ standard deviations per generation (median 1.58×10^{-3}) on timescales of tens to thousands of generations.

9.2.23 Pliocene Rodent Cosomys primus

Cosomys primus is a microtine rodent of the family Cricetidae. Deborah Lich (1990) studied statistical samples of *C. primus* from a Pliocene stratigraphic sequence to see whether they changed through time. The samples came from the Glenns Ferry Formation on the Snake River Plain near Hagerman, Idaho, in the western United States.

Two radiometric ages in the Hagerman sequence indicate that Lich's fossil samples range in age from about 3.345 to 3.181 Ma (spanning an interval of about 164 kyr). Lower tooth row length for *C. primus* is about 6.6 mm (Zakrzewski, 1969, figure 7e), which translates to a body weight of 100 g (Freudenthal and Martín-Suárez, 2013). This indicates, in turn, a generation time of about one year (Equation 3.2).

The LRI log-rate-interval temporal scaling slope for the *C. primus* time series as a whole is −0.780, which is significantly different from expectation for a direc-tional time series (0.000) and for a random time series (−0.500). The LRI slope is not significantly different from expectation for a stationary time series, reinforcing Lich's interpretation of a lineage in stasis.

Base rates for the three *C. primus* tooth measurements range from $h_{4.398} = 3.20 \times 10^{-6}$ to $h_{3.921} = 9.09 \times 10^{-5}$ standard deviations per generation (median 2.40×10^{-5}), on timescales of thousands to tens of thousands of generations.

9.2.24 *Eocene Primate* Cantius

Cantius is an early Eocene fossil primate with an exceptional stratigraphic record in the Clarks Fork Basin of northwestern Wyoming. William Clyde and Philip Gingerich studied a total of 451 lower first molars (M_1) of *Cantius* from 40 successive stratigraphic intervals, and 121 upper first molars (M^1) from 18 intervals. The Clarks Fork Basin record of *Cantius* evolution spans approximately 1.8 million years of geological time (Figure. 9.8). Clyde and Gingerich (1994) found *Cantius* to exhibit directional change in tooth size while simultaneously exhibiting stasis in overall tooth shape.

The Clarks Fork Basin stratigraphic section of interest includes the Paleocene–Eocene thermal maximum, when *Cantius* first appeared, which is now calibrated at 56 Ma (Vandenberghe et al., 2012). The highest interval producing *Cantius* is in the Wa-5 biozone, just below magnetochron 24N (Gingerich, 2010), dated at approximately 54 Ma (Vandenberghe et al., 2012). Tooth size in *Cantius* can be used to estimate body weight (Gingerich et al., 1982, figure 5), which can be used in turn to estimate generation time (Equation 3.2).

LRI log-rate-interval temporal scaling of all 1,803 rates of change in tooth size yields a slope of -0.534, which is close to and not significantly different from the slope expected for random change (-0.500). The observed slope is significantly different from slopes expected for directional change (0.000) and stasis (-1.000). This overall random-change result differs from the directional change of M_1 length and width found by Clyde and Gingerich (1994) because here more rates are pooled, and the regression model is slightly different.

Log-rate-interval temporal scaling of all 2,159 rates of change in tooth shape yields a slope of -1.079, which is not significantly different from -1.000 expected for stasis. It is significantly different from slopes expected for random change (-0.500) and directional change (0.000), corroborating Clyde and Gingerich's finding of stasis in tooth shape.

There are 90 base rates for tooth size change in *Cantius*, that range from $h_{4.527} = 1.84 \times 10^{-6}$ to $h_{3.803} = 9.78 \times 10^{-4}$ standard deviations per generation (median 2.82×10^{-5}), on timescales of about 6 to more than 140 thousand generations.

There are 161 base rates for tooth-shape change in *Cantius*, and these range from $h_{5.072} = 3.08 \times 10^{-7}$ to $h_{4.069} = 1.20 \times 10^{-3}$ standard deviations per

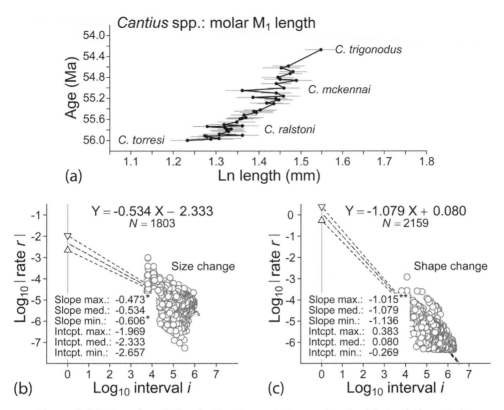

Figure 9.8 Rates of evolution in the size and shape of molar M_1 teeth for a 1.6-million-year lineage of the early Eocene mammal *Cantius* documented in the Clarks Fork Basin of Wyoming (Clyde and Gingerich, 1994). (a) Time-calibrated stratophenetic graph shows change in length of the first lower molar from *Cantius torresi* to *C. trigonodus*. Independent variable time is on the vertical or *y* axis. Horizontal lines span ±1 standard deviation. (b) Log-rate-interval LRI plot shows that rates of change for tooth size scale with a slope of −0.534. This is close to −0.500 expected for random change but significantly different from stasis (−1.000) and directional change (0.000). Size in the *Cantius* lineage is increasing but increasing more randomly than directionally. (c) LRI plot shows that rates of change for tooth shape scale with a slope of −1.079, which is close to and not significantly different from the slope (−1.000) expected for stasis. Size can change in an evolutionary lineage while shape remains stationary.

generation (median 8.76×10^{-6}), on timescales of about six thousand to more than 251 thousand generations.

9.2.25 *Pleistocene-to-Recent Island Snail* Mandarina

Mandarina chichijimana is a land snail endemic to the western Pacific island of Chichijima in the Ogasawara Archipelago lying 1,000 kilometers south of Tokyo in Japan. Here Satoshi Chiba documented a 40,000-year record of late Pleistocene to Recent *M. chichijimana.*

Chiba (1996) provided a suite of 10 shell measurements for 14 successive radiocarbon-dated fossil samples. Several Recent samples were measured as well, and he chose Recent sample BCL5 for comparison with the fossil samples. The generation time for *Mandarina* is minimally 1.5 years (Chiba, 2007), and two years is a more likely estimate.

Measurements provided by Chiba (1996) yield a total of 1,036 nonzero rates. Taken together these have a temporal scaling slope of -0.571. This is not significantly different from -0.500 expected for a random time series, but the observed slope is significantly different from expectation for both stasis (-1.000) and directional change (0.000). Phyletic evolution in *M. chichijimana* is, as Chiba (1996, p. 185) concluded, "essentially irregular."

One hundred thirty-nine base rates for size change in shells of *M. chichijimana* range from $h_{3.243} = 7.89 \times 10^{-6}$ to $h_{2.332} = 9.84 \times 10^{-3}$ standard deviations per generation (median 3.74×10^{-4}), on timescales of 215 to just over 3,000 generations.

9.2.26 *Paleocene-Eocene Mammal* Ectocion

In 1943, George Gaylord Simpson introduced the idea of a "chronocline" as the temporal equivalent of a geographic cline. He viewed a geographic cline as a "sequence of contemporaneous populations arrayed geograpically and intergrading in progressively changing morphological characters" where the "gradation is normally maintained by genetic exchange" (Simpson, 1943, p. 174). Thus a chronocline for Simpson was a sequence of sympatric populations arrayed temporally and intergrading in progressively changing morphological characters where the gradation is normally maintained by genetic exchange.

The example Simpson (1943) chose to illustrate a chronocline was an evolutionary lineage of the late Paleocene and early Eocene mammal *Ectocion.* Simpson arranged the specimens he had available by geological age and by the length of their first lower molar (M_1), and then combined what he regarded as ascending "vertical species" *E. parvus*, *E. ralstonensis*, and *E. osbornianus* into a chronocline of increasing size. We know more about the stratigraphic distribution of the species now, and have much larger samples of each (Figure 9.9a). Ironically, what Simpson viewed as a progressive increase in size we now know is a stationary time series with the smallest species, *E. parvus*, not at the beginning but in the middle of the series. *E. parvus* is restricted to the narrow stratigraphic interval of the Paleocene–Eocene thermal maximum (PETM) when many species were dwarfed in response to a global greenhouse warming event.

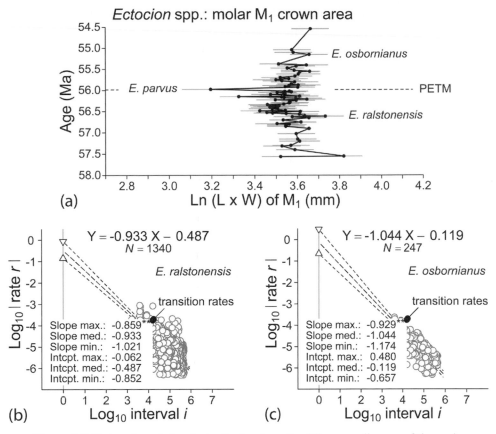

Figure 9.9 Rates of evolution in tooth size for a 3-million-year lineage of the early Eocene mammal *Ectocion* spanning the Paleocene–Eocene thermal maximum (PETM) in the Clarks Fork Basin of Wyoming (Gingerich, 2003). (a) Time-calibrated stratophenetic graph shows change in crown area of the first lower molar from *Ectocion ralstonensis* to *E. parvus* and then *E. osbornianus*. Independent variable time is on the vertical or *y* axis. Horizontal lines span ± 1 standard deviation. (b) Log-rate-interval LRI plot for change in tooth size within the *E. ralstonensis* lineage. In this species rates scale with a slope of -0.933, which is close to and not significantly different from -1.000 expected for stasis. (c) LRI plot for change in tooth size within the *E. osbornianus* lineage. Here rates scale with a slope of -1.044, which is close to and not significantly different from the slope (-1.000) expected for stasis. Note that the transition rates from *E. ralstonensis* to *E. parvus* and from *E. parvus* to *E. osbornianus* (filled circles in panels b and c) lie within the distribution of rates for each of the longer-surviving species. Figure is modified from Gingerich (2003), reproduced by permission of the Geological Society of America

The *Ectocion* specimens summarized in Figure 9.9a were previously analyzed in Gingerich (1989, 2003). Here they are compared in four groups: (1) within the late Paleocene lineage *E. ralstonensis*, (2) during the transition from latest *E. ralstonensis* to *E. parvus*, (3) during the transition from *E. parvus* to earliest

E. osbornianus, and (4) within the early Eocene lineage *E. osbornianus*. Body weight is estimated from tooth size using the ungulate regression of Legendre (1989, table 1), and generation time is estimated from body weight using Equation 3.2.

A log-rate-interval LRI plot of rates within the *E. ralstonensis* lineage yields a temporal scaling slope of -0.933 (Figure 9.9b), which is close to and not significantly different from -1.000 expected for a lineage in stasis. This slope is significantly different from expectation for both random and directional change. The rates for the transition from *E. ralstonensis* to *E. parvus* and from *E. parvus* to *E. osbornianus* (filled circles in Figure 9.9b) lie within the distribution of rates for change within *E. ralstonensis*.

A log-rate-interval LRI plot of rates within the *E. osbornianus* lineage yields a temporal scaling slope of -1.044 (Figure. 9.9c), which is again close to and not significantly different from -1.000 expected for a lineage in stasis. This slope too is significantly different from expectation for random and directional change. Both transition rates (filled circles in Figure 9.9c) lie within or near the distribution of rates for change within *E. osbornianus*.

Aaron Wood and co-authors (2007) analyzed multivariate shape in relation to time in the *Ectocion* lineage. They found no change in tooth shape, consistent with a lack of change in size. The only indication of a chronocline in *Ectocion* is in the dwarfing and recovery of *E. parvus* during and after the PETM.

There are 76 base rates in the *Ectocion* time series, and these range from $h_{3.914} = 3.93 \times 10^{-7}$ to $h_{3.591} = 8.81 \times 10^{-4}$ standard deviations per generation (median 8.25×10^{-5}), on timescales of about two thousand to 160 thousand generations.

9.2.27 *Miocene Stickleback* Gasterosteus doryssus *Revisited*

Michael Bell and colleagues returned to the Miocene *Gasterosteus doryssus* fossil site in Nevada described above (Figure 9.6) to resample the 88 to 108-thousand-year interval with the rapid excursion in pelvic scores. New samples were taken at 250-year intervals that enabled tracing of pelvic scores, dorsal spine counts, and predorsal pterygiophore counts in much finer detail (Bell et al., 2006). The rapid excursion in pelvic score stands out in the new record (Figure 9.10) as it did in the old.

The new record was studied in lineage segments comprising samples 1–17 preceding the pelvic score excursion, samples 17–19 in the pelvic score excursion, and, for most traits, samples 19–78 during the subsequent recovery interval. Rates for pelvic scores were calculated using samples 1–17, 17–19, and 29–78, omitting samples 19 through 28 from the latter because of their redundancy and, in most cases, lack of variance.

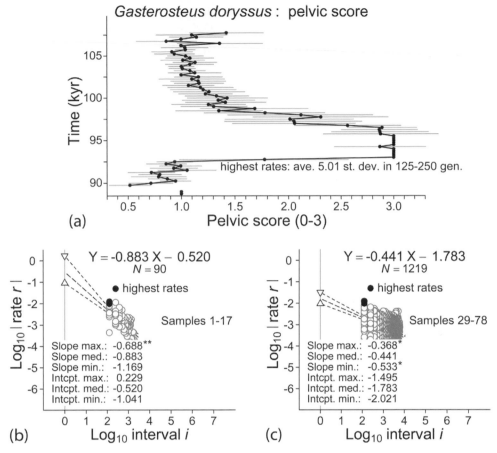

(a)

(b)

(c)

Figure 9.10 Rates of evolution in pelvic scores for a 17-thousand-year lineage of Miocene *Gasterosteus doryssus* fish from the Truckee Formation of western Nevada (Bell et al. 2006). This is a part of the record illustrated in Figure 9.6, sampled more finely. (a) Time-calibrated stratophenetic graph of pelvic score change in *G. doryssus*. Independent variable time is on the vertical or *y* axis. Horizontal lines span ±1 standard deviation. (b) Log-rate-interval LRI plot shows that rates of change in the first 17 samples scale with a slope of −0.883. This is close to the slope expected for a stationary time series (−1.000) and significantly different from expectation for a random or directional time series. (c) LRI plot shows that rates of change in the last 50 samples scale with a slope of −0.441. This is close to the slope expected for a random time series (−0.500) and significantly different from the slopes expected for stasis (−1.000) and directional change (0.000). Note that the highest rates, from samples 17 to 19 (represented by filled circles in panels b and c) are substantially higher than rates for the preceding and following series of samples, but these are not too high to be explained by directional selection at step rates observed in field studies compiled in Chapter 8.

As before, the highest rates of change in pelvic scores were found during the rapid excursion in the 500-year interval from 92.5 to 93.0 kyr (Figure 9.10). The average change was 5.01 standard deviations in an interval of 125 to 250 generations, and the rates range from lows of 0.010 and 0.012 standard deviations per generation on a timescale of 125 generations to a high of 0.049 standard deviations per generation on a timescale of 250 generations. This rate is the highest for *G. doryssus* on this timescale, but it is not instantaneous.

Rates of change in pelvic scores for the sample 1–17 lineage segment are shown graphically in Figure 9.10b. The LRI temporal scaling slope is −0.883, which is not significantly different from −1.000 expected for stasis. This slope is significantly different from those expected for random (−0.500) and directional change (0.000). Rates of change in pelvic scores for the sample 29–78 lineage segment are shown in Figure 9.10c. Here the LRI temporal scaling slope is lower, −0.441, which is not significantly different from −0.500 expected for random change. This slope is significantly different from those expected for stasis (−1.000) and directional change (0.000).

Sixty-two base rates of change for pelvic scores range from $h_{2.097} = 7.89 \times 10^{-5}$ to $h_{2.097} = 1.24 \times 10^{-2}$ standard deviations per generation (median 3.13×10^{-3}), all on a timescale of 125 generations. The other two traits yielded an additional 140 base rates that range from $h_{2.097} = 6.44 \times 10^{-5}$ to $h_{2.097} = 1.72 \times 10^{-2}$ standard deviations per generation (median 3.02×10^{-3}) on a timescale of 125 generations.

9.2.28 Paleocene-Eocene Mammal Haplomylus

Haplomylus is a small herbivorous mammal common in upper Paleocene and lower Eocene strata of the Clarks Fork and Bighorn basins in Wyoming, U.S.A. (Figure 9.11a). One lineage of two species, *H. palustris* and *H. simpsoni*, is known from the late Paleocene, and a second lineage of two species, *H. speirianus* and *H. scottianus*, is known from the early Eocene. Both lineages increased in tooth size and body size through time and each illustrates Simpson's concept of a chronocline, discussed above in relation to *Ectocion*, and here there is substantial change in each lineage through time.

The biggest change in the fossil record of *Haplomylus* happened during the onset of the Paleocene–Eocene thermal maximum (PETM) when the relatively large, latest Paleocene species *H. simpsoni* was abruptly replaced by the dwarfed earliest Eocene species *H. zalmouti*. *Haplomylus* evolution has been illustrated by Gingerich (1976; 1994), but the full story was not evident until discovery of *H. zalmouti* (Gingerich and Smith, 2006).

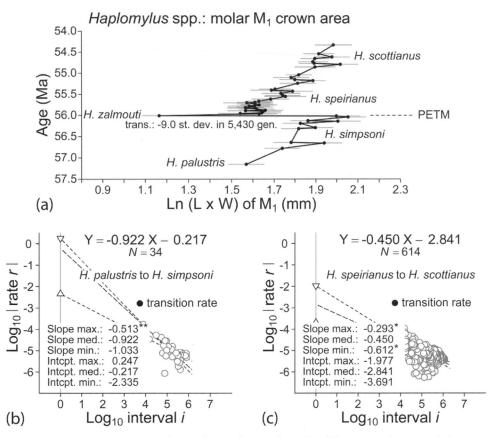

(a)

(b)

(c)

Figure 9.11 Rates of evolution in tooth size for a 3-million-year lineage of the early Eocene mammal *Haplomylus* documented in the Clarks Fork Basin of Wyoming (Gingerich and Smith, 2006). (a) Time-calibrated stratophenetic graph shows change in crown area of the first lower molar from *Haplomylus palustris* to *H. simpsoni* and from *H. speirianus* to *H. scottianus*. Independent variable time is on the vertical or *y* axis. Horizontal lines span ±1 standard deviation. (b) Log-rate-interval LRI plot shows that rates of change for tooth size in *H. palustris* and *H. simpsoni* scale with a slope of −0.922. The slope is not significantly different from −1.000 expected for stasis, but it is significantly different from expectation for random change (−0.500) and directional change (0.000). (c) LRI plot shows that rates of change for tooth size in *H. speirianus* and *H. scottianus* scale with a slope of −0.450. The slope is close to and not significantly different from the slope expected for random change. It is significantly different from expectation for stasis and directional change. Here as in the *Cantius* lineage of Figure 9.8, size is increasing but increasing randomly with a directional component.

The *Haplomylus* specimens are analyzed in three groups: (1) within the late Paleocene lineage of *H. palustris* and *H. simpsoni*, (2) during the transition from latest *H. simpsoni* to *H. zalmouti*, and (3) within the early Eocene lineage of *H. speirianus* and *H. scottianus*. Body weight is estimated from tooth size using the ungulate regression of Legendre (1989, table 1), and generation time is estimated from body weight using Equation 3.2.

A log-rate-interval LRI plot for rates in the *H. palustris* to *H. simpsoni* lineage yields a temporal scaling slope of −0.922 (Figure 9.10b), close to that expected for stasis (−1.000), with a confidence interval that just excludes −0.500 expected for random change. Change in this lineage looks both random and directional, but the appearance of directionality is due largely to the single early sample of *H. palustris*. The highest rate is that for the transition from latest Paleocene *H. simpsoni* to earliest Eocene *H. zalmouti* (filled circle in Figure 9.10b). This lies outside the distribution of rates for change in the *H. palustris* to *H. simpsoni* lineage, but it is not too fast to be explained by directional selection.

A log-rate-interval LRI plot for rates in the *H. speirianus* to *H. scottianus* lineage yields a temporal scaling slope of −0.450 (Figure 9.10c), which is close to and not significantly different from −0.500 expected for random change. The observed slope is significantly different from expectation for both stasis and directional change but represents randomness with a directional component. The *H. simpsoni* to *H. zalmouti* transition rate (filled circle in Figure 9.10c) again lies well outside the distribution of rates for change in the *H. speirianus* to *H. scottianus* lineage.

There are 43 base rates in the *Haplomylus* time series, and these range from $h_{4.681} = 2.88 \times 10^{-6}$ to $h_{3.735} = 1.65 \times 10^{-3}$ standard deviations per generation (median 2.47×10^{-5}), on timescales between 5 thousand and 264 thousand generations.

9.2.29 Cenozoic Ostracod Poseidonamicus

Ostracoda are a class of minute shrimp-like representatives of the phylum Crustacea covered by a carapace of left and right hinged valves. *Poseidonamicus* is an Eocene-to-Recent deep-sea ostracod with a distinct pattern of reticulation on the surface of the carapace. Gene Hunt and Kaustuv Roy measured valve lengths in 19 species of *Poseidonamicus* to quantify changes in body size through time. Their growth is indeterminate, but there are a limited number of instars, and instars can be identified on the basis of size and other characteristics.

Fifteen species measured by Hunt and Roy (2006) are represented by two or more samples of known geological age. These can be used to calculate rates of change within the species. Coefficients of variation average 0.03 (Hunt and Roy, 2006), and generation time is assumed to be five years (Hunt, 2007).

A total of 376 nonzero rates of change can be calculated from within-species sample means. Taken together, these scale with a log-rate-interval LRI slope of −0.635. This is on the stationary side of expectation for random time series while differing significantly from expectation for stationary (-1.000), random (-0.500), and directional series (0.000).

Within the full sample, there are 76 base rates that range from $h_{6.215} = 2.95 \times 10^{-8}$ to $h_{4.415} = 2.52 \times 10^{-4}$ standard deviations per generation (median 6.50×10^{-6}), on timescales ranging from two thousand to nearly three million generations.

9.2.30 *Pleistocene-to-Recent Freshwater Diatom* Stephanodiscus

Stephanodiscus is a small centric diatom that is abundant and ecologically important in many freshwater ecosystems. Edward Theriot and colleagues traced the late Pleistocene to Recent evolution of one particular species, *S. yellowstonensis*, which is endemic to Yellowstone Lake in northwestern Wyoming, U.S.A. Theriot and colleagues recognized two phases in this evolutionary history (Figure 9.12a): (1) some 3–4 thousand years of change, from about 13.7 Ka to 10 Ka, during the evolution of *S. yellowstonensis* from the ancestral species *S. niagarae* and (2) a 10,000-year interval of relative stability from 10 Ka to the present within *S. yellowstonensis* itself.

The Yellowstone *Stephanodiscus* were sampled from a core retrieved from the central basin of Yellowstone Lake. This was calibrated radiometrically using carbon-14. Pollen from the same core facilitated environmental interpretation. Three morphological traits of *Stephanodiscus* were measured or counted: (1) valve diameter, (2) number of costae per valve, and (3) number of spines per valve. A fourth trait calculated from these proved informative: (4) the number of spines present relative to valve diameter, calculated as the slope of spine counts in a sample regressed on valve diameters in the sample.

Yellowstone Lake is a high mountain lake, and *Stephanodiscus* is only able to reproduce during the summer. According to Theriot (personal communication), cells divide about once every two days during summer, for a maximum of 45 clonal generations per year. During the Pleistocene when the water remained colder cell division probably took place about every five days, for a maximum of about 18 clonal generations per year. *Stephanodiscus* is thought to reproduce sexually about once in four years, and four years is used as an estimate of generation time.

Here, following Theriot et al. (2006), the *Stephanodiscus* time series was divided into two phases, one preceding 10 Ka (dashed line in Figure 9.12a), and the second following 10 Ka. Successive values for the fourth trait listed above, quantification of the number of spines present relative to valve diameter, are shown in Figure 9.12a. Rates in the initial transitional phase, from 13.7 to 10.0 Ka, have a log-rate-interval LRI temporal scaling slope of -0.609 (Figure 9.12b). This slope is not significantly different from expectation for random change (-0.500), but it is

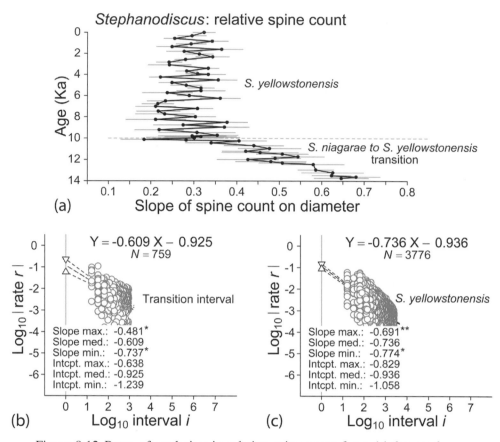

Figure 9.12 Rates of evolution in relative spine count for a 14-thousand-year lineage of the late Pleistocene to Recent diatom *Stephanodiscus yellowstonensis* documented in Yellowstone Lake, northwestern Wyoming (Theriot et al. 2006). (a) Time-calibrated stratophenetic graph of relative spine count for *S. yellowstonensis*. Independent variable time is on the vertical or *y* axis. Horizontal lines span ±1 standard deviation. (b) Log-rate-interval LRI plot shows that rates of change in the first 20 samples, the transition interval, scale with a slope of −0.609. This is close to the slope expected for a random time series (−0.500) and significantly different from expectation for a stationary or directional time series. (c) LRI plot shows that rates of change in the last 44 samples scale with a slope of −0.736. This lies between slopes expected for random change (−0.500) and stasis (−1.000), indicating a pattern combining the two.

significantly different from expectation for both stasis (−1.000) and directional change (0.000). Rates after 10.0 Ka have an LRI slope of −0.736 (Figure 9.12c), which lies between expectation for a random time series (−0.500) and stasis (−1.000). The post-10 Ka LRI slope is significantly different from expectation for stasis and random change but represents a combination of both.

The total number of nonzero rates for all four traits combined, within the intervals before and after 10 Ka, is 4,535. There are 247 base rates in this set, and these base rates range from $h_{1.809} = 5.34 \times 10^{-5}$ to $h_{1.207} = 9.72 \times 10^{-2}$ standard deviations per generation (median 8.68×10^{-3}), on timescales ranging from some 16 to 129 generations.

9.2.31 Miocene-to-Recent Bryozoan **Metrarabdotos**

Metrarabdotos is a genus in the order Cheilostomata of the phylum Bryozoa. It is a colonial marine invertebrate with polypides inhabiting small mineralized box-shaped zooids of definitive size. Alan Cheetham spent much of his career analyzing tropical American *Metrarabdotos* from an evolutionary perspective. The complexity and definitive growth of *Metrarabdotos* zooids make them advantageous for evolutionary study.

The principal study of *Metrarabdotos* is by Cheetham et al. (2007), in which 22 species were analyzed stratophenetically, cladistically, and morphometrically. Pairs of sister lineages were then compared in terms of within-species variability and between-species distances, quantified in Mahalanobis D multivariate standard deviation units, all based on discriminant functions involving variable sets of morphological traits. Four of these sets of sister lineages are illustrated in Figures 9.13a–d. The first species in each pair has its within-species discriminant scores normalized to a zero mean, and the derived species in each pair has its within-species discriminant scores normalized to the distance between the species in a more inclusive analysis. This procedure explains why Mahalanobis D values for *M. boldi* in Figure 9.13a differ slightly from those in Figure 9.13b.

The generation time for *Metrarabdotos* is seemingly about 10 years; if rapidly growing colonies of *Stylopoma* reach reproductive maturity in one year, then the average age of reproduction for slowly growing *Metrarabdotos* is likely to be near the upper end of the 1–10-year range proposed by Cheetham and Jackson (1995).

There are 725 differences between Mahalanobis D values for within-species (within-lineage) comparisons in Figures 9.13a–d. These differences are shown as open circles in the log-difference-interval LDI plot of Figure 9.13e. Temporal scaling of the differences yields a slope of −0.008, which is almost exactly the slope expected for stationary time series (0.000). There are four differences (dashed lines) separating initial representatives in each lineage pair in Figures 9.13a–d. These differences are shown as filled circles in Figure 9.13e. Each between-lineage difference is greater than all within-lineage differences for the corresponding time interval.

There are similarly 725 rates between Mahalanobis D values for within-species (within-lineage) comparisons in Figures 9.13a–d. These rates are shown as open

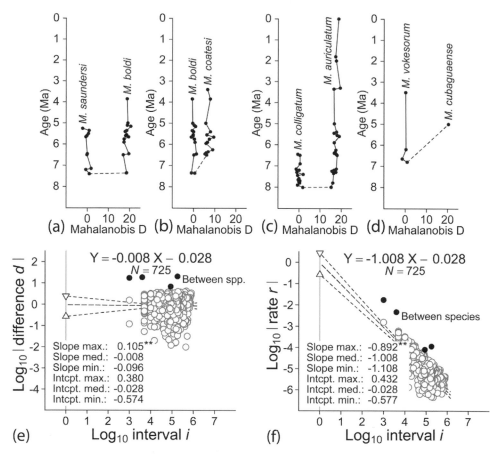

Figure 9.13 Rates of evolution within and between lineages of Miocene to Recent *Metrarabdotos* bryozoans from tropical American oceans (Cheetham et al., 2007). (a) Time-calibrated stratophenetic graphs are shown for (a) *M. saundersi* and *M. boldi* species pair, (b) *M. boldi* and *M. coatesi* species pair, (c) *M. colligatum* and *M. auriculatum* species pair, and (d) *M. vokesorum* and *M. cubaguaense* species pair. (e) Log-difference-interval LDI plot for all within-lineage differences showing that these scale with a slope of −0.008. This is almost exactly the scaling expected for stationary time series (0.000) and significantly different from the slopes expected for a random or directional time series. Note that differences between species lineages (solid circles) are greater than differences within lineages. (f) Log-rate-interval LRI plot for all within-lineage rates showing that these scale a slope of −1.008. This is again almost exactly the slope expected for a stationary time series (−1.000) and significantly different from the slopes expected for a random or directional time series. Note that rates between species lineages (filled circles) are higher than rates within lineages sampled on the same scale of time, but they are not too high to be explained by directional selection (see Chapter 11.2.2).

circles in the log-rate-interval LRI plot of Figure 9.13f. Temporal scaling of the rates yields a slope of -1.008, which is again almost exactly the slope expected for stationary time series (-1.000). There are four rates (slopes of dashed lines) separating initial representatives in each lineage pair in Figures 9.13a–d. These rates are shown as filled circles in Figure 9.13f. Each between-lineage rate is greater than all within-lineage rates for the corresponding time interval.

The rates within lineages include 90 base rates, which range from $h_{5.431} = 6.82 \times 10^{-7}$ to $h_{3.000} = 1.54 \times 10^{-3}$ standard deviations per generation (median 8.50×10^{-5}), on timescales of one thousand to 270 thousand generations. The four rates between lineages are all base rates, and these range from $h_{4.935} = 8.02 \times 10^{-5}$ to $h_{3.000} = 1.72 \times 10^{-2}$ standard deviations per generation (median 2.38×10^{-3}), on timescales of one thousand to 180 thousand generations.

9.2.32 *Eocene Dawn Horse* Hyracotherium *at Its First Appearance*

The Paleocene–Eocene Thermal Maximum, or PETM, is a global greenhouse warming event that defines the end of the Paleocene epoch of geological time and the beginning of the Eocene at about 56 Ma. Evolution in response to the event was often both rapid and substantial. A study by Ross Secord and colleagues provided information that enables calculation of rates for the dawn horse *Hyracotherium* (or "*Sifrhippus*") during onset of the PETM and during recovery from the greenhouse event.

Secord et al. (2012) published tooth size measurements and body weight estimates for 44 specimens of *Hyracotherium sandrae* and *H. grangeri* collected from 24 stratigraphic levels within and following the PETM. Tooth size was measured as ln M_1 molar crown area (ln L $\times$ W). This has a weighted standard deviation of 0.082. Body weights are those of Secord et al. (2012), and generation times are estimated from the body weights (Equation 3.2).

Twenty-two nonzero base rates can be calculated from the 24 successive samples of *Hyracotherium* known from the onset through recovery of the PETM. These range from $h_{4.138} = 6.65 \times 10^{-6}$ to $h_{3.007} = 1.56 \times 10^{-3}$ standard deviations per generation (median 3.17×10^{-4}), on timescales of some 700 to 24,000 generations.

9.2.33 *Late Miocene Murine Rodents*

Murine rodents, including forms related to the modern mouse genus *Mus*, are well represented the Siwalik fossil record of late Miocene age in Pakistan. Screen-washing yielded large samples of murine teeth from 31 stratigraphic levels in a 7.55- million-year sequence ranging in age from 14.05 to 6.5 Ma. Yuri Kimura and colleagues made a detailed morphometric study of this record to quantify the evolution of size and shape in murine first upper molars (M^1; Kimura et al. 2015). Here rates are quantified for lineages of sequential samples for five of the genera involved.

The genera analyzed are *Antemus*, *Progonomys*, *Karnimata*, *Mus*, and *Parapelomys*. Traits studied are tooth length, tooth width, four measures of shape involving tooth cusps (van Dam's index, anterostyle ratio, anterostyle angle, protocone–entocone angle), and two measures of shape involving the tooth outline (PC-I and PC-II). Generation time was estimated from body weight using Equation 3.2, with body weight estimated from tooth size (using information for the genera studied here in Freudenthal and Martín-Suárez, 2013).

The two measures of size yield 436 rates with an LRI temporal scaling slope of -0.804 lying between, and significantly different from, the slopes expected for stasis (-1.000) and random change (-0.500). In contrast, the six measures of shape yield 1,269 rates with an LRI slope of -0.996, which is almost exactly the slope expected for stasis (-1.000). It is significantly different from slopes expected for random (-0.500) and directional change (0.000). Here, as we have seen previously, size may change significantly within lineages while shape is seemingly more constrained. These are generalizations about change in the generic lineages as a whole that may of course obscure stasis or change on finer scales of time.

Seventy base rates of change in size range from $h_{5.957} = 1.12 \times 10^{-8}$ to $h_{4.646} = 2.44 \times 10^{-5}$ standard deviations per generation (median 3.25×10^{-6}). A total of 202 rates of change in shape range from $h_{5.870} = 2.10 \times 10^{-8}$ to $h_{4.786} = 8.08 \times 10^{-5}$ standard deviations per generation (median 1.95×10^{-6}). These base rates span timescales of about 28,000 to 1.3 million generations.

9.2.34 *Pleistocene-to-Recent Freshwater Diatom* Cyclostephanos

Cyclostephanos, like *Stephanodiscus* analyzed above, is a small centric diatom that is abundant and ecologically important in many freshwater ecosystems. Trisha Spanbauer and colleagues analyzed a 312-thousand-year time series of valve diameters in *Cyclostephanos andinus* from a core in Lake Titicaca on the border of Bolivia and Peru (Spanbauer et al., 2018). Three intervals of stasis were identified in the *C. andinus* time series, running from 312 to 220 Ka (lineage segment 1), from 220 to 70 Ka (lineage segment 2), and from 70 Ka to virtually the present (lineage segment 3). These were separated by punctuation events at 220 Ka and 70 Ka (Figure 9.14a).

A total of 266 samples yielding *C. andinus* were analyzed, averaging nearly one sample per thousand years of core. Trisha Spanbauer provided original measurements for valve diameters. Spanbauer et al. (2018) used a well-established age

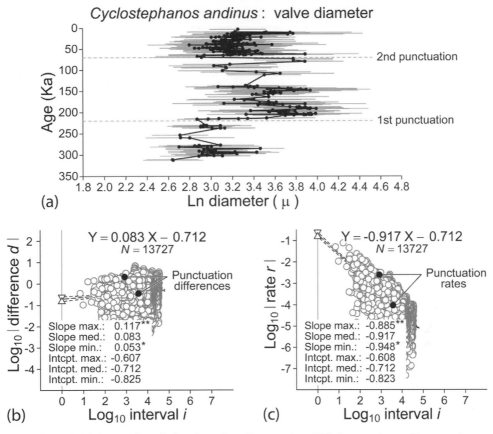

(a)

(b)

(c)

Figure 9.14 Rates of evolution for valve diameter in a 300-thousand-year lineage of the Pleistocene to Recent diatom *Cyclostephanos andinus* in Lake Titicaca (Spanbauer et al., 2018). (a) Time-calibrated stratophenetic graph of valve diameter for *C. andinus*. Independent variable time is on the vertical or *y* axis. Horizontal lines span ±1 standard deviation. (b) Log-difference-interval LDI plot shows that differences between samples within lineage segments and the two "punctuation" differences between segments (filled circles) scale with a slope of 0.083. This is close to the slope expected for a stationary time series (0.000) but it is significantly different from 0.000 and from expectation for random (0.500) or directional change (1.000). (c) Log-rate-interval LRI plot shows that rates of change between samples within lineage segments and the two punctuation rates between segments (filled circles) scale with a slope of −0.917. This is close to the slope expected for a stationary time series (−1.000) but significantly different from −1.000 and from expectation for random (−0.500) or directional change (0.000). Both analyses indicate that the time series is stationary with a random component.

model for the core based on radiocarbon ages, uranium-series dating of aragonite laminae, and tuning to the Vostok CO_2 record. The life history of *Cyclostephanos*, like that of *Stephanodiscus*, is complicated, and here again four years is used as an estimate of generation time.

The first analysis of *C. andinus* involved log-rate-interval LRI scaling of rates in the three lineage segments of Figure 9.14a. These yielded slopes of -0.924, -0.831, and -0.962, respectively. Only the first, lineage segment 1, had a confidence interval wide enough to include -1.000 expected for stasis. Lineage segments 2 and 3 had narrower confidence intervals that excluded stasis. All had confidence intervals too narrow to include random (-0.500) and directional change (0.000). Thus lineage segment 1 is reasonably interpreted as a segment in stasis. Lineage segments 2 and 3 exhibit a combination of stasis and random change.

In a second analysis all differences for lineage segment 1, the first punctuation, lineage segment 2, the second punctuation, and segment 3 were combined on an LDI plot (Figure 9.14b) and all corresponding rates were combined on an LRI plot (Figure 9.14c). The slope for the LDI plot is close to but significantly different from 0.000 expected for stasis. Similarly, the slope for the LRI plot is close to but significantly different from -1.000 expected for stasis. Both differ significantly from expectation for random or directional change. The full *C. andinus* time series in Figure 9.14a is interpreted as a lineage combining stasis and random change.

Note that the differences associated with the first and second punctuations (solid circles in Figure 9.14b) are embedded within the distribution of differences for the lineage segments themselves. Similarly, the rates associated with the two punctuations (solid circles in Figure 9.14c) are embedded within the distribution of rates for the lineage segments themselves. There is seemingly no difference between the punctuation events and background change in the lineage as a whole.

There are a total of 265 base rates in the *C. andinus* lineage. These range from $h_{2.671} = 2.09 \times 10^{-6}$ to $h_{1.208} = 768 \times 10^{-2}$ standard deviations per generation (median 2.21×10^{-3}), on timescales of 7 to about 6,400 generations.

9.3 Discussion

The 34 paleontological studies of fossils reviewed in this chapter yielded no step rates on a timescale of one generation. The fossil record is different in this respect from laboratory selection experiments reviewed in Chapter 7 and from field studies reviewed in Chapter 8. To learn about rates of change on the generation-to-generation timescale of evolution by natural selection we must look to laboratory experiments and field studies: The fossil record tells us nothing about these. Paleontological studies in this chapter do provide 3,251 independent base rates of change on timescales longer than a generation, and 44,510 net rates compounded from the base rates.

9.3.1 Rates of Evolution in Fossil Studies

Rates of evolution in field and fossil studies are summarized in Table 9.1 and compared in Figure 9.15. Field rates compiled in Chapter 8 include 814 step rates on a timescale of one generation, and 3,627 independent base rates ontimescales longer than a generation. These are compared in two ways in Figure 9.15: (1) base and step rates are compared as histograms on a log rate axis (Figure 9.15a) and (2) base and step rates are compared as bivariate scatters on an LRI log-rate-interval plot, where rates are matched to their associated intervals (Figure 9.15c). Step rates are the shaded rates in Figures 9.15a and 9.15c.

Fossil rates compiled here in Chapter 9 include no step rates but do include 3,251 independent base rates. For ease of comparison these are plotted like the field rates in Figure 9.15: (1) as a simple histogram on a log-rate axis (Figure 9.15b) and (2) as a bivariate scatter on an LRI log-rate-interval plot, where again rates are matched to their associated intervals (Figure 9.15d).

Field-derived base rates and step rates together (Figure 9.15a) span a range from near zero to 5.412 standard deviations per generation (-4.476 to $+0.733$ on a $\log_{10}$ scale), with a joint median rate of 0.006 standard deviations per generation (-2.218 on a $\log_{10}$ scale). In contrast, fossil-derived base rates (Figure 9.15b) span a range from near zero to 0.097 standard deviations per generation (-7.950 to -1.012 on a $\log_{10}$ scale), with a median base rate of $h_{3.814} = 0.00007$ standard deviations per generation (-4.164 on a $\log_{10}$ scale).

A median rate of 0.00007 standard deviations per generation for the base rates in fossil studies is almost two orders of magnitude less than the median of 0.004 standard deviations per generation for base rates in field studies. And a median of 0.00007 is miniscule compared to the median of 0.15 standard deviations per generation for step rates in field studies. Any direct comparison of fossil base rates with field base rates or of field base rates with field step rates is bound to be misleading.

To understand this it is necessary to compare field base rates with field step rates in the context of their corresponding interval-length denominators. Field-derived base rates and step rates together have denominators spanning a range from 1 to 580 generations (abscissa in Figure 9.15c; ranging from 0 to 2.763 on a $\log_{10}$ scale). The rates themselves span a range, as we saw in Figure 9.15a, from near zero to 5.412 standard deviations per generation (ordinate in Figure 9.15c; ranging from -4.476 to $+0.733$ on a $\log_{10}$ scale). However, the LRI plot for field rates in Figure 9.15c shows clearly that the rates and intervals covary and are not independent: Rates decrease systematically as interval lengths increase, and rates increase systematically as interval lengths decrease. The rates are dependent to some extent on their denominators. We can estimate the strength and effect of this covariation

Table 9.1 *Summary statistics for intervals, differences, step rates and base rates in the field and fossil studies analyzed in Chapters 8 and 9. Differences and rates here are differences and rates for unsigned absolute values.*

	Observed values			Log_{10} values		
	Minimum	Maximum	Median	Minimum	Maximum	Median
Field studies						
Single-generation step rates ($N = 814$)						
Intervals (gen.)	1	1	1	0.0000	0.0000	0.0000
Differences (s.d.)	0.0007	5.4124	0.1534	−3.1568	0.7334	−0.8142
Rates (s.d./gen.)	0.0007	5.4124	0.1534	−3.1568	0.7334	−0.8142
Multi-generation base rates ($N = 3627$)						
Intervals (gen.)	1	580	111	0.0000	2.7634	2.0453
Differences (s.d.)	0.0004	5.9784	0.4188	−3.4346	0.7766	−0.3780
Rates (s.d./gen.)	0.0000	1.0884	0.0044	−4.4764	0.0368	−2.3529
Step rates plus base rates ($N = 4441$)						
Intervals (gen.)	1	580	111	0.0000	2.7634	2.0453
Differences (s.d.)	0.0004	5.9784	0.3642	−3.4346	0.7766	−0.4386
Rates (s.d./gen.)	0.0000	5.4124	0.0061	−4.4764	0.7334	−2.2179
Fossil studies						
Multi-generation base rates ($N = 3251$)						
Intervals (gen.)	7	9355693	6519	0.8487	6.9711	3.8141
Differences (s.d.)	0.0010	39.2197	0.5448	−3.0086	1.5935	−0.2638
Rates (s.d./gen.)	0.0000	0.0972	0.0001	−7.9495	−1.0122	−4.1641

Intervals are in generations (*gen.*); differences are in standard deviations (*s.d.*); and rates are in standard deviations per generation (*s.d./gen.*).

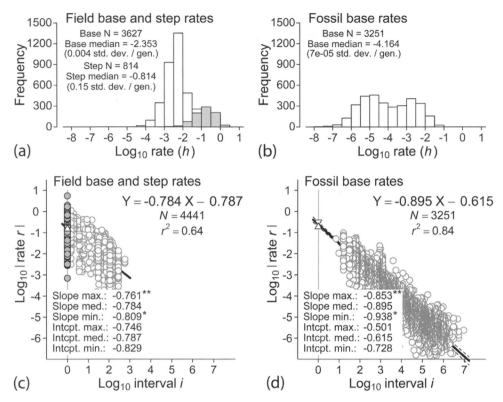

Figure 9.15 Comparison of 3,627 base and 814 step rates from field studies with 3,251 base rates from fossil studies. Fossil studies yielded no generation-to-generation step rates. (a) Histogram of base rates (open bars) and step rates (shaded bars) in field studies. (b) Histogram of base rates (open bars) in fossil studies. (c) Log-rate-interval plot comparing rates from field studies to their interval-length denominators. Shaded circles here correspond to the shaded histogram for step rates in panel a; open circles correspond to open bars in panel a. (d) Log-rate-interval plot comparing rates from fossil studies to their interval-length denominators. Open circles correspond to open bars in panel b. Note the similar temporal-scaling slopes and intercepts for field and fossil rates in panels c and d.

by calculating a coefficient of determination, r^2, for a simple linear regression of rate on interval. The resulting $r^2 = 0.64$ for field rates indicates that 64% of the variation in rates is explained by the variation in interval lengths.

We see the same effect even more clearly in the fossil rates. Fossil-derived base rates span an interval range from 7 to nearly 10 million generations (abscissa in Figure 9.15d; ranging from 0.849 to 6.971 on a $\log_{10}$ scale). The rates themselves span a range, as we saw in Figure 9.15b, from near zero to 0.097 standard deviations per generation (ordinate in Figure 9.15d; ranging from -7.950 to

-1.012 on a $\log_{10}$ scale). Here again, the LRI plot for fossil rates in Figure 9.15d shows clearly that the rates and intervals covary and are not independent: Again rates decrease systematically as interval lengths increase, and the rates increase systematically as interval lengths decrease. The rates are dependent on their denominators, and we can again estimate the strength and effect of this covariation by calculating a coefficient of determination, r^2, for a simple linear regression of rate on interval. The resulting $r^2 = 0.84$ indicates that 84% of the variation in fossil base rates is explained by the variation in interval lengths.

Field and fossil rates are complementary, and it is important to recognize the similarities in their LRI distributions. The slopes for the LRI distributions in Figure 9.15c–d are similar, at -0.784 and -0.895, respectively. Each slope is significantly different from those expected for directional change (0.000), random change (-0.500), and stasis (-1.000). Both of the observed slopes lie between the values expected for random change and stasis. Within this range, the slope observed for field rates is closer to expectation for random change, and the slope observed for fossil rates is closer to expectation for stasis. This complements r^2 values in illustrating that the fossil rates are more dependent on interval length.

The intercepts for the LRI distributions in Figure 9.15c–d are also similar, at -0.787 and -0.615, respectively. Both are close to the median value of -0.814 for the field-study step rates. If we were to construct histograms of residual values for the regressions shown in Figure 9.15c–d, these histograms would be larger but very similar in location and dispersion to the shaded histogram of step rates in Figure 9.15a. All are telling us basically the same thing about evolution on a generation-to-generation timescale.

The step rates we observe in field studies, with a median of -0.814 standard deviations per generation on a timescale of one generation, are representative of change on the generation-to-generation timescale of the evolutionary process. Base rates from field studies that average 0.004 standard deviations per generation are averaged over timescales ranging from 1 to 580 generations (ranging from 0 to 2.763 on a $\log_{10}$ scale). Base rates from fossil studies that average 7×10^{-5} standard deviations per generation are averaged over timescales ranging from seven generations to nearly 10 million generations (ranging from 0.849 to 6.971 on a $\log_{10}$ scale). Base rates from field and fossil studies averaged over any range of timescales are worth little in representing the evolutionary process.

In the absence of any direct evidence of change on a generation-to-generation timescale, we could use field or fossil base rates, in aggregate, to extrapolate and predict rates on a per-generation timescale, projecting these to the intercept of an LRI plot. Fortunately the extrapolation is not necessary because we have a good sample of step rates from field studies. These are the rates that represent the step-by-step, generation-by-generation process of evolution, and these are the highest of the rates we observe. Empirical evidence shows that evolution by natural selection is fast.

What constitutes rapid change in a fossil study? This is again a comparative question of how a given rate relates to expectation. And it is an empirical question because the rate expected depends on what has been observed before. The LRI distribution of haldane rates and intervals in Figure 9.15d represents the set of rates observed to date, indexed by the corresponding interval length. A rate for a given interval length that is higher than the line fit to the points in Figure 9.15d is a rate that is higher than expectation. Similarly, a rate for a given interval length that is lower than the line fit to the points is a rate that is lower than expectation.

9.3.2 Explanation for Inverse Scaling of Rates and Intervals

In the previous section we estimated the strength of the dependence of evolutionary rates on their interval lengths or rate denominators by calculating a coefficient of determination, r^2, for a simple linear regression of log rate on log interval. The resulting $r^2 = 0.84$ indicated that 84% of the variation in fossil base rates is explained by the variation in interval lengths. Here we look at this in a different way to explain the inverse scaling.

The shaded symbols in Figures 9.15a and 9.15c that represent step rates cover rates that range from a minimum of -3.157 to a maximum of 0.733 on a $\log_{10}$ scale. Thus the total range of field step rates spans 3.890 orders of magnitude. This is the range of step rates we can expect in nature independent of any effect of interval length. In contrast, the minimum interval length in Figures 9.15c and 9.15d is 0 on a $\log_{10}$ scale. The maximum interval length in Figures 9.15c and 9.15d is 6.971 on a $\log_{10}$ scale. Thus the total range in interval lengths spans 6.971 orders of magnitude. Rounding, the range of interval lengths is three orders of magnitude greater than the range of step rates. The former, step rates, which reflect the limits of morphological change, are much more constrained than the latter, interval lengths, which reflect the length of geological time. And this is not the "fullness" of geological time, which would be greater, but the length of geological time over which the rates here are calculated. When rate numerators are constrained to span a range three orders of magnitude smaller than the range of their denominators, inverse scaling of long-term rates is inevitable.

Another way to think about this was introduced by Gingerich (1983) and illustrated by Gingerich (2001). One of the smallest living mammals is the least shrew weighing about 3 or $e^{1.1}$g, and the largest living mammal is the blue whale weighing about 100 metric tonnes or $e^{18.4}$g. The standard deviation of body weight in mammals is about 0.15 units on a natural logarithmic scale. Hence the largest and smallest mammals living today differ by approximately 100 standard deviations, which is 10^3 0.1-standard-deviation units. These are physiological limits and mammals have never been much smaller or much larger.

If we assume that the generation time for an average living mammal is on the order of one year, and the Cenozoic history of the modern orders of mammals, shrews and whales included, goes back 55–65 million years, this is an interval that is conservatively about 10 million or 10^7 generations. In this example the evolutionary history of the group exceeds the variation in morphology by four orders of magnitude. If we were modeling the history on a game board, the board would have to be *four orders of magnitude* longer in time than its width in morphology, and in a forward-modeling exercise the constraint on morphology would soon lead to inverse scaling of long-term rates and interval lengths.

9.3.3 Proportion of Zero Rates in Fossil Studies

Distributions of rates, like those plotted on a $\log_{10}$ scale in Figure 9.15d, have an upper limit dictated by the standard-deviation width of a normal curve, and rates greater than 2.0 standard deviations per generation (0.30 on a $\log_{10}$ scale) are extremely rare. None of the 3,251 base rates in Figure 9.15d is greater than 2.0.

The numbers of rates expected, observed, and zero for each of the 34 fossil studies analyzed here are listed in Table 9.2. One study has a high number missing (48). This is the study with the highest proportion missing, but the proportion missing is only 6.7%. Overall, the number of missing rates is 0.52% of the total expected. Here again, addition of a small number of zero rates has a negligible effect on our estimation of rates that are representative in nature.

9.3.4 Three-Fold Classifications of Evolutionary Rates

George Gaylord Simpson (1944; 1953) proposed a three-fold classification of change and rates of change based on study of the fossil record. The core of the classification was what Simpson considered normal or standard change or *horotely*, and standard rates or *horotelic* rates of change (Simpson, 1944, p. 133; from Greek *horos*, "standard," and *telos*, "consummation"). Simpson called change and rates of change that are slow in comparison to standard *bradytely* or *bradytelic*. Change and rates of change that are fast in comparison to the standard Simpson called *tachytely* or *tachytelic* (Van Valen, 1974, called these "epistandard"). The words horotely, bradytely, and tachytely thus have no absolute meaning but represent (1) whatever is standard change, (2) change less rapid than standard, and (3) change more rapid than standard.

Simpson's three-fold division of rates was originally based on taxonomic longevity and on rates of taxonomic turnover, but horotely, bradytely, and tachytely have subsequently been generalized to characterize evolution more broadly. One problem with use of such relative terms can be illustrated by the step rates shaded in Figure 9.15a. Step rates are faster than standard if standard refers to field rates as a whole, and thus step rates can reasonably be called tachytelic. However, step rates are themselves the standard for rates on a timescale of one generation: step rates in this sense are horotelic, and everything else is bradytelic.

A second three-fold classification of change and rates of change was proposed by Kurtén (1959; 1963), who distinguished what he called "A" rates, "B" rates, and "C" rates. Kurtén's "A" rates were originally defined in terms of Haldane's darwin units and considered to span a range from about 3.7 to 43 darwins. When the same rates are recalculated in haldanes, the range is about 0.0003 to 0.005 haldanes (-2.2 to -3.4 on a $\log_{10}$ scale). These "A" rates span the mode on the left side of the open histogram in Figure 9.15a and the mode on the right side of the open histogram in Figure 9.15b. Kurtén's "B" rates, recalculated, span a range from about 0.000001 to 0.0002 haldanes (5.9–3.8 on a $\log_{10}$ scale). These "B" rates span much of the mode on the left side of the histogram in Figure 9.15b. Kurtén envisioned "C" rates as being even slower.

Kurtén considered "A" rates to correspond closely to the tachytelic rates of Simpson (1944). "A-rates . . . are very fast, they have short temporal duration, and they often though not always lead to the rise of new taxa." (Kurtén 1959, p. 213). Kurtén considered his "B" rates to conform well to Simpson's definition of horotelic rates but also included "C" rates in the horotelic range: "B" rates being standard for the Quaternary, and "C" rates standard for the Tertiary. The Quaternary of Kurtén included what is here called Pleistocene and Holocene, while the Tertiary of Kurtén included all of the Paleogene and Neogene.

Now we come to the same problem we had before. If "A" rates range from 0.0003 to 0.005 haldanes (-2.2 to -3.4 on a $\log_{10}$ scale) and "A" rates are tachytelic, then what are the shaded step rates in Figure 9.15a? These have a modal value of 0.15 haldanes (-0.814 on a $\log_{10}$ scale) and are hence faster than tachytelic. Quantification of rates avoids these semantic problems.

9.3.5 *"Cope's Rule" and Paired Differences in Fossil Studies*

The rates compiled here have a bearing on a phenomenon (or illusion) widely discussed in paleontology: "Cope's Law" or "Cope's Rule" stating that animals tend to evolve toward larger size. Edward Drinker Cope, to the extent that he is the author of Cope's Law, included this in a discussion of his "law of the unspecialized." Cope emphasized evolution *from* small size rather than *toward* large size

Table 9.2 *Number of rates expected and observed for time series in the fossil studies analyzed here.*

Study	Author	Year	Text section	Taxon	Trait	Expected	Observed	Zero	Zero %
1	Sewertzoff	1931	9.2.01	Equidae	Shoulder hgt.	1	1	0	0.0
2	Simpson	1944	9.2.02	Equidae	Molar size and shape	24	24	0	0.0
3	Colbert	1948	9.2.03	Dinosauria	Body length	8	8	0	0.0
4	Haldane	1949	9.2.04	*Homo* spp.	Opisthion index	1	1	0	0.0
5	Bader	1955	9.2.05	Merycoidodonts	Various	191	191	0	0.0
6	Kurtén	1955	9.2.06	Ursidae	M^1 length	15	15	0	0.0
7	Kurtén	1959	9.2.07	Ursidae	M_2 length	55	55	0	0.0
8	Kurtén	1959	9.2.08	Postglacial mammals	Various	33	33	0	0.0
9	Ziegler	1966	9.2.09	*Eocoelia* spp.	Various	234	234	0	0.0
10	Maglio	1973	9.2.10	Elephantidae	Molar traits	270	267	3	1.1
11	Kellogg	1975	9.2.11	*Pseudocubus vema*	Thorax wid.	561	561	0	0.0
12	Gingerich	1976	9.2.12	*Hyopsodus* spp.	Tooth size	318	318	0	0.0
13	Malmgren and Kennett	1981	9.2.13	*Globorotalia* spp.	Various	4,423	4,399	24	0.5
14	McDonald	1981	9.2.14	*Bison* spp.	Various	339	339	0	0.0
15	Raup and Crick	1981	9.2.15	*Kosmoceras* spp.	Shell diameter	1,989	1,984	5	0.3
16	Malmgren et al.	1983	9.2.16	*Globorotalia* spp.	Various	3,770	3,723	47	1.2
17	Hayami	1984	9.2.17	*Cryptopecten versiculosus*	Various	90	90	0	0.0
18	Bell	1985	9.2.18	*Gasterosteus doryssus*	Various	1,236	1,219	17	1.4
19	Stanley and Yang	1987	9.2.19	Bivalve Mollusca	Size	93	93	0	0.0

20	MacFadden	1988	9.2.20	Equidae	Various	145	143	2	1.4
21	Lister	1989	9.2.21	*Cervus elaphus*	Various	10	10	0	0.0
22	Forstén	1990	9.2.22	*Equus germanicus*	Tooth traits	714	666	48	6.7
23	Lich	1990	9.2.23	*Cosomys primus*	Tooth size	135	129	6	4.4
24	Clyde and Gingerich	1994	9.2.24	*Cantius* spp.	Tooth size and shape	3,997	3,962	35	0.9
25	Chiba	1996	9.2.25	*Mandarina chichijimana*	Shell size	1,050	1,036	14	1.3
26	Gingerich	2003	9.2.26	*Ectocion* spp.	Tooth size	1,590	1,589	1	0.1
27	Bell et al.	2006	9.2.27	*Gasterosteus doryssus*	Various	5,041	5,020	21	0.4
28	Gingerich and Smith	2006	9.2.28	*Haplomylus* spp.	Tooth size	652	650	2	0.3
29	Hunt and Roy	2006	9.2.29	*Poseidonamicus* spp.	Valve length	382	376	6	1.6
30	Theriot et al.	2006	9.2.30	*Stephanodiscus yellowstonensis*	Various	4,544	4,535	9	0.2
31	Cheetham et al.	2007	9.2.31	*Metrarabdotos* spp.	Mahalanobis D	729	729	0	0.0
32	Secord et al.	2012	9.2.32	*Hyracotherium* spp.	Tooth size	23	22	1	4.3
33	Kimura et al.	2015	9.2.33	*Murinae*	Tooth size and shape	1,715	1,705	10	0.6
34	Spanbauer et al.	2018	9.2.34	*Cyclostephanos andinus*	Valve diameter	13,727	13,727	0	0.0
	Totals					48,105	47,854	251	0.52

and, ironically, credited this to an observation by his fellow American and some-times bitter rival Othniel Charles Marsh (Cope, 1896, pp. 172–174).

Cope's Law was first called a "law" by Charles Depéret in 1907. Depéret (1907, p. 87) wrote of a "Loi féconde, dont la formule précise parait bien appartenir à Cope" [fertile law, whose formulation seems attributable to Cope]. Depéret, like Cope, emphasized that major groups of mammals began as "types de petit taille et de chétive puissance" [species of small size and little power].

Cope's Law was reborn in a review article by Bernard Rensch (1943), as a rule rather than a law, who referred to the "Gaudry–Copeshe Regel der Größenzunahme in den Stammesreihen" [Gaudry-Cope Rule of size increase in phyletic lineages]. When Rensch discussed the rule again, in the 1947 first edition of his *Neuere Probleme der Abstammungslehre*, it was simply the "Copesche Regel" or "Cope's Rule." Rensch cited both Cope (1896) and Gaudry (1896) in the 1947 bibliography but gave no explanation for dropping Gaudry's name from the rule.

Rensch (1947, p. 198) wrote of "die im ganzen Tierreich am weitesten verbrei-tete Form der Orthogenese ... die zukzessive Größensteigerung in den Stammes-reihen, wie sie als Copesche Regel geläufig ist" [the most widespread form of orthogenesis in the whole animal kingdom: gradual increase in the size of phyletic lines familiar as Cope's Rule]. Rensch identified Cope's Rule with orthogenesis, the idea that evolution is irreversible, and in this connection altered the rule to evolution toward larger size. A rule is usually viewed as something softer than a law, but Rensch's reformulation was in fact more rigid.

Norman Newell (1949) broadened the scope of Cope's Rule to include the fossil record of evolution in diverse groups of invertebrates. He attributed increasing size to a perceived statistical advantage of superior bulk for individuals within and between species. Newell then concluded that "increase in the mean size of individ-uals during evolution is a widespread phenomenon, affecting many ... groups throughout the animal kingdom ... it may well be the most important of all invertebrate trends" (Newell, 1949, p. 122).

Steven Stanley tempered enthusiasm for Cope's Rule, noting that "reliance solely on inherent advantages of larger size to explain the high incidence of size increase suffers from deficiencies resembling some of those attached to the dis-carded concept of orthogenesis" (Stanley, 1973, p. 1). In summarizing, Stanley (1973, p. 22) emphasized three important points: (1) "Cope's Rule ... cannot be explained by intrinsic advantages of larger size"; (2) "skewing [of size distribu-tions] occurs very rapidly because possible increments of size change ... are a direct function of body size, so that early spreading of the range proceeds more rapidly in a positive direction than in a negative direction"; and (3) returning to Cope's and Depéret's original formulation – "Cope's Rule is more fruitfully viewed as describing evolution from small size rather than toward large size."

Cope's Rule is rarely tested by comparing the relative frequencies of change toward larger or smaller size. However we regard the rule, we can ask whether there is a predominance of change from smaller to larger size, a predominance of change from larger to smaller size, or simply a statistical tie with change in both directions? The differences and corresponding rates compiled in this chapter are signed, positive for an increase in size and negative for a decrease, and the signs can be used to test for predominance.

A comparison of positive and negative rates is shown in Table 9.3. Positive rates that increase from smaller to larger size support Cope's Rule. Negative rates that decrease from larger to smaller size contradict Cope's Rule. There are no single-generation step rates. The only rates that matter for the test are the independent base rates, where 1,313 are positive and 1,365 are negative. This difference in positive and negative rates is not significant, and there is no evidence in this test-of-relative-frequencies to support Cope's Rule. Jablonski (1997) came to a similar conclusion after a study of Cretaceous mollusks. Cope's Rule seems to be an illusion.

The net rates in Table 9.3 seem to have a significant excess of negatives over positives (17,668 positives, 21,247 negatives), but these, by virtue of their compounding, are not statistically independent. Thus the test supporting significance is inappropriate in this case. The same caveat applies to the totals for all rates considered together.

Why is the assumption of evolution toward larger size so pervasive? This almost certainly reflects a failure to reason proportionally. Size on an arithmetic scale, however measured, yields a number greater than zero, and nothing can be smaller than zero. Thus a shrew with a 10 g body weight or a clam with a 10 cm shell diameter cannot lose more than 10 of these size units. Each, in theory, can gain hundreds or thousands of units, yielding the skewing mentioned by Stanley (1973) that was cited above. If we reason proportionally then 10 g and 10 cm are observations on geometric scales — scales of ... 0.1, 1, 10, 100, 1000 ...; or ... 2.5, 5, 10, 20, 40 ... — scales with no beginning and no end. There are physiological limits to be sure, but a proportional geometric scale no longer has the asymmetry that favors the illusion of evolution toward larger size.

Acceptance of Cope's Rule depends largely on failure to log-transform axes when graphing distributions of size. Stephen Jay Gould (1988) explained evolutionary trends in size, like those embodied in Cope's Rule, as an artifact of increasing variance; but changing variance by itself will not create a false trend. False trends result from failure to reason proportionally, introducing asymmetry when variance is increasing. There is not a single logarithmic axis in Gould's (1988) examples of trends explained by a change in variance. Cope's Rule is an asymmetry enhanced if not created by a collective failure to reason proportionally.

Table 9.3 *Positive and negative rates in fossil studies as a test of Cope's Rule that evolution proceeds from smaller to larger size*

Rates from fossil studies	Positive: increasing size	Negative: decreasing size	Difference: positive or negative	Total rates	Minimum as a test value	Cumulative binomial probability	Conclusion
Single-generation step rates	—	—	—	—	—	—	—
Multigeneration base rates	**1,313**	**1,365**	**−52**	**2,678**	**1,313**	**0.153**	**Not significant**
Compound net rates	17,668	21,247	−3,579	38,916	17,668	0.000	Not independent
All rates together	18,981	22,612	−3,631	41,594	18,981	0.000	Not independent

Test of significance is based on the cumulative density (probability) in a symmetrical binomial distribution with the number of trials equal to the test value.

9.4 Summary

1. Analysis of 34 paleontological studies in the fossil record yields no step rates quantifying change from one generation to the next.
2. The 34 paleontological studies compiled here yield 3,251 independent base rates of change on timescales longer than a generation, and 44,510 net rates compounded from these. The median base rate is $h_{3.814} = 0.00007$ standard deviation per generation (-4.164 on a $\log_{10}$ scale).
3. Paleontological studies that yield no step rates cannot inform our understanding of generation-to-generation change by natural selection; however, temporal scaling of base rates found in the fossil studies yields an LRI intercept and residuals consistent with the step rates found in selection experiments and field studies.
4. A median rate of 0.00007 standard deviations per generation for base rates in fossil studies is almost two orders of magnitude less than the median of 0.004 standard deviations per generation for base rates in field studies, and both are miniscule compared to the median of 0.15 standard deviations per generation for step rates in field studies. Some 84% of the variance in evolutionary rates is determined by variance in interval lengths (rate denominators), meaning rates in the fossil record cannot be compared to rates in field studies without temporal scaling.
5. The log-rate-interval LRI slope for base rates in fossil studies lies between expectation for a stationary time series (-1.000) and expectation for a random time series (-0.500). It is significantly different from both but closer to the slope expected for stasis. Stasis predominates in the fossil record but change is evident too.
6. Rapid change in a fossil study is change lying above a line fit to an empirical LRI distribution of base rates and their corresponding intervals: change lying above $Y = -0.895 \cdot X - 0.615$, where Y is $\log_{10}$ of the rate r and X is $\log_{10}$ of the corresponding interval i (Figure 9.15d).
7. When rate numerators are constrained to a range three orders of magnitude smaller than the range of their denominators, inverse scaling is inevitable.
8. Empirically, the proportion of zero rates in fossil studies is 0.52%, and zero rates have a negligible effect on the rate statistics presented here.

10

A Quantitative Synthesis

> Let us now see whether the several facts and rules relating to the
> geological succession of organic beings, better accord with the common
> view of the immutability of species, or with that of their slow and
> gradual modification, through descent and natural selection.
>
> *Charles Darwin, 1859,* Origin of Species, *p. 312*

In the introductory chapter we asked how we can reconcile the Lamarck–Darwin thesis of slow and gradual, little-by-little, step-by-step evolutionary change through time with the Lyell–Linnaeus antithesis of no-change "equilibrium" supported by paleontologists for more than a century. The first step toward reconciliation comes in identifying what the thesis and antithesis share so we can focus on what distinguishes them. The second step considers the distinguishing characteristics to see whether and how they can be reconciled.

10.1 The Lamarck–Darwin Thesis and the Lyell–Linnaeus Antithesis

We can take Charles Darwin's (1859) conjectural "accompanying diagram" of species changing through evolutionary time as a graphic representation of the Lamarck–Darwin thesis. This is the only illustration that Darwin included in any of his six editions of the *Origin of Species*, and it appeared in each edition.

Similarly, we can take Niles Eldredge and Stephen Jay Gould's (1972) conjectural summary in their figure 5–10 as a graphic representation of the Linnaeus–Lamarck antithesis of species-level evolution through time.

The two graphics, one by Darwin and the other by Eldredge and Gould, were quantified and compared in terms of their relative rates in Figure 1.3 in the Introduction.

10.1.1 What the Thesis and Antithesis Share

In Figure 1.3 we saw that the feature shared on Darwin's diagram and on Eldredge and Gould's diagram is a dominant central mode of stasis representing negligible

270

change through time. Both diagrams have a preponderance of lineages exhibiting negligible deviation from vertical when the independent variable, time, is plotted on the *y*-axis and morphology is plotted on the *x*-axis. Stasis is what the Lamarck–Darwin thesis and the Linnaean–Lyellian antithesis share.

Stasis underpins everything in evolution, and this has been true since the Modern Synthesis of the 1940s and 1950s when long-term unidirectional orthogenesis lost its last supporters. Disagreements about stasis remain, but the debate is now about the causes of stasis. Does stasis represent some form of inherent homeostasis within species, however that might be explained, or does stasis reflect the dominance of stabilizing selection in assemblages of interacting species?

10.1.2 Differences in the Thesis and Antithesis

In Figure 1.3 we saw too that the principle feature distinguishing Darwin's diagram from that of Eldredge and Gould is the distribution of rates of change about the central mode. Rates in Darwin's illustration are more or less normally distributed about the central stasis mode, while rates in Eldredge and Gould's diagram are distinctly trichotomous, with positive and negative punctuation modes removed from, and distinctly different from, the stasis mode. Negative and positive change are continuous with stasis in Darwin's diagram, and change takes place at broadly variable rates. Change is discontinuous in Eldredge and Gould's diagram, and any variation within the flanking modes of negative and positive change spans a narrow range.

Variable rates of change forming a single central distribution are the distinguishing feature of Darwin's diagram and, by extension, the distinguishing feature of the Lamarck–Darwin thesis. Discrete "change"and "no change" are the distinguishing features of Eldredge and Gould's diagram and, by extension, the distinguishing features of the Lyell–Linnaeus antithesis.

10.1.3 A Simple Resolution

A simple resolution of differences distinguishing Darwin's diagram from that of Eldredge and Gould was shown in Figures 8.10a and 8.10b, where one empirical histogram is shown for laboratory step rates and another is shown for field step rates. Each includes a positive half and also the positive half reflected to represent the corresponding negative rates. All of the rates together cluster near and vary about a central stasis mode that is exactly our expectation for the continuous distribution of rates from Darwin's diagram in Figure 1.3. There is no hint of the discontinuous distribution of rates to be expected from the Eldredge–Gould diagram.

The empirical distribution of step rates in Figure 8.10 resembles the distribution of rates in Darwin's diagram and clearly favors the Lamarck–Darwin thesis of step-by-step evolutionary change through time. The empirical distribution of step rates in Figure 8.10 appears, on the face of it, to refute the Eldredge–Gould expectation and the Lyell–Linnaeus antithesis of dichotomous unexplained origins and no-change equilibrium. However, a full resolution may be deeper than this.

10.2 Source for a Dichotomous Distribution of Rates

The simple resolution just given is part of the story, however it does not explain how or why Eldredge and Gould and so many other supporters of Lyell–Linnaeus no-change "equilibrium" came to believe that the fossil record displays a dichotomous distribution of rates. No one practices paleontology without thinking, indeed knowing, that the fossil record is important. The fossil record is after all the empirical evidence we have, recorded in strata, that life has a history — and a long history. Further, it gives us a very detailed outline, through geological time, for large and important parts of the history. Why should we not believe that the fossil record tells us even more: how evolution works? Darwin, after all, acknowledged father of natural selection, spent his formative years as a geologist and paleontologist.

Here we have to be careful. Darwin was a geologist and paleontologist, yes, and from geology and paleontology he knew there was a long history of life to explain. But Darwin was a naturalist more broadly and his development of natural selection came from his experience as a botanist, zoologist, biogeographer, and ecologist. This is clear in the *Origin of Species* when he worried so about the fossil record, writing: "Geology assuredly does not reveal any ... finely graduated organic chain; and this, perhaps, is the most obvious and gravest objection which can be urged against my theory" (Darwin, 1859, p. 280). The "theory" part of Darwin's writing came from his broader experience as a naturalist, and it was not narrowly derived from his experience in paleontology and geology. The fossil record is not what gave Darwin insight into natural selection.

So when we study the fossil record we have to respect it but be careful about what we take from it. There is no reason to think that the fossil record should be taken literally, in all aspects on all scales of space and time. When we study rates of change in the fossil record, we find the rates to be very low. An empirical example derived from Chapter 9 is shown in the left-hand histogram of Figure 10.1a. There the median rate is the nearly infinitesimal 0.000007 standard deviations per generation for rates on timescales of 10,000 generations or more. Darwin believed evolution was slow: should we?

Very little of the fossil record is known in sufficient detail to resolve time on scales as fine as 10,000 generations, and a rate of 0.000007 standard deviations per generation on this and longer time scales will not explain much of the change we

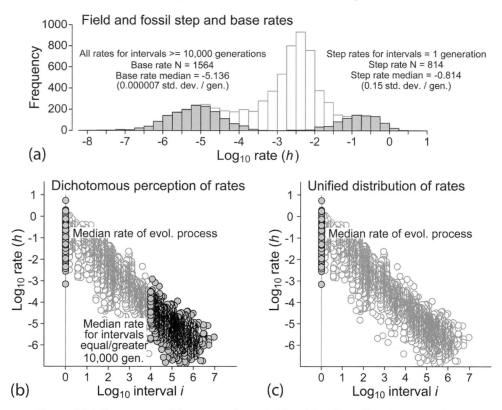

Figure 10.1 Step rates and base rates from field and fossil studies. (a) comparison of a distribution of base rates on timescales of 10,000 or more generations visible in the fossil record (shaded histogram at left) and a distribution of step rates on timescales of a single generation recorded in field studies (shaded histogram at right). Base rates on intermediate scales of time fill the gap between the shaded histograms (open histogram in background). (b) The same rates on an LRI log-rate-interval plot showing that the dichotomous distribution of rates visible in field and fossil studies, with very different medians, results from deficient sampling of the intermediate range of intervals. (c) Unified distribution of rates showing a continuous pattern of decline from step rates found in field studies to lower and lower base rates as these are averaged on longer and longer scales of time.

know has taken place on shorter timescales. So if we take the slow rates we see in the fossil record literally, and add more rapid rates not seen in the fossil record to explain what we cannot explain any other way, we naturally develop a dichotomous perception of rates. When we are impressed by the long-lived species that seem to dominate the fossil record but change little, and then postulate shorter-lived species to fill the gaps we cannot explain any other way, we reinforce our dichotomous perception of rates. Both are then based, in part, on a literal reading of what we can see and not see in the fossil record.

Georges Cuvier's "formes fixes" and "révolutions subites" reflect what he saw in the fossil record and what he could not explain any other way. Francis Galton's "stability of types" and "changes in jerks" is another of many examples in Table 1.1 reflecting a biologist's perception of a dichotomous distribution of rates.

10.3 A Unified Perception of Rates and Change

The left-hand histogram of rates visible to a paleontologist is not the only histogram in Figure 10.1a. The right-hand histogram in the figure is an equally empirical distribution of rates with a median of 0.15 standard deviations per generation, now on a timescale of a single generation, based on the step rates found in Chapter 8. Step rates are, as we saw in Chapter 9, completely invisible in the fossil record. We can think of them as the rates that explain the change that paleontologists cannot see — the change we cannot explain without such rapid rates.

We know from Chapters 8 and 9 that the left-hand histogram of Figure 10.1a, showing rates that represent change visible in the fossil record, and the right-hand histogram of rates in the same figure, invisible to a paleontologist, are connected by a continuous spectrum of empirical rates of intermediate value. These are shown by the broader central histogram visible in the background.

Rates in the two shaded histograms of Figure 10.1a are portrayed on an LRI log-rate-interval plot in Figure 10.1b, again with shaded symbols. The LRI plot further emphasizes their dichotomous distribution. The medians of the distributions of rates are different as before, at 0.000007 versus 0.15 standard deviations per generation, a difference of more than four orders of magnitude. The distributions of rates barely overlap. The distributions of rates and intervals in Figure 10.1b are dichotomous too in the scales of time they represent, and these again differ by about four orders of magnitude.

We know from Chapters 8 and 9 that the lower-right scatter of points in Figure 10.1b, showing rates that represent change visible in the fossil record, and the upper-left scatter of rates in the same figure, invisible to a paleontologist, are connected by a continuous spectrum of quantified rates of intermediate value. These are again shown by the broader central scatter visible in the background.

The background rates in Figure 10.1b that unite the shaded step rates from field studies with the shaded long-term base rates from fossil studies are no less real than the shaded rates, and they represent change that is no less real. Arbitrary separation of evolutionary rates on timescales visible and invisible to a paleontologist is a limitation rather than a distinction of importance: All rates and intervals are important for what they tell us about rates and about their dependence on interval length.

10.4 Evolution in an Environmental Context

Most lineage turnover in the history of life has occurred in pulses, nearly synchronous across diverse groups of organisms, and in predictable synchrony with changes in the physical environment.

Vrba 1985, p. 232

Explanations for rapid change and stasis in terms of rates and interval lengths are not an explanation for their common coupling in a larger pattern. Faunal turnover in the fossil record, with the appearance of new associations of taxa, led Allison Palmer (1965) to propose the concept of "biomeres" for associations seemingly coming from somewhere outside the study area. Geerat Vermeij (1977) wrote of long-term equilibria interrupted by destabilizing revolutions. Peter Williamson (1981) described long intervals of stasis in lacustrine mollusks interrupted by simultaneous speciation events affecting all lineages, with the interruptions related to lake level. Elisabeth Vrba (1985) introduced what she called a "turnover-pulse hypothesis" to explain faunal turnover in African antelopes in response to global climatic forcing of speciation. Carlton Brett and Gordon Baird (1995) found a pattern of persistent biotas, terminated by events of abrupt breakdown and restructuring. In their view, without episodic perturbations and collapse of stable ecosystems there would have been little change in the history of life.

High rates of evolution mean that taxa integrate and faunas stabilize rapidly, and both remain stable until perturbed by changes in the physical or biotic environment around them. Environmental change happens on a broad range of temporal and spatial scales. These range from extraterrestrial bolide impacts that affect the whole earth more or less instantaneously to long-term declines in individual species that affect local communities. Ironically, between perturbations, long-term patterns of integrated faunal stasis are promoted and enhanced by high rates of evolutionary change and adaptation, generation-to-generation, in the species that comprise the faunas. It seems clear that stasis in a biota reflects a dominance of stabilizing selection in assemblages of interacting species.

10.5 Quantitative Synthesis

In the nineteenth century Jean-Baptiste Lamarck and Charles Darwin proposed and defended a radical thesis of slow, gradual, *peu-à-peu*, step-by-step evolutionary change through time. Lamarck and Darwin were then confronted by contemporaries promoting an older antithetical worldview that is here associated with Charles Lyell and Carl Linnaeus. Discussion and debate between the sides continue still, strongly influenced by paleontologists' perception of unseen change followed by stasis in the fossil record.

Figures 10.1a and 10.1b show that paleontologists, in some sense, cannot be blamed for our dichotomous perception of rates and its carryover to theories of "quantum evolution" and "punctuated equilibria." Practical difficulties of fossil sampling and time resolution mean that most paleontologists necessarily study evolution on timescales of 10,000 generations or longer: long intervals for which average rates are very low. The rates are so low that they cannot explain evolutionary changes that we know must have happened on shorter timescales. Near-zero rates of evolution are justified by measurements on long timescales in the fossil record, and at the same time change is known that paleontologists are forced to imagine at higher rates. Experimental and field biologists see the higher rates, seemingly reinforcing the dichotomy.

It is only when we study rates on a full range of timescales, combining rates from field studies with rates from fossil studies, that we see the unified distribution displayed in the background of Figures 10.1a and 10.1b. The unified distribution tells us that the dichotomy so easily imagined of slow rates and fast rates is an artifact of sampling. The distribution is dichotomous because it is sampled dichotomously.

We can take the now-unified distribution in Figure 10.1b a step farther and again consider the timescale of evolution by natural selection. We saw in Chapter 6 that natural selection is a process that takes place from one generation to the next, over and over, but always from one generation to the next. We were careful to distinguish and record rates on this timescale as step rates. We found that step rates from experimental laboratory selection are similar to step rates from field studies, so the latter are really what we expect for change by natural selection. No step rates are known from fossil studies, and this tells us again that what we see in the fossil record cannot inform our understanding of the process of change by natural selection.

The step rates from field studies are shown as shaded symbols in Figure 10.1c. These by themselves span some four orders of magnitude, all on a timescale of one generation. What happens when step rates are applied not once in the span of a generation but twice in a span of two generations? The signs are free to vary by selection or by chance, and the rates themselves may vary too. The result after two generations will be the average of what happened in each of the generations individually. If the rates in two succeeding generations are the same, then a rate averaged over the two can be the same, but a rate averaged over multiple generations will soon be averaged to a lower rate. When organisms face the selection that is inescapable near or at any competitive or physiological limit, discussed in Chapter 9, further reversal and downward averaging of rates is inevitable.

The unified distribution of rates in Figure 10.1c shows this pattern of decline from the initial one-generation step rates to lower and lower base rates on longer and longer scales of time. When we are able to sample the resulting distribution continuously, there is no dichotomous pattern of "change" versus "no change" like that so often advocated by paleontologists. The rates, and the changes they represent, are all parts of one continuous distribution. People enjoy discussion and debate, and these will no doubt continue. However, in the full scale of time, the Lamarck–Darwin thesis of step-by-step evolutionary change and the Lyell–Linnaeus antithesis of "change" and "no change" (or simply "no change") are unified in a quantitative synthesis. Both are compromised to some degree. Evolution on the generational timescale of the evolutionary process is gradual but not slow, and paleontologists working on timescales of 10,000 generations or more can never see change that is fast.

Lamarck seemingly never commented on tempo, but Darwin was wrong in claiming evolution to be slow: A step rate averaging 0.15 standard deviations per generation is not slow in any sense. Evolution appears to be slow when we cannot see it in detail and rates are averaged down over thousands and millions of generations. Field studies indicate that step rates in each generation are high, so high that faunas and floras may equilibrate rapidly between environmental perturbations and then persist with little change. High rates and rapid evolutionary change are the key to a quantitative synthesis that unites and explains what we see in the past and in the present.

10.6 Summary

1. The Lamarck–Darwin thesis of slow and gradual, step-by-step evolutionary change through time and the Lyell–Linnaeus antithesis of more rapid change and no change both share a dominant central stasis mode representing negligible change through time.
2. The Lamarck–Darwin thesis and Lyell–Linnaeus antithesis differ in the distribution of rates about the central stasis mode. Slow step-by-step evolutionary change, positive and negative, leads to a single central distribution of rates. More rapid changes imperceptible on the timescales generally studied by paleontologists create what are portrayed to be distinct secondary modes, positive and negative, flanking the central stasis mode.
3. Quantification of field rates and fossil rates on all scales of time provides a unified perception of rates and change.
4. Long-term patterns of integrated faunal and floral stasis are promoted and enhanced by high rates of evolutionary change and adaptation, generation to generation, in the species that comprise the faunas and floras.

5. The Lamarck–Darwin thesis of step-by-step evolutionary change through time, and the Lyell–Linnaeus antithesis of "change" and "no change" (or simply "no change") are unified when rates are studied on all scales of time. Lamarck and Darwin were right about the gradual nature of evolutionary change.

6. Darwin was wrong in claiming evolution to be slow. Rates are slow when averaged down over thousands and millions of generations, but the step rates averaging 0.15 standard deviations per generation that animate evolutionary change are fast by any measure.

11

Retrospective on Punctuated Equilibria

> One of the crowning achievements of paleontology, and of surpassing importance in the development of evolutionary theory, has been the discovery of innumerable graded morphological series of fossils showing progressive change as we ascend the geological scale of time.
>
> *Norman D. Newell,* Evolution, *1949, p. 103*

Punctuated equilibria is the name Niles Eldredge and Stephen Jay Gould gave to their personal refutation of twentieth-century evolutionary paleontology and also, obliquely, Norman Newell. Newell, quoted in the epigraph, was professor of geology at Columbia University and curator of invertebrate paleontology at the American Museum of Natural History. Newell was the PhD advisor for Eldredge and for Gould. The two introduced punctuated equilibria in a book-chapter manifesto titled "Punctuated equilibria: an alternative to phyletic gradualism" (Eldredge and Gould, 1972).

The Eldredge and Gould study warrants rereading, as does a classic punctuated-equilibria study of the bryozoan *Metrarabdotos* by Alan Cheetham and colleagues. A review of these two studies is followed by discussion of objections Stephen Jay Gould and other paleontologists have raised in opposition to temporal scaling of evolutionary rates.

11.1 Rereading Eldredge and Gould (1972)

It is interesting to read the Eldredge and Gould (1972) manifesto again, some 45 years after it appeared. When published it frustrated some paleontologists and seemingly pleased many others. The battle joined was the long-standing conflict between Lamarck–Darwin progressives who envision species changing through time and traditionalists who believe that species are fixed in form (Table 1.1). What was frustrating about punctuated equilibria as an interpretation of the fossil record was not only prolongation of this long-standing conflict, but Eldredge and Gould's ideas on how the conflict might be resolved.

11.1.1 Denigration of Evidence and Imperfection of the Fossil Record

Eldredge and Gould began by arguing that "All observation is colored by theory and expectation" (Eldredge and Gould, 1972, p. 85). Then, "Science progresses more by the introduction of new world-views . . . than by the steady accumulation of information" (p. 86). But are measurements of a tooth or a shell and establishment of a fossil's position in a sequence of strata really colored by theory and expectation? Do we record information about fossils because they exist, like pebbles in a gravel-pit, or do we record information about fossils to test theories? A blanket claim that all observation is colored by theory and expectation is nothing but a denigration of observation, and an assault, in advance, on information that might contradict claims to come later in the manifesto. Denigration of observation may appeal to people who are not involved in making observations, but observations and information are critical for progress in science — critical for testing alternative theories about change.

Eldredge and Gould (1972, p.87) then claimed that "Charles Darwin viewed the fossil record more as an embarrassment than as an aid to his theory." Darwin never said this. What he wrote was:

The number of intermediate varieties, which have formerly existed on the earth, [must] be truly enormous. Why then is not every geological formation and every stratum full of such intermediate links? Geology assuredly does not reveal any such finely graduated organic chain; and this, perhaps, is the most obvious and gravest objection which can be urged against my theory. The explanation lies, as I believe, in the extreme imperfection of the geological record.

Darwin, 1859, Origin of Species*, p. 280*

History being what it is, should we expect the fossil record to be perfect? Darwin was writing in 1859, when epochs were still being added to the geological timescale. His *Origin* was a challenge to search for "intermediate varieties," and as Chapter 9 shows, many have been documented in the subsequent century and a half.

Eldredge and Gould (1972, p. 90) went on to write: "Under the influence of phyletic gradualism, the rarity of transitional series remains as our persistent bugbear. From the reputable claims of a Cuvier or an Agassiz to the jibes of modern cranks and fundamentalists, it has stood as the bulwark of anti-evolutionist arguments." Are we to accept the claims and experience of nineteenth-century anti-evolutionists like Cuvier and Agassiz as reputable and representative of twentieth- and twenty-first-century paleontology? As Chapter 9 shows, the fossil record known now is not the record known to Cuvier and Agassiz, nor is it the record known to Darwin. There are numerous species connected by the finest graduated steps. Generations of paleontologists made transitional series common before 1972, and the field and laboratory tools available now make it possible to document transitions in ways unthinkable to Cuvier and Agassiz.

11.1.2 Empirical Tests of Theory

Hypothetico-deductive science starts with a conjecture or hypothesis and proceeds by testing the hypothesis with evidence. If refuted by evidence, the hypothesis is falsified. If corroborated by evidence, the hypothesis lives on to be tested by new evidence, or to be evaluated in terms of relative likelihood against new hypotheses. Testing and comparison of hypotheses is always carried out in the light of evidence. Thus it is odd to read (Eldredge and Gould, 1972, p. 91): "If we doubt phyletic gradualism, we should not seek to 'disprove' it 'in the rocks.' We should bring a new picture from elsewhere and see if it provides a more adequate interpretation of fossil evidence." What is the imagined difference between "in the rocks" and "fossil evidence"? This is a distinction without a difference. Phyletic gradualism and any "new picture" must be tested by fossil evidence "in the rocks."

11.1.3 Allopatric Speciation

The "new world view" introduced by Eldredge and Gould to guide thinking about change in the fossil record was "allopatric speciation," the idea that speciation has a geographic component. "We contend that a notion developed elsewhere, the theory of allopatric speciation, supplies a more satisfactory picture for the ordering of paleontological data" (Eldedge and Gould, 1972, p. 86). But allopatric speciation is a theory to explain how genetic differences arise through geographic isolation. Genetic differences inhibit interbreeding and may lead to separation of populations within a species. Genetic separation is a precondition for morphological differentiation, but it does not explain how morphological change takes place during any subsequent differentiation. Change visible in the fossil record is morphological change and the underlying genetic differences are unknowable.

Allopatric speciation does not enhance or constrain change visible in the fossil record. Eldredge and Gould admit as much when they contradict themselves later (1972, p. 93) and acknowledge that "mechanisms of speciation can be studied directly only with experimental and field techniques applied to living organisms." How then is allopatric speciation to be applied to the fossil record?

Eldredge and Gould claimed (1972, p. 94): "As a consequence of the allopatric theory, new fossil species do not originate in the place where their ancestors lived. It is extremely improbable that we shall be able to trace the gradual splitting of a lineage merely by following a certain species up through the local rock column." However, if speciation is splitting of a lineage, then it is likely, by chance, that one half of new fossil species *did* originate where their ancestors lived. If splitting is not involved, then it is possible that an even larger proportion of species originated where their ancestors lived.

Eldredge and Gould continued (1972, pp. 94–95): "Since selection always maintains an equilibrium between populations and their local environment, the morphological features that distinguish the descendent species from its ancestor are present close after, if not actually prior to, the onset of genetic isolation." This is a nonsequitur, implying that the onset of genetic isolation between populations somehow alters the local environment and then, anticipating or through selection, gives rise to the morphological features that distinguish species. Species respond to natural selection and environmental conditions in the broadest sense, and species change in response to environmental change — or they do not. It does not necessarily follow that "Most evolutionary changes in morphology occur in a short period of time relative to the total duration of species," and it would be more accurate to say that most evolutionary changes in morphology occur when the environment of a species changes. If environmental changes are episodic, then speciation and evolutionary changes in morphology may be episodic as well, but if environmental changes are persistent, then changes in morphology should be persistent as well.

To a skeptic, invocation of allopatric speciation in this way seems a transparent attempt to justify a conclusion that does not stand on its own evidence. Several case studies of local stratigraphic sections in Chapter 9 show a descendant species separated from its ancestor by a sharp morphological break, but others show more continuous gradual change. Some breaks in the fossil record express evolution in response to rapid environmental change, but many simply reflect discontinuities in the accumulation of sedimentary strata and their enclosed fossils.

11.1.4 *Eldredge and Gould's Perception of Phyletic Gradualism*

Eldredge and Gould (1972, p. 97) adopted an unorthodox caricature of phyletic gradualism as an alternative to their theory of punctuated equilibria. They assumed that phyletic gradualism is unidirectional orthoselection (their word) "involving constant adjustment to unidirectional change in the physical environment." This is not what "gradual" means. Natural selection is involved, but there is nothing unidirectional or "ortho" about it. Gradual change is change by gradation, step-by-step, on a generation-to-generation timescale or tracking some coarser succession of base intervals. Gradual change is "constant" in the sense that it happens every generation, but it is not constant in any other sense.

Phyletic gradualism is the name given to patterns like those in Figures 9.2–9.6 and 9.8–9.14. In most instances successive samples overlap each other in time and are statistically indistinguishable from each other in form. Many, perhaps most, gradual patterns involve some change, but others like *Ectocion ralstoni* and *E. osbornianus* in Figure 9.9 and the species of *Metrarabdotos* in Figure 9.13 are stationary in a statistical sense.

There is no reason to assume, as Eldredge and Gould do, that gradual change under selection precludes a genetic, ecological, and geographic process of speciation. Phyletic gradualism is like punctuated equilibria in being a theory about how change takes place before, during, and after speciation, not a theory about the proliferation of species. Both are equally compatible with allopatric speciation.

11.1.5 Examples of Punctuated Equilibria: Poecilozonites

One way to absorb a newly proposed theory is through examples given to illustrate the new idea. Eldredge and Gould (1972) gave two examples to illustrate their understanding of allopatric speciation and punctuated equilibria. The first example was a study of evolution in the Bermuda land snail *Poecilozonites bermudensis* during the last 300,000 years of the Pleistocene. This was originally published by Gould (1969). The second example was a study of the trilobite *Phacops rana* from Middle Devonian strata of North America. This was originally published by Eldredge (1971; 1972).

Gould (1969) made an exhaustive multivariate study of shell form in late Pleistocene and Recent land snails of the genus *Poecilozonites* in Bermuda. The objective was development and interpretation of a coherent pattern of morphological variation and evolutionary change through time. Gould recognized the advantage of endemism in an island setting when he compared his goal for *Poecilozonites* to classical studies of Galápagos finches.

The principal figure Gould chose to illustrate his findings concerning speciation (subspeciation in this case) was figure 21 in the 1969 study, reprinted as figure 5–5 in Eldredge and Gould (1972). This shows the pattern of change in what Gould called the "lower aperture eccentricity" — a ratio of two width measurements for the shell aperture. Many other variables were measured in the 1969 study and summary data were included in appendix tables. A generalized interpretation of the pattern of change inferred from figures 21 and 5–5 was published on the page facing figure 21 in Gould (1969) and again facing figure 5–5 in Eldredge and Gould (1972).

Anyone who has reached Chapter 11 in this book will want to know what rates might be involved as *Poecilozonites* changed through time. Here we encounter some unwelcome ambiguity. Gould's figure 21 shows how the mean of sample means changed through time, not how the sample means themselves changed, nor how the original measurements changed. Fortunately, all of the sample means are listed in Gould's appendix (1969, pp. 520–521), and his figure 21 can be redrawn with all of the sample means (Figure 11.1). Gould's sample means are the open and filled circles in Figure 11.1, which has the same axes as Gould's original figure 21. The one change is to the *x*-axis, which has been logged and reversed to increase from left to right.

Solid circles in Figure 11.1 are means for samples of *Poecilozonites bermudensis zonatus*, the stem lineage that gave rise to other derived subspecies that have sample means represented by open circles: *P. b. fasolti*, *P. b. sieglindae*, and *P. b. bermudensis*. The origin of *P. b. fasolti* preceded any known sample of *P. b. zonatus* so we cannot say anything about it. The two arrows in Figure 11.1 indicate the origin of *P. b. sieglindae* from *P. b. zonatus*, and the origin of *P. b. bermudensis* from *P. b. zonatus*. The two arrows mark what Eldredge and Gould (1972) interpreted as punctuated speciation events.

When we plot multiple sample means for each subspecies, we see clear change between subspecies but also some change within subspecies. What the change means in each instance depends critically on within-sample variation in the eccentricity ratio for each sample of each subspecies — variance numbers that were not provided. Gould wrote in the preface to his work on *Poecilozonites*:

It is easy, after a work is completed, to see its shortcomings. My quantitative comparisons among samples, for example, are based only upon differences among means. While variances were used to test the statistical significance of mean differences for all major conclusions, this use of variance as a tool does not exploit its full potentiality.

Gould, 1969, p. 409

Variances and standard deviations are necessary to scale any and all differences between samples.

Here the weighted standard deviation of sample means is used to estimate the variation in all populations. This is possibly a poor substitute but, lacking variances, it is the best we can do. Normal curves in Figure 11.1 show the dispersion expected for each species in each time interval based on the weighted standard deviation of sample means. Since the quantity measured is a ratio, the actual dispersion could be less than or greater than shown – which might focus or diffuse any interpretation. As it stands, Figure 11.1 supports Gould's idea of a stem lineage giving rise to successive branches.

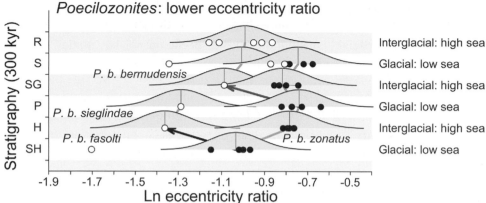

Figure 11.1 Stratigraphic record for the extant Bermuda land snail *Poecilozonites bermudensis* and its late Pleistocene subspecies documented by Gould (1969). Closed circles are sample means for the "lower eccentricity" ratio in the principal stem lineage *P. b. zonatus*. Open circles are sample means for subspecies that branched from the stem lineage (*P. b. fasolti*, *P. b. sieglindae*, and *P. b. bermudensis*). Arrows show the origin of the *P. b. sieglindae* and *P. b. bermudensis* branches. Normal curves show the dispersion expected for each subspecies in each time interval, based on the weighted standard deviation of sample means (individual measurements were not published, nor were sample standard deviations); actual dispersion may be greater or less than shown. Eccentricity is shown on the *x*-axis, where the scale is logged and runs from smaller to larger values (opposite the direction in Gould's plots). Time, based on stratigraphic position, is shown on the *y*-axis. Note that successive stratigraphic units reflect an alternation of high and low sea stands. The entire stratigraphic interval represents about 300,000 years of evolutionary time. Abbreviation of stratigraphic formation names: *SH*, Shore Hills; *H*, Harrington; *P*, Pembroke; *SG*, St. George's; *S*, Southhampton; and *R*, Recent or living samples.

What rates are required to effect the changes seen in Figure 11.1? There are four base rates for the *P. b. zonatus* lineage (filled circles). These range from $h_{4.699} = 7.64 \times 10^{-6}$ to $h_{4.699} = 4.32 \times 10^{-5}$ on a timescale of 50,000 generations. The punctuation rate leading to *P. b. sieglindae* (lower arrow) is $h_{4.699} = 5.69 \times 10^{-5}$ on a timescale of 50,000 generations. The punctuation rate leading to *P. b. bermudensis* (upper arrow) is $h_{4.699} = 6.01 \times 10^{-5}$ on a timescale of 50,000 generations. Three additional base rates for *P. b. sieglindae* and *P. b. bermudensis* range from $h_{4.699} = 2.69 \times 10^{-6}$ to $h_{4.699} = 1.39 \times 10^{-5}$ on a timescale of 50,000 generations. The two "punctuation" rates are above average for their timescale, but all fall comfortably astride the regression and within the range of fossil base rates shown in Figure 9.15d.

The evolutionary history that Gould developed for *Poecilozonites* included "branching of peripheral isolates by paedomorphosis in *P. bermudensis*, and zigzag fluctuations in the central stock of *P. b. zonatus* ... The major temporal variations of morphology in the *P. b. zonatus* stock are adaptive in nature" (Gould, 1969, p. 491). The history of *Poecilozonites* is slightly complicated (Figure 11.1), but there was no special "punctuation" in the speciation nor any static "equilibrium" in Gould's original interpretation.

11.1.6 Examples of Punctuated Equilibria: Phacops

The second example Eldredge and Gould (1972) gave to illustrate allopatric speciation and punctuated equilibria came from a study of the Middle Devonian trilobite *Phacops rana* published by Eldredge (1971; 1972). Eldredge studied *Phacops* through an interval of approximately 7 million years (ca. 390 Ma to 383 Ma) in the Eifelian and Givetian geological stages of eastern and central North America (Cazenovia, Tioughnioga, and Taghanic in New York State terminology).

Eldredge was able to trace a lineage of *Phacops rana* through Middle Devonian strata in marginal seas of the Appalachian Basin flanking the Acadian Mountains, and he was able to show that change in *P. rana* in the Appalachian Basin presaged what was found later in the epeiric sea of the Michigan and Illinois basins farther to the west. The history is summarized in Figure 11.2. The most interesting change in *P. rana* through time was reduction in the number of dorsoventral columns or "files" of eye lenses. These are discrete and easy to count.

Phacops rana was seemingly an immigrant to North America, with a close relative in the Eifelian of Germany. The earliest North American form had 18 files of eye lenses. A subspecies with this number, *Phacops rana milleri*, dispersed to the western epeiric sea during the Cazenovian interval of the Middle Devonian. At this time the number of eye-lens files decreased in the marginal sea from 18 to 17. *Phacops rana rana* dispersed to the epeiric sea in the early Tioughniogan interval, while *P. r. rana* persisted in the marginal sea. Finally, early in the Taghanic interval the number of eye-lens files decreased again from 17 to 17, 16, and 15, and a subspecies, *P. r. norwoodensis*, dispersed to the epeiric sea.

Phacops rana in the marginal sea was the stem lineage from which auxiliary subspecies evolved and dispersed. Thus it deserves special attention. Six geological formations that yield *P. rana* are known from the marginal sea (Eldredge, 1972). These are, in order, the Cardiff, Stafford, Centerfield, Ludlowville, Windom, and Tully formations. The first formation has *Phacops* with 18 and 17 eye-lens files, the next four formations all have *Phacops* with 17 eye-lens files, and the last formation has *Phacops* with 17, 16, and 15 eye-lens files. Eldredge (1972) gives this information but no sample counts indicating how many specimens were studied in each sample, no sample means, and no sample standard deviations. Thus again we are forced to improvise.

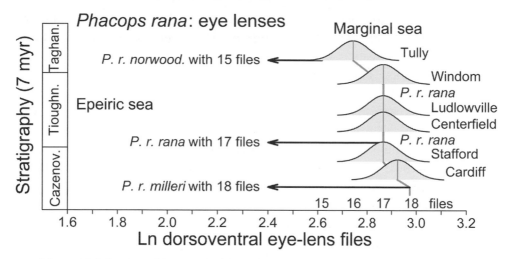

Figure 11.2 Stratigraphic record of the Middle Devonian trilobite *Phacops rana* in eastern North America documented by Eldredge (1971; 1972). Evolutionary change in the eastern marginal sea of the Appalachian Basin gave rise to successive subspecies in the epeiric sea farther west in the North American continental interior. Normal curves show the dispersion expected for each subspecies in each time interval based on the information provided by Eldredge (1972; individual measurements are not available, nor are sample statistics): Actual dispersion may be greater or less than shown. The principal change observed in the marginal sea was reduction in the number of dorsoventral files of eye lenses from 18 alone, to both 17 and 18, to 17 alone, and finally to 15, 16, and 17 in the Tully Formation. Populations with 18, 17, and 15 eye-lens files dispersed from the marginal sea to populate the epeiric sea. The entire stratigraphic interval represents about 7 million years of evolutionary time. Rates of change for eye-lens number in the marginal sea are typical of rates in the fossil record on comparable scales of time. Formations yielding marginal-sea *Phacops rana* are, in sequence: Cardiff, Stafford, Centerfield, Ludlowville, Windom, and Tully.

We can be reasonably certain that means in the marginal sea changed from 17.5 to 17.0 to 16.0 eye-lens files (2.862, 2.833, and 2.771 on a natural-log scale). If we assume that the Tully sample with a mean of 16 included tails with 15 and 17, then the standard deviation must have been about one quarter of the difference between ln 15 and ln 17, or about 0.031 on a natural log scale. This dispersion makes it reasonable to capture two morphs (18 and 17) in the earliest sample, one morph (17) in the next four samples, and three morphs (17, 16, and 15) in the last sample. This value, 0.031, is the standard deviation used to draw the normal curves in Figure 11.2.

The entire sequence shown in Figure 11.2 spanned about seven million years of geological time. This is the interval between the 390 Ma Tioga Ash at the base of the Cazenovian in Pennsylvania (Roden et al., 1990), and the end of the Middle Devonian at 383 Ma (Gradstein et al. 2012). Ages within the interval are not known with precision, but it is reasonable to assume that formations yielding *P. rana* are separated by about one million years. The generation time for *P. rana* is not known, but 10 years is a reasonable possibility.

Combining these numbers, we can calculate a rate of $h_{5.000} = 9.13 \times 10^{-6}$ standard deviations per generation on a timescale of 100,000 generations for the transition from early *Phacops rana* with 18 eye lenses to *P. r. rana* with 17 eye lenses. We can calculate a rate of $h_{5.000} = 4.00 \times 10^{-5}$ standard deviations per generation on a timescale of 100,000 generations for the transition from late *P. r. rana* with 17 eye lenses to *P. r. norwoodensis* with 15 eye lenses. These two "punctuation" rates fall comfortably within the range of fossil base rates — for their interval length — shown in Figure 9.15d.

The history of *Phacops* developed by Eldredge (1972), like that of *Poecilozonites*, is slightly complicated (Figure 11.2), but there was no special "punctuation" in the speciation, nor any statistically justified "equilibrium" in Eldredge's original interpretation.

11.1.7 Conclusions from Rereading Eldredge and Gould (1972)

We conclude this rereading of Eldredge and Gould (1972) by emphasizing that there is nothing unusual about the patterns of change or the rates required for evolution of the late Pleistocene land snail *Poecilozonites bermudensis* or the Middle Devonian trilobite *Phacops rana*, the two examples presented to illustrate punctuated equilibria. Inadequate statistics were provided as background in both studies, but, piecing together what we can, the calculated rates are within the range one would expect for changes in the fossil record on the timescales involved. Some changes are more rapid than others, but there are no outlying "punctuated" rates involved in these studies; and, with the exception of the 17-file eye lenses in successive samples of *Phacops rana rana*, where no variance is reported, there is no real stasis either.

Each study is interesting in involving a more stable stem lineage that gave rise to two or more geographically separated subspecies. There is a geographic component in both of the histories. However the rates and timescales for the *Poecilozonites* and *Phacops* fossil studies are very different from the rates and timescales typically involved in the geographic or allopatric speciation familiar to biologists.

11.2 Punctuated Equilibria and the Bryozoan *Metrarabdotos*

The textbook example of punctuated equilibria is not the evolution of subspecies in the land snail *Poecilozonites bermudensis*, nor is it the evolution of subspecies in the trilobite *Phacops rana*. It is instead the development of multiple species in the Caribbean Neogene bryozoan genus *Metrarabdotos* studied by Alan Cheetham (1986). A later updated and more comprehensive treatment was published by Cheetham et al. (2007). The Cheetham et al. (2007) study was introduced in Section 9.2.31 of Chapter 9, where four sets of species lineages were analyzed (Figure. 9.13). These are the four sets of lineages illustrated in figure 10 of Cheetham et al. (2007). Alan Cheetham provided the ages and Mahalanobis D values that he used to construct this figure. More lineages were illustrated and analyzed in Cheetham (1986), but sample ages and Mahalanobis D values were not published. The Cheetham et al. (2007) sets of species lineages analyzed here seem to be representative of *Metrarabdotos* in general.

11.2.1 Lineages in Stasis

Metrarabdotos species lineages changed little through time in comparison to the distances between their earliest representatives (Figure 9.13a–d). Cheetham (1986), following Charlesworth (1984), developed a test of punctuated versus gradual evolutionary patterns that compares rates of change within species lineages to rates of change between them. By this test, Cheetham (1986) found rates of change between species lineages to be significantly greater than rates within them for all nine of his case studies. Some of the discreteness of the species lineages may result from the use of discriminant analysis to separate predetermined groups, but here we are dependent on the original assembly of samples and treatment of measurements.

No intervals were published for lineages as a whole, nor were they published for transitions between lineages. Thus it is not possible to convert Cheetham's (1986, table 5) rates to haldane units for comparison with the *Metrarabdotos* rates compiled in Chapter 9. This is an oversight for another reason: Cheetham's within-lineage rates were calculated for the full duration of each lineage, and this interval was in every case much longer than any corresponding between-lineage interval. As we have seen in previous chapters, rates calculated over intervals of different lengths cannot be compared directly. Cheetham was correct, that differences between species of *Metrarabdotos* are greater than differences within species, as we found too in Figure 9.13e. However, this does not mean that all rates between species are necessarily greater than rates within them (Figure 9.13f).

The LRI temporal scaling slope for rates within species lineages of *Metrarabdotos* is -1.008, which is virtually the -1.000 expectation for lineages in stasis. Given the stasis within lineages, the question of interest is how change takes place in *Metrarabdotos*. What are the rates required to explain the change between lineages?

11.2.2 Change between Lineages

Figure 11.3 is a four-part figure illustrating the initial stages of speciation in the four species pairs of Cheetham et al. (2007). Parts a–d in Figure 11.3 correspond to parts a–d in Figure 9.13 and to panels 1–4 of Cheetham's (2007) figure 10. For each pair of lineages, samples of the ancestral lineage are shown as filled circles, and samples of the descendant lineage are shown as open circles. The vertical axis for each part of Figure 11.3 has been moved to a common time of

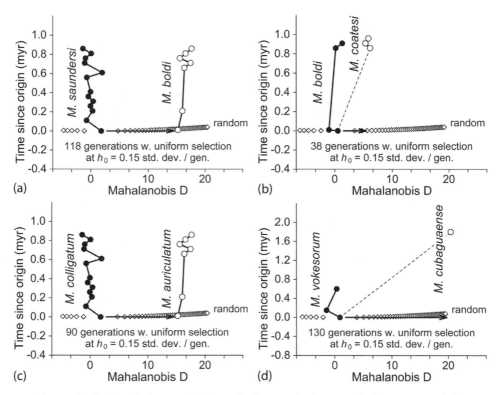

Figure 11.3 Detailed examination of change during speciation proposed by Cheetham et al. (2007) for paired *Metrarabdotos* species lineages. (a) Change from *M. saundersi* to *M. boldi*: 17.590 *D* in 4,000 generations. (b) Change from *M. boldi* to *M. coatesi*: 5.608 *D* in 86,000 generations. (c) Change from *M. colligatum* to *M. auriculatum*: 13.395 *D* in 1,000 generations. (d) Change from *M. vokesorum* to *M. cubaguaense*: 19.486 *D* in 180,000 generations (time-scale compressed). All are scaled to start at time 0 on the y-axis. Solid arrows show the Mahalanobis *D* multivariate standard deviation distance separating initial representatives for each species pair. Open diamonds enclose a 95% confidence envelope for random diffusion from the origin at a step rate of 0.15 standard deviations per generation. All changes can be explained by directional selection at this rate, and all except *M. colligatum* to *M. auriculatum* can be explained by random diffusion at this rate.

origin corresponding to the geological age of the first sample of the ancestral species.

The ancestral species *M. saundersi* in Figure 11.3a is first known from 7.40 Ma and the earliest descendant species sample for *M. boldi* is known from 7.36 Ma. The time difference is 40,000 years. With a 10-year generation time, this is an interval of 4,000 generations. The Mahalanobis D distance or difference between the morphology of the earliest ancestor and the morphology of the earliest descendant is 17.590 multivariate standard deviation units. Combining the interval and the difference yields the rate found in Section 9.2.31 of $h_{3.602} = 4.40 \times 10^{-3}$ on a timescale of 4,000 generations. This is a high rate, but it falls within the spectrum of base rates for fossil studies on a 4,000-generation timescale (Figure 9.15d).

We can also ask how long it would take natural selection to effect the change between *M. saundersi* and *M. boldi* when operating at the median step rate for field studies of $h_0 = 0.15$ standard deviations per generation (Figure 9.15a). Simple division of the 17.590 standard deviation difference between initial samples by the 0.15 standard deviation per generation rate indicates that 118 generations would be required (solid arrow in Figure 11.3a) — an interval much shorter than the 4,000-generation interval available.

Finally, we can ask whether the observed difference of 17.590 standard deviations separating *M. saundersi* and *M. boldi* can be explained by purely random change at a step rate of 0.15 standard deviations per generation. We can test this by comparing the observed difference between the species with the two-tailed 95% confidence envelope for random change calculated as $D.sd = 1.96 \cdot 0.15 \cdot \sqrt{T.g}$. When time in generations $(T.g) = 4,000$, then $D.sd = 18.594$, so the observed difference of 17.590 standard deviations lies within the confidence envelope for random change (open diamonds in Figure. 11.3a). The rate for the transition from *M. saundersi* to *M. boldi* is high, but the transition is easily explained by natural selection at the median step rate for the process and it can also be explained, plausibly, by random genetic drift. The step rate of 0.15 standard deviations per generation is an empirical average, with (unknown) average heritabilities and effective population sizes built into it.

The remaining three species pairs, *M. boldi* and *M. coatesi*, *M. colligatum* and *M. auriculatum*, and *M. vokesorum* and *M. cubaguaense*, are analyzed in the same way in Figures 11.3b, c, and d. The *M. boldi–M. coatesi* transition requires only 38 generations of uniform selection, and it falls well within the confidence envelope for random change. The *M. vokesorum* and *M. cubaguaense* transition requires 130 generations of uniform selection, and it too falls within the confidence envelope for random change.

The *M. colligatum–M. auriculatum* transition is the one most similar to the *M. saundersi–M. boldi* transition. The two species appeared at 8.01 and 8.00 Ma, a difference of 10,000 years and 1,000 generations. Division of the 13.395 standard deviation difference between initial samples by the 0.15 standard deviation per generation rate indicates that 90 generations of directional selection would be required to explain the difference (solid arrow in Figure 11.3c) — an interval much shorter than the 1,000-generation interval available. For time in generations $T.g = 1,000$, then $D.sd = 9.297$ and the observed difference of 13.395 standard deviations is outside the confidence envelope for random change (open diamonds in Figure 11.3c). Directional selection is required because the change observed cannot be explained by random change alone.

The *Metrabdotos* species lineages described by Cheetham et al. (2007) are lineages in stasis. Species pairs diverged rapidly, yielding punctuated patterns, but the divergences are easily explained by directional selection at the median step rate for field studies documented in Chapter 8. Punctuated equilibria is a description of a pattern, like other descriptions listed in Table 1.1, but there is nothing about any pattern of change in *Metrarabdotos* that precludes explanation in terms of ordinary Darwinian natural selection. Empirical step rates are high and change by natural selection is much faster than many paleontologists appreciate.

11.3 Objections to Temporal Scaling of Evolutionary Rates

Various colleagues defending punctuated equilibria have objected to the temporal scaling framework employed in earlier publications to extract information from evolutionary time series. Stephen Jay Gould was first and foremost among the critics, and others have followed. Some response is necessary.

No one has identified a simpler, more straightforward way to recover: (1) the temporal scaling slopes of differences and rates necessary for efficient recognition of stasis, random change, and directional change in an evolutionary time series and (2) step rates appropriate for representing evolution as a process in our thinking and in our genetic models.

11.3.1 Gould and His Psychological and Mathematical Artifacts

Gingerich (1983) has shown that many of our orthodox comparisons for evolutionary rates between different groups of organisms are wrongly stated as a result of measurement over different temporal scales... Nevertheless, Gingerich's quantitative apparatus for illustrating this phenomenon is an artifact of human psychology, and his chosen mode of mathematical plotting is not a property of the empirical world. He bases his conclusion on the continuous and straight array of data, with slope of -1.0, for a plot of the natural logarithm of the evolutionary rate as the ordinate and the natural logarithm of the measurement interval on the abscissa.

Stephen Jay Gould, Science, *1984, p. 994*

If the LRI log-rate-interval plots used throughout this book were psychological and mathematical artifacts in 1983, as Stephen Jay Gould (1984) titled his rejoinder, then one might question their efficacy today. However, Gould misrepresented plotting rates against their denominators as plotting time against its own reciprocal. He then misinterpreted an average difference in rate numerators as a constant difference, and asked how such invariance can be a property of the world. How indeed?

The point of Gould's comment became clear when he argued: "Finally, the artificial character of Gingerich's curve refutes his general conclusion that its smoothly linear character demonstrates that 'microevolution and macroevolution are different manifestations of a common underlying process'" (Gould, 1984, p. 995). Gould, it seems, feared that macroevolution on long timescales might, at its heart, depend on the process that explains microevolution from one generation to the next.

At this point we have seen enough LDI log-difference-interval and LRI log-rate-interval plots to understand their empirical basis and how these plots represent evolutionary time series. We have seen LDI and LRI plots representing a spectrum of case studies from simple to complex, involving experimental selection, natural selection in the field, and natural selection in the fossil record. The plots tell us things we cannot learn any other way.

11.3.2 Bookstein and the Existence of Evolutionary Rates

Evolutionary rates exist only when the hypothesis of symmetric random walk can be refuted. This hypothesis, not the assumption of an evolutionary rate equal to zero, is the appropriate null model for studies of tempo and mode.

Fred Bookstein, Paleobiology, *1987, p. 446*

Fred Bookstein's first sentence is false. The second is true. The problem with Bookstein's logic is that random walks have rates. As we saw in Chapter 4, every random walk has a step rate (the step rate that Bookstein elusively called a "speed"), the change per step to which a sign, positive or negative, is applied. In addition, random walks have net rates that may be positive or negative. The net rates increase in increasing multiples of the number of steps in the random walk. Random walks have rates, and so do the alternatives to be compared to random walks. Nonrandom no-net-change stasis is one alternative to a random walk, and nonrandom cumulative directional change is the other alternative.

Bookstein also wrote that "for the notion of a rate to exist, the ratio of evolutionary 'distance traveled' to geological 'travel time' must have a limit as travel time becomes shorter. If this limit does not exist, then ... we will compute higher and higher 'rates' ... as the temporal scale of comparison becomes shorter" (Bookstein, 1987, p. 446). This emphasis on limits from mathematical calculus, on "a derivative with respect to time," is inappropriate for empirical comparisons of "distance" and "travel time," as Bookstein himself indicated in an earlier discussion of random walks (Bookstein et al., 1978, p. 121).

The rate that Bookstein caricatured in quotation marks in the previous paragraph is a ratio and a rate on the timescale of the rate denominator — whether or not any rate limit exists. It is a rate on the particular timescale of its calculation and not a rate on any other timescale — which is why every rate must be indexed by its timescale. The dependence of a rate on its timescale, inherent in fractal processes such as random walks, would be a problem if there were no natural timescale for comparison of rates. The natural timescale is provided by the step-to-step timescale of the random walk, and by the generation-to-generation timescale of evolution by natural selection.

We can end on a cautionary note about random walks: Random walks can be made to fit any time series if the time series can be subdivided arbitrarily and step rates ("speeds") calculated to fit each subdivision — as Bookstein (1987) did for *Globorotalia tumida* studied by Malmgren et al. (1983). Our interest should not be the meaningless exercise of reducing every evolutionary lineage to a series of random walks. Temporal scaling of rates and intervals enables us to identify the lineages and parts of lineages that are not changing randomly. Temporal scaling showed that changes in size within the species *Globorotalia plesiotumida* and *G. tumida* were significantly less than expected if random. Temporal scaling also showed that change in size between the species was significantly more than expected if random (Chapter 9.2.16; Figure 9).

11.3.3 Sheets, Mitchell, and Spurious Self-Correlation

> Since the inverse dependence of evolutionary rate on interval...[is] consistent with the phenomenon of spurious self-correlation, it is not possible to interpret the inverse dependence of rate on interval as indicating anything other than the low correlation of change with interval...We should focus our attention on the distribution of evolutionary change over time, rather than rate, as change is not subject to the problem of spurious self-correlation.
>
> (*Sheets and Mitchell, 2001, p. 443*)

David Sheets and Charles Mitchell (2001) interpreted the inverse dependence of evolutionary rate on interval as "spurious self-correlation." In the preceding chapters we have seen enough LDI log-difference-interval graphs and LRI log-rate-interval graphs plotted side by side to confirm that evolutionary differences and rates are inverses with respect to time. Long-term evolutionary change is largely independent of interval, so the inverse, rate, is largely dependent. Sheets and Mitchell are correct to focus attention on the distribution of change over time, but we are also interested in rates. Plotting differences versus their corresponding intervals and rates versus their corresponding intervals give us equivalent information about a time series. The rate of greatest interest, the step rate given by the intercept on an LDI plot, is the same as the step rate given by the intercept on an LRI plot. LDI slopes and LRI slopes, as we have seen, are complementary.

11.3.4 Hunt and "Biased Random Walks"

In Chapter 4 we used i_0, d_0, and r_0 to represent the step interval, step displacement, and step rate for each step of a random walk. In evolutionary studies the step interval is a constant equal to the generation time, and the step displacement is a normally distributed random variable with mean $\mu = 0$, standard deviation σ, and variance σ^2. The step rate for each interval, each time step, is then the ratio of the step displacement, in σ units, to the step interval (which rate, for a unit interval, is just the step displacement). The walk itself has coordinates defined by the increasing sum of successive intervals, and by the corresponding increasing or decreasing sum of successive displacements. The walk is symmetric in the sense that the sign of each displacement is random, giving each step an equal chance for a positive or negative displacement.

Gene Hunt (2004) developed an unusual formulation for a random walk. He considered the "step distribution" for a random walk to have a mean μ_{step} and variance σ^2_{step}, with no restriction on the step mean. If the mean is zero, the resulting random walk will be symmetric and unbiased. If the mean is positive or negative, the resulting random walk will be biased positively or negatively, and hence directional. The long-term average rate, on the timescale of the entire time series, is given by μ_{step}. Rates on shorter time scales are erroneously assumed to be simple fractions of this.

Randomness and directionality are both built into Hunt's random walk model, and a purely random random walk is considered to be a special case of the general model. This means that careful attention is required when reading to recognize and distinguish an "unbiased random walk" from a larger set of "biased random walks." Hunt's unusual formulation for a random walk required the complicated extension he developed to model stasis as an evolutionary mode. For Hunt the step variance, σ^2_{step}, is also a measure of evolutionary rate (in a quadratic reference frame).

Hunt used "step distribution" to refer to the statistical distribution of displacements in a time step, with no restriction on time steps. He started by assuming that time steps are generations, but relaxed this to allow a time step to be "a generation, a year, a million years, etc." (Hunt 2004, p. 428). Later he wrote:

> In order for these equations to apply to real paleontological data, it is necessary to relate time in the model, which is composed of discrete steps at which evolutionary changes occur, to the actual time scale by which fossil ages are measured. In other words, do the evolutionary steps occur every year, every thousand years, or at some other frequency? This question is unanswerable, but fortunately unimportant. As long as consistency is maintained within an analysis, results are unaffected by how frequently (in real time) model steps occur. *Altering the assumed duration of steps amounts to changing the units in which time is measured, which scales parameter estimates accordingly* (e.g., parameter estimates are one thousand times larger when steps occur every Kyr rather than every year). Because overall statistical inference is otherwise unchanged by this scaling, researchers can just measure time in whatever units are convenient.
>
> *Hunt, 2006, p. 584; emphasis added*

Two things are wrong here: (1) The timescale of the model must match the timscale of the process being modeled. Evolution by natural selection and by random drift, two processes, take place on a timescale of generations, not "every year, every thousand years, or at some other frequency." Models that represent these processes are necessarily models on a timescale of generations. (2) Hunt proposed that "altering the assumed duration of steps ... scales parameter estimates accordingly" — which is only true when change is purely directional. A review of the experimental, field, and fossil case studies in Chapters 7–9 shows that evolutionary change is rarely, if ever, purely directional.

Finally, to follow up on the first of these points, Hunt's (2006) $\mu_{\text{step}} = 5.10$ standard deviations per million years for M_1 molar length in *Cantius* is a single long-term rate close to the long-term haldane rate of $h_{5.716} = 10^{-5.408}$ or 3.91 standard deviations per million years (or million generations) published by Clyde and Gingerich (1994, figure 4). Neither of these rates on million-year timescales is a rate of interest for modeling evolution or random drift. Clyde and Gingerich extrapolated the temporal scaling of a large sample of rates for M_1 in *Cantius* to find an intercept of -3.458. They estimated the corresponding step rate as $h_0 = 10^{-3.458}$, or 3.48×10^{-4} standard deviations per generation on a timescale of one generation. The Clyde–Gingerich estimate from extrapolation is near the minimum step rate observed in field studies (Table 9.1). The median step rate for empirical field studies in Table 9.1, $h_0 = 0.1534$, rounded to $h_0 = 0.15$, provides a better estimate, without extrapolation, for rates of selection and drift to be used in evolutionary models.

11.4 Summary

1. The theory of punctuated equilibria promoted by Niles Eldredge and Stephen Jay Gould (1972) rekindled the long-standing debate between Lamarck–Darwin gradualists who envision species changing through time and Lyell–Linnaeus traditionalists who regard species as fixed in form (Table 1.1).

2. The history of *Poecilozonites* developed by Gould (1969; Figure 11.1) is slightly complicated, but there was no special "punctuation" in the speciation nor any static "equilibrium" in Gould's original interpretation.

3. The history of *Phacops* developed by Eldredge (1972; Figure 11.2) is similarly slightly complicated, but there was no special "punctuation" in the speciation, nor any statistically justified "equilibrium" in Eldredge's original interpretation.

4. The *Metrabdotos* species lineages described by Cheetham et al. (2007; Figure 11.3) are lineages in stasis. Species pairs diverged rapidly, yielding punctuated patterns, but the divergences can all be explained by directional selection at the median step rate for field studies documented here in Chapter 8.

5. Study of evolutionary rates in relation to their denominators provides a simple and straightforward method for recovering important information: (1) temporal scaling slopes necessary for efficient recognition of stasis, random change, and directional change and (2) empirical step rates representing evolution as a process.

6. Empirical step rates are high, and change by natural selection is much faster than many paleontologists appreciate.

12

Genetic Models

The widest disparity which has so far developed in the field of population genetics is that which separates those who accept from those who reject the theory of "drift" or "non-adaptive radiation," as it has been called by its author, Professor Sewall Wright of Chicago.

R. A. Fisher and E. B. Ford, Heredity, *1950, p. 117*

There are many good biologists who think that natural selection is almost, or in fact is, irrelevant to evolution. Because many are molecular biologists, their view might be called the "molecular" view, but there is broad overlap with both the "random" and "constraint" views.

John Endler, Natural Selection in the Wild, *1986, p. 240*

Sewall Wright (1931) is famous for developing the theory that subdivision of a population into small isolated or semi-isolated groups is important for evolution because it promotes random changes in gene frequencies. Later Ronald A. Fisher and E. B. "Henry" Ford, responding to Wright, made the observation highlighted in the epigraph. Fisher and Ford (1950) reported that large as well as small populations experience changes in gene frequencies beyond any effect attributable to population size, and considered the observation fatal for the "Sewall Wright effect." The Sewall Wright effect has survived, and genetic drift remains a central idea in modern evolutionary theory. Here we consider genetic models that have been used to study both drift and selection in long-term evolution documented in the fossil record.

12.1 Natural Selection and Random Drift in Phenotypic Evolution

Russell Lande (1976) used population-genetic methods to develop models for the evolution of an average phenotype in a population, first by natural selection, and then by random genetic drift.

298

12.1.1 Minimum Selection in Observed Evolutionary Events

Lande first calculated the natural selection required to produce rates of evolution observed in quantitative (polygenic) characters changing through geological time. Two assumptions were stated (Lande, 1976, p. 319), and a third was implicit. The stated assumptions were: (1) change has a genetic basis and (2) random change — random genetic drift — is not involved. The implicit assumption, unstated, was: (3) the modeled change is unidirectional. The model was based on truncation selection because, as Lande wrote, this requires the minimum selective mortality for a given change in phenotype.

Lande then calculated the rate of change corresponding to weak selection on a normally distributed character (Lande's equation 11):

$$\Delta \bar{z}(t) = \pm \frac{h^2 \cdot \sigma}{\sqrt{2\pi}} e^{-b^2/2} \qquad (12.1)$$

The quantities $\Delta \bar{z}(t)$, h^2, and σ for the character of interest are, respectively, the rate of change per generation in average phenotypic value, the heritability, and the phenotypic standard deviation. The quantity related to selection, b, is the truncation point expressed in standard deviations from the average phenotype. The sign of the rate is opposite that of b, which means that the rate is positive when the truncation point b is less than the average phenotype, and vice versa. To take an example, if h^2 = 0.5 and b = 1, then $\Delta \bar{z}(t) = -0.121 \cdot \sigma$.

Lande reasoned that h^2 and σ^2 can be regarded as constant for polygenic characters. He then calculated the total change z after t generations as the product of t times the rate per generation:

$$z = t \cdot \Delta \bar{z}(t) \qquad (12.2)$$

Here $\Delta \bar{z}(t)$ too is regarded as constant. Substituting z/t for $\Delta \bar{z}(t)$ in Equation 12.1, multiplying both sides by t, and dividing by σ yields:

$$\frac{|z|}{\sigma} = \frac{h^2 \cdot t}{\sqrt{2\pi}} e^{-b^2/2} \qquad (12.3)$$

Symbols $|z|/\sigma$ and t are the numerator and denominator of the haldane rate units we have been using throughout the book, the former ln-transformed and expressed in units of phenotypic standard deviation, and the latter expressed in generations. Here the rate is a constant that is dependent on b and h^2:

$$\frac{|z|/\sigma}{t} = \frac{h^2}{\sqrt{2\pi}} \cdot e^{-b^2/2} \qquad (12.4)$$

Finally, Equation 12.4 can be solved for the truncation point b:

$$b = \pm\sqrt{-2 \cdot \ln\left(\sqrt{2\pi} \cdot \frac{|z|/\sigma}{h^2 \cdot t}\right)} \qquad (12.5)$$

The first example Lande used to illustrate evolution by natural selection was the example of Simpson's horses (Simpson, 1944, 1953) that was introduced here in Section 9.2.2. Lande (1976, table 1) used ages of 50 Ma and 30 Ma for *Hyracotherium borealis* and *Mesohippus bairdi*, respectively, and a generation time of two years. Ln paracone heights for the two species are 1.54 and 2.12, with standard deviations of 0.062 and 0.048, respectively. Heritability was assumed to be $h^2 = 0.5$. These numbers yielded, by Lande's calculation, morphological change of $z/\sigma = 10.6$ standard deviations in $t = 10 \times 10^6$ generations. The quotient, 1.06×10^{-6} standard deviations per generation, would be smaller than any step rate found here, and smaller by five orders of magnitude than the median step rate $h_0 = 0.153$ observed in the field studies reviewed in Chapter 8 and summarized in Table 9.1.

Lande substituted $z/\sigma = 10.6$, $t = 10 \times 10^6$, and $h^2 = 0.5$ into Equation 12.5 to find the value $b = \pm 4.929$. The complement of the normal-curve distribution function for $b = 4.929$ indicates that truncation corresponds to a tail proportion of 4.11×10^{-7}. This was interpreted as the minimum selective mortality, 4 deaths per 10 million individuals, necessary to explain change at the observed rate. Selection other than truncation would elevate the mortality. Lande (1976, p. 328) concluded that "extremely weak" selection can explain the change observed. His conclusion is not wrong, in light of the assumptions, but it is also not quite right.

Two flags were raised while reviewing Lande's treatment of paracone heights in *H. borealis* and *M. bairdi*. The first was raised by calling attention to the implicit assumption that the modeled change was unidirectional, a casual commitment reminiscent of nineteenth-century orthogenesis. The second flag was raised when elevating Equation 12.2 to numbered status after Lande buried it in his text. None of the 100 or so multi-generation experimental, field, and fossil studies reviewed in Chapters 7–9 was purely directional (unidirectional), which is the assumption required to make Equation 12.2 work. Even controlled directional selection experiments, with LRI temporal scaling slopes averaging -0.412 (Figure 7.7d), do not come close to the LRI zero-slope directionality assumed in Equation 12.2. There is too much complexity in environmental change and too much back-and-forth alternation in the direction of selection to model long-term evolution as a simple unidirectional process.

It is interesting to compare Lande's minimal estimate of natural selection in an observed evolutionary event (4 deaths per 10 million individuals in each generation) to the minimum number of generations of ordinary selection that would explain the same event. For example, the change of $z/\sigma = 10.6$ in paracone height found in horse evolution can be explained by just 70 generations of unidirectional natural selection at the median step rate for field studies (0.153 standard deviations per generation; Table 9.1). Similar numbers emerge for Lande's other examples.

Lande concluded that extremely weak natural selection can explain the changes observed in Simpson's horses, Bader's oreodonts, or Gingerich's *Hyopsodus*. This would carry more weight if the model represented how evolution works. Nothing evolves in a straight line for long. Further, natural selection is cumulative, as a sum of values in each generation irrespective of changes in direction. A realistic model would capture more of the generation-to-generation fractal roughness inherent in evolutionary change.

12.1.2 Random Genetic Drift in Phenotypic Evolution

Lande (1976) concluded that extremely weak natural selection can explain changes in phenotypes observed in the fossil record. From this it was logical to question whether any selection was required at all. Genetic drift due to random survival depends critically on the sizes of samples drawn at random from a source population. Drift is probable in small samples and less probable in large samples, so population size has an inverse effect. Any effect of randomly sampling a gene pool is lost if it is not heritable, so heritability has a direct effect.

We have already seen that a two-tailed 95% confidence interval for random change in a phenotype is given by $s_{(\alpha \leq 0.05)} = \pm 1.96 \cdot \sigma \cdot \sqrt{t}$ (Equation 4.5), where σ is the step rate, and the standard normal deviate ± 1.96 is the value, in standard deviation units, that encloses 95% of the area under a normal curve. The genetic-drift equivalent is:

$$z/\sigma = \pm 1.96 \cdot \sqrt{\frac{h^2}{N_e}} \cdot \sqrt{t} \qquad (12.6)$$

Here z/σ represents the two-tailed 95% confidence interval in phenotypic standard deviation units, ± 1.96 is the standard normal deviate, h^2 is the heritability, N_e is the effective population size, and t is time in generations. The middle term on the right side, $\sqrt{(h^2 / N_e)}$, is now the step rate in standard deviations per generation that will generate a confidence interval $\pm z/\sigma$ units wide after t generations. The function itself is quadratic, and the trace of z/σ for increasing t is parabolic.

Squaring and rearranging terms yields Lande's equation 19:

$$N^* = \frac{(1.96)^2 \cdot h^2 \cdot t}{(z/\sigma)^2} \qquad (12.7)$$

N^* is the limiting effective population size that will explain a change of z/σ or more in t generations by chance, with no assistance from selection. If the population size is larger, drift cannot explain the change.

In the transition from *Hyracotherium borealis* to *Mesohippus bairdi* the limiting population size that would explain a change of $z/\sigma = 10.6$ standard deviation units of paracone height by chance is $N^* = 1.71 \times 10^5$ (not 2×10^3 as reported). A smaller population might explain more change, but a larger population cannot explain the observed change of $z/\sigma = 10.6$ by drift. If we now set $z/\sigma = 10.6$ and $t = 10 \times 10^6$ and solve for the middle term in Equation 12.6, we find $h_0 = 1.71 \times 10^{-3}$ standard deviations per generation as the step rate that will generate the observed change in the time available.

A step rate of $h_0 = 1.71 \times 10^{-3}$ standard deviations per generation would be at the low end of step rates, and smaller by about two orders of magnitude than the median step rate $h_0 = 0.153$ observed in the field studies reviewed in Chapter 8 and summarized in Table 9.1. The observed change in paracone height for Simpson's horses is $z/\sigma = 10.6$ standard deviation units — which lies at the margin of drift at a rate of $h_0 = 1.71 \times 10^{-3}$. Figure 12.1, drawn to scale, illustrates this. The drift

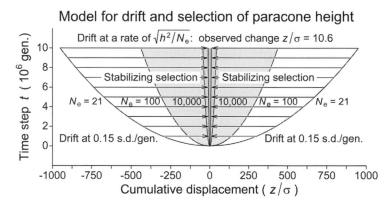

Figure 12.1 Model for drift and selection of paracone height in the transition from *Hyracotherium borealis* to *Mesohippus bairdi* studied by Simpson (1944) and Lande (1976). Heritability is assumed to be constant at $h^2 = 0.5$. The narrowest central parabola (darkly shaded) represents drift at a step rate of $h_0 = \sqrt{(h^2 / N_e)}$ standard deviations per generation calculated to explain the observed change of $z/\sigma = 10.6$ standard deviations in the 10×10^6 generation transition. Effective population size N_e is limited to $N^* \leq 1.71 \times 10^5$. Drift at $h_0 = 0.15$ standard deviations per generation, the median step rate found in field studies (Table 9.1), would yield a drift envelope corresponding to $N_e = 21$ that is wider by nearly two orders of magnitude. Stabilizing selection (arrows) constrains change to a much narrower range than would be expected from the median step rate found in field studies. Similar models can be constructed for the other traits and transitions studied by Lande (1976).

envelope for observed change is the more darkly shaded central parabola in Figure 12.1.

In Chapter 8 we saw that step rates in field studies are normally much higher (faster). Random drift at a step rate of 0.153 standard deviations per generation would enclose an envelope much wider than the change of $z/\sigma = 10.6$ we observed. The only plausible explanation for the narrowness of the envelope of change and plausible drift for paracone height in Figure 12.1 is stabilizing selection, and here the amount of stabilizing selection (represented by arrows) is some two orders of magnitude greater than the drift modeled by Lande (1976). Smaller upper limits of effective population size (lines with N_e values in Figure 12.1) would allow more change to be explained by drift. However, for paracone height in this example, drift at a rate of $h_0 = 0.153$ would correspond to an unrealistic 10-million-generation-long average N_e of about 21 breeding horses.

The empirical step rates that biologists observe in field studies today are much higher than any long-interval substitutes that paleontologists calculate from the fossil record. Higher step rates mean that the potential for random drift is also much greater than most authors estimate. Drift does not begin to equal its potential — which is evidence for the near ubiquity of natural selection in the form of stabilizing selection.

12.2 Random Drift and Mutation

Michael Lynch (1990) addressed evolutionary change and random drift in a completely different way. He considered how the phenotypic change we measure in the fossil record is related to the change in morphology expected from the accumulation of polygenic mutation and related drift. Lynch, like Lande, focused on the evolution of mammals. We begin with Lynch's model and then examine the rates of polygenic mutation and morphological change that underlie the model's interpretation.

12.2.1 Rates k and Δ

Lynch (1990) defined σ_m^2 as a rate, the rate of input of genetic variance in a generation due to mutation. He reasoned that the change in genetic variance after t generations will then be $\sigma_m^2 \cdot t$. If mutation and drift are the only forces shaping populations in a lineage, then the phenotypic difference between them after t generations will also be $\sigma_m^2 \cdot t$. In the case of populations diverging from a common ancestor, t is the sum of generations in the two divergent lineages.

The input of phenotypic variance due to mutation and drift in a generation, σ_m^2, is necessarily some fraction k of the whole within-population phenotypic variance

σ^2 (ln z) in that generation. Given $\sigma_m^2 = k \cdot \sigma^2$ (ln z), multiplying both sides by t and rearranging terms yields:

$$k \cdot t = \frac{\sigma_m^2 \cdot t}{\sigma^2(\ln z)} \quad \text{and} \quad k = \left[\frac{\sigma_m^2 \cdot t}{\sigma^2(\ln z)}\right]/t \qquad (12.8)$$

The numerator is the variance between groups and the denominator is the variance within groups. In the right-hand form of Equation 12.8, k is the rate of change of the ratio of these two variances. We can estimate k experimentally from the within- and between-population phenotypic variances of highly inbred, genetically uniform laboratory populations mutating and drifting through time, giving us an expectation for the rate of selection-free neutral divergence in nature.

Changing notation, Lynch (1990) introduced Δ as the equivalent rate of change in the variance ratio for populations diverging in nature:

$$\Delta \cdot t = \frac{\text{var}_B(\ln z)}{\text{var}_W(\ln z)} \quad \text{and} \quad \Delta = \left[\frac{\text{var}_B(\ln z)}{\text{var}_W(\ln z)}\right]/t \qquad (12.9)$$

$\Delta \cdot t$ is a dimensionless ratio of the between-species and within-species variances for a trait, and Δ itself is conceived to be the rate of change in this ratio through time.

Lynch (1990) gave equations for calculating var_W (ln z) and var_B (ln z) from means and standard deviations (SD) for two-population analyses, as follows:

$$\text{var}_W(\ln z) = \frac{(n_1 - 1) \cdot [\text{SD}(\ln z_1)]^2 + (n_2 - 1) \cdot [\text{SD}(\ln z_2)]^2}{n_1 + n_2 - 2} \qquad (12.10)$$

$$\text{var}_B(\ln z) = \left[\frac{n_1 \cdot n_2 \cdot (\overline{\ln z_1} - \overline{\ln z_2})^2}{n_1 + n_2} - \text{var}_W(\ln z)\right]/n_0 \qquad (12.11)$$

where n_1 and n_2 are the sizes of the two populations, and $n_0 = n_1 + n_2 - [(n_1^2 + n_2^2)/(n_1 + n_2)]$. The bars over ln z_1 and ln z_2 indicate means calculated after natural-log transformation. These means of logged values can be approximated as the log of untransformed values minus one half the square of the coefficient of variation for the untransformed values. The standard deviation of logged values can be approximated as the coefficient of variation of unlogged values.

Estimates for the rate k calculated by Lynch (1988) fall in a range from about 10^{-4} to 10^{-2}. Lynch (1990) noted that skeletal characteristics in mammals average about $k = 10^{-2}$, which he called Δ_{max}, but he also considered k (or Δ_{min}) $= 10^{-4}$. As we shall see, the value $k = 10^{-2}$ is an average of base rates and underestimates the step rate of interest.

12.2.2 Temporal Scaling of Δ Rates

The final figure in Lynch (1990, figure 3) is an LDI log-difference-interval plot for $\Delta \cdot t$ and t in the mammalian case studies that he reviewed. This is redrawn with full axes in Figure 12.2a. Lynch showed that the observed total divergence $\Delta \cdot t$ in each study is related to the corresponding interval t, and he considered a least-squares regression with slope 0.52 and intercept −2.174 to provide a good approximation. Lynch (1990, p. 736) interpreted his figure 3 to indicate that "the cumulative divergence of cranial morphology in mammalian lineages is proportional to the square root of divergence time in generations," a reference to the slope of 0.52.

Figure 12.2a is even more interesting when transformed to an LRI log-rate-interval plot (Figure 12.2b). The LRI plot contains the same information as Figure 12.2a and the same information as Lynch's (1990) figures 2 and 3 but shows how Δ rates change systematically with interval length. The slope in Figure 12.2b is the complement of the previous slope (−0.48 = 0.52 − 1.00), and the intercept is the same. The shaded area in the background of Figure 12.2b shows the rates $k_{min} = 10^{-4}$ and $k = 10^{-2}$ that Lynch labeled Δ_{min} and Δ_{max}.

What do the slopes and intercepts in Figures 12.2a and 12.2b tell us about phenotypic evolution in the taxa involved? First, the slopes, 0.52 and −0.48, are almost exactly the slopes expected for random or "square root" temporal scaling.

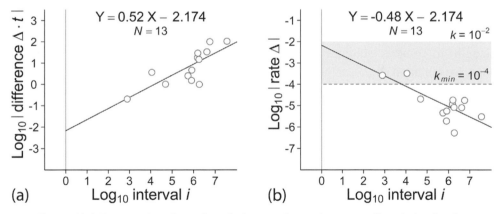

(a) (b)

Figure 12.2 Temporal scaling of evolutionary change in mammalian skeletal traits calculated by Lynch (1990). (a) LDI log-difference-interval plot of $\Delta \cdot t$ versus t. Lynch calculated a slope of 0.52 and an intercept of −2.174 that he reported in the form $10^{-2.174} = 0.0067$. (b) Corresponding LRI log-rate-interval plot for Δ versus t, with a slope of 0.48 and an intercept of −2.174. Note that the slopes are close to the null expectation for each, 0.500 and −0.500 respectively, ruling out any simplistic assumption of directional change. Both plots yield the same step rate $\Delta_0 = -2.174$, which as a $\log_{10}$ value is close to the average of $k = 10^{-2}$ that Lynch calculated for polygenic mutation and random drift (and well above Lynch's $k_{min} = 10^{-4}$).

These are not the slopes expected for either stasis or directional change. As we saw above in Section 12.1.1, any implicit assumption that the underlying change was unidirectional is refuted. Our model for change in Δ through time should be a neutral-model scaling with a slope of -0.500.

Second, the intercepts of -2.174 in both panels indicate that the LDI and LRI plots are telling us the same thing about Δ_0 as a rate: $\Delta_0 = 10^{-2.174}$ on a timescale of one generation. This is close to the rate $k = 10^{-2}$ that Lynch called Δ_{max}. On the face of it we might expect k to explain Δ_0, but we shall see that this is unlikely.

12.2.3 Rates of Polygenic Mutation

Lynch (1988) compiled a set of variance ratios V_m/V_E for morphological change due to polygenic mutation in mammals. These are all based on inbred strains of laboratory mice, *Mus musculus*, and the divergence of sublines over time due to random genetic drift. Two studies based on measurement of continuous traits were analyzed, one by Bailey (1959) and the other by Festing (1973). Each is re-examined here.

Bailey (1959) studied rates of divergence in two inbred strains of mice. Two sublines of one strain were allowed to diverge for an average of nine generations (18 generations of separation) and two sublines of another strain were allowed to diverge for an average of 46.5 generations (93 generations of separation). Bailey measured four cranial traits (three distances between landmarks on the skull base, height of the foramen magnum) and two postcranial traits (ulna length, ilium length) to compare sublines in each strain. Lynch (1988) calculated the variance due to mutation, V_m, as the between-group variance in Bailey's sublines. He calculated the environmental variance, V_E, from $V_E + 4 \cdot V_m$, estimated as the sum of the variances Bailey provided between individuals within litters and between litters within mothers. Lynch (1990) considered k to be the ratio V_m/V_E independent of the generations involved.

Log_{10} values of k from the Bailey study are shown on the LRI plot in Figure 12.3a, where a range of intercepts provides a range of estimates for k_0 as a step rate. The median of the Bailey k values is 0.004 (yielding, when logged, an intercept of -2.38). A line fit to the logged k values ($Y = -0.71 \cdot X - 1.33$) has an intercept of -1.33. The intermediate intercepts of -1.69 and -1.61 are calculated from the median value of k within each experiment and a slope of -0.500 expected for random change (as in Figure 12.2b). From this exercise it appears that 0.020 and 0.025 may provide the best estimates for k from Bailey's mouse measurements. These are minimally five times greater than the unscaled median of 0.004.

Festing (1973) provided statistics for 13 traits measured on the mandibles of inbred mice. An LRI plot for Festing's Mahalanobis D values for the entire sample

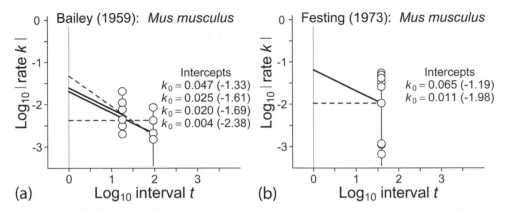

Figure 12.3 Temporal scaling of rates of phenotypic change due to polygenic mutation and drift in inbred laboratory mice. (a) Open circles are rates for six traits in sublines of inbred mice studied by Bailey (1959). Sublines were studied after separations averaging 18 and 93 generations, respectively. Horizontal dashed line represents change at the median rate for both lines with no scaling. Oblique dashed line is fit to all the rates. Solid lines are fit the median for each subline with a slope of -0.500. (b) Open circles are rates for 13 traits in sublines of inbred mice studied by Festing (1973). Sublines were studied after separations averaging 39 generations. Horizontal dashed line represents change at the median rate with no scaling. Solid line is fit the median rate, with a slope of -0.500. Slopes of -0.500 are the slopes expected for random change (compare Figure 12.2). Note that rates of polygenic mutation per generation on a timescale of one generation are substantially higher than rates per generation averaged over many generations.

of 20 sublines has a slope of -0.403 and an intercept of -0.410. The latter, exponentiated, indicates a high step rate of $h_0 = 0.389$ standard deviations per generation on a timescale of one generation. This is a rate representing the morphological change expected from mutation and drift that is more than double the median step rate $h_0 = 0.153$ observed in field studies.

Lynch (1988, p. 142) restricted analysis of rates to the nine Festing sublines of strain C57BL/Gr, for which the average two-way divergence time was about 39 generations. He calculated the variance due to mutation, V_m, for each trait as one half of the slope of a regression of between-subline variance, $var(\bar{z})$, on divergence time. The sum $V_E + 4 \cdot V_m$ for each trait was estimated as the square of the within-subline standard deviations in Festing's table 1. From these numbers Lynch (1988) calculated a mean variance ratio, V_m / V_E of 0.023. The median is 0.011.

Log$_{10}$ values of k from the Festing study are shown on the LRI plot in Figure 12.3b, where two intercepts provide two estimates for k_0 as a step rate. The median of the Festing k values is 0.011 (yielding, when logged, an intercept of -1.98). The alternative intercept of -1.19 is calculated from the median value of k and a slope of -0.500 expected for random change (as in Figure 12.2b). From this exercise it appears that 0.065 is the best estimate for k_0 from Festing's mouse measurements. This is again some five times greater than the unscaled median k of 0.011.

12.2.4 Paracone Height in Horses Revisited

Lynch (1990) reported the average value of Δ to be about 2×10^{-5} for paracone heights and ectoloph lengths in Simpson's horses, the horses studied by Lande (1976) and analyzed above in Sections 9.2.2 and 12.1.1. This empirical Δ is one fifth the value of Lynch's $k = 10^{-4}$ and only 1/500 the value of $k = 10^{-2}$. On the face of it the observed drift for tooth measurements in horses is less than our minimal expectation based on experimental studies and much less than our general expectation.

The example in Figure 12.1 shows drift and selection for paracone height in the transition from *Hyracotherium borealis* to *Mesohippus bairdi*, quantified following Lande (1976). The same example is reanalyzed here. Following Lynch (Equations 12.9–12.11), the Δ value calculated for paracone height in the transition from *H. borealis* to *M. bairdi* is $\Delta_{7.000} = 5.71 \times 10^{-5}$. This is not the step rate. From Equation 12.6: $\Delta_{7.000} \cdot t = \pm 1.96 \cdot \Delta_0 \cdot \sqrt{t}$. Dividing both sides by 1.96 and $\sqrt{t}$ and rearranging terms yields Δ_0:

$$\Delta_0 = \pm \frac{\Delta_{7.000}}{1.96} \cdot \sqrt{t} \qquad (12.12)$$

The Δ_0 value calculated for paracone height in the transition from *H. borealis* to *M. bairdi* is $\Delta_0 = 0.0921$, and the $\Delta_{7.000} \cdot t$ value expected after 10×10^6 generations is 571.

The three expectations for drift at experimental rates k_0 are shown graphically in Figure 12.4, where they are represented by nested parabolas in darker to lighter shades of gray. The outer unshaded parabola represents the change in paracone height observed in the transition from *Hyracotherium borealis* to *Mesohippus bairdi*, with $\Delta_0 = 0.0921$ and $\Delta_{7.000} \cdot t = 571$. Observed change exceeds the change expected by drift at any of the experimental rates.

Figure 12.4, where observed change exceeds all expectations for drift, is the opposite of Figure 12.1, where, for the same trait in the same two taxa, observed change is less than all realistic expectations for drift. The high step rate of $h_0 = 0.389$ standard deviations per generation for drift implicit in Festing's (1973) analysis of inbred mice offers a clue to resolution of these opposites in favor of Figure 12.1 and drift. The Festing rate of $h_0 = 0.389$ is more than twice the median rate of $h_0 = 0.153$ standard deviations per generation found in field studies and would make the drift envelope in Figure 12.1 even wider.

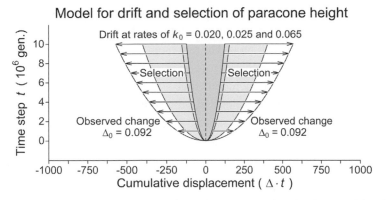

Figure 12.4 Model for drift and selection of paracone height in the transition from *Hyracotherium borealis* to *Mesohippus bairdi* studied by Simpson (1944), Lande (1976), and Lynch (1990). Parabolic drift envelopes, representing step rates of $k_0 = 0.020$, 0.025, and 0.065 (darker to lighter shading), cannot explain the observed cumulative change of $\Delta \cdot t = 571$ variance units in the 10×10^6 generation transition. Directional selection (arrows) is required. Similar models can be constructed for the other traits and transitions studied by Lynch (1990), but k_0 is difficult to determine and may be underestimated.

Lynch (1988, p. 145) mentioned that "Selection is always a potential source of bias in estimates of V_m. It is difficult to quantify the magnitude of such bias but it will usually be in the downward direction. A conflict between stabilizing natural selection and directional artificial selection will result in underestimates of V_m/V_E derived from long-term selection experiments. Stabilizing selection will also reduce the rate of divergence of 'unselected' lines." Later he argued (Lynch 1990, p. 736) that "there is no compelling evidence that the estimates of Δ [Δ_{min} or k_{min} and Δ_{max} or k_{max}] are biased in the downward direction." Now, however, it appears that k_0 and Δ_0 may be seriously and systematically underestimated.

12.3 Discussion

Lande (1976, p. 333) concluded that "The minimum mortality rates needed to explain observed rates of evolution in tooth characters of Tertiary mammals are very small, typically about one selective death per million individuals per generation ... These results support the contention that random genetic drift may play a significant role in phenotypic evolution." The idea that one selective death per million individuals will explain long-term change in the fossil record assumes, unrealistically, that change is always directional, and it assumes, again unrealistically, that a long-term base rate calculated over a million or so generations represents what happens on a timescale of one generation.

Genetic models represent change on a timescale of one generation, iterated over and over to represent change on longer scales of time. It makes no sense to use long-term rates from the fossil record to represent what happens in evolution on the timescale of drift and natural selection.

Lynch (1990, p. 736) concluded that "the diversification of mammalian morphology was not caused solely by the interaction of polygenic mutation and random genetic drift. It is too slow for that ... The divergence rate [is] well beyond the neutral expectation, stabilizing selection has predominated in the long term, preventing a potentially much greater diversification of mammals than we see today." Lynch may have been wrong about k_0 and Δ_0, but this quotation is an appropriate interpretation of Figure 12.1 — and a fitting conclusion for the chapter.

Rates on a generational timescale show that random drift has long-term evolutionary potential far beyond what we see. Estes and Arnold (2007) recognized this and explained it with reference to an underlying adaptive landscape. It is also possible that adaptive zones are nothing more than a space between constraining physiological limits, where species divide and interact and in the process create the fitness landscape. High step rates of change in morphology and high speciation rates (not discussed here) are the key to understanding why the limits are reached rapidly and net rates slow down — pending the next perturbation.

Some authors find stabilizing selection difficult to detect because they are trying to measure its effect when it is weakest near a fitness optimum (Hoekstra et al., 2002; Haller and Hendry, 2014). It makes more sense to quantify stabilizing selection by comparing observed evolutionary change to the change expected from random drift. If the observed change is greater, then the difference is attributable to directional selection. If the observed change is less, then the difference is a measure of stabilizing selection.

Rereading the epigraphs here, R. A. Fisher's comprehension of drift and natural selection appears deeper than Sewall Wright's. Random drift, the entropy in the system, is sufficiently powerful that evolution is mostly about constraint and stabilizing selection. Directional selection is possible too, at net rates more rapid than expected for drift.

12.4 Summary

1. Change in evolution is sometimes assumed to be purely unidirectional, but unidirectional change is not found in directional selection experiments, field studies, or fossil studies, and it should not be assumed in evolution. Nothing evolves in a straight line for long.

2. Genetic models are based on generation-to-generation change and the models mislead us when long-term net rates of evolution are substituted for the generation-scale step rates required.
3. The empirical step rates that biologists observe in field studies today are much higher than any long-interval substitutes that paleontologists calculate from the fossil record.
4. High step rates mean that the potential for random drift is also much greater than most authors estimate.
5. Random drift does not begin to equal its potential in evolution — which is evidence for the near ubiquity of natural selection in the form of stabilizing selection (with some directional change included too).
6. Directional and stabilizing selection can be quantified by comparing change observed in evolution to the change expected from random drift. If the observed change is greater, then the difference is attributable to directional selection. If the observed change is less, then the difference is a measure of stabilizing selection.

13

Independent Contrasts: Phylogeny's Influence on Phenotypes

> Species are part of a hierarchically structured phylogeny, and thus cannot be regarded for statistical purposes as if drawn independently from the same distribution...There is one case in which the problem does not arise. That is when the characters respond essentially instantaneously to natural selection in the current environment, so that phylogenetic inertia is essentially absent.
>
> *Joseph Felsenstein,* American Naturalist, *1985, pp. 1, 6*

In 1985 Joseph Felsenstein published an influential and widely cited study showing how phylogeny may affect quantitative comparisons in biology. The statistical procedures generally used to make comparisons among taxa, to make functional interpretations, and to make predictions all assume that species are sampled independently from a common distribution. However, species are related to each other in a hierarchically structured phylogeny, and statistical independence may be compromised when some species are more closely related than others. Felsenstein (1985) dealt with the problem by introducing what he called "independent contrasts" — independent of phylogeny — and proposed that we compare contrasts between species rather than compare species directly.

Felsenstein (1985) noted that species being parts of a hierarchically structured phylogeny will not be a problem when "characters respond essentially instantaneously to natural selection ... so that phylogenetic inertia is essentially absent" (epigraph). Phylogenetic inertia refers to phylogenetic dependence and constraint. Here we pursue the idea of independent contrasts in light of our understanding of the rate of response of species to natural selection, and the rate of response of species to "random walk" regimes of randomly changing sign. What is the time span of phylogeny's influence on phenotypic characteristics? When should we be concerned about the statistical independence of traits for species we know are related phylogenetically?

13.1 Random Walks and Brownian Diffusion

"Brownian motion" is a key element of Felsenstein's independent contrasts solution to phylogeny's influence on phenotypes. Brownian motion was originally a mathematical model to describe the collective movement of particles colliding in a fluid. It is also commonly used to describe the statistical behavior of a collection of random walks evolving independently from a common source or ancestor, in which case the expectation is one of Brownian diffusion through time.

In an evolutionary random walk a phenotypic trait changes by some displacement or difference d_0, a random variable, positively or negatively, in each time step t of the walk (Chapter 4). Time steps are discrete with a constant interval value i_0, and the indexed displacement d_i associated with each time step t_i is independent of all displacements that came before. The randomness of a random walk lies in the sign, positive or negative, assigned with equal probability to each successive d_i. Each step t_i of a random walk has an associated displacement $\pm d_i$ and step rate $r_i = d_i / t_i$.

The expected behavior of a random walk over time can be studied by forward modeling, generating many walks with the same parameters. Cumulative displacements are normally distributed with an expected value (mean) of zero, variance proportional to the number of steps, and standard deviation proportional to the square root of the number of steps (Figure 4.4b). Following Mandelbrot (1983, p. 240), random walks have an expected fractal dimension D = 1.5. The fractal dimension of an individual random walk is given by D = 2 − | m |, where m is the slope on an LDI log-rate-difference plot or by D = 1 + | m | where m is the slope on an LRI log-rate-interval plot. This is useful when interpreting time series because $m \approx \pm 0.5$ and D $\approx$ 1.5 are fingerprints of a random walk. As we have seen before, a random walk does not have a single rate but has a rate (or distribution of rates) for each interval length, and no rate has meaning independent of its timescale.

The power of Brownian random diffusion as a model for evolutionary change lies in the generality of its application, and in the simplicity of its parameters: with an expected value (mean) of zero, variance increasing in proportion to time, and standard deviation increasing in proportion to the square root of time. This generality and simplicity is also a liability. Brownian diffusion can be tailored to fit every history. Some might consider this an advantage; however, a model that explains everything often explains nothing. The possibility of representing every history by Brownian diffusion is probably one reason that random drift is assumed to be so common in evolution. However, a realistic evolutionary model must: (1) match the timescale being modeled and (2) employ rates that are representative of evolution on this timescale.

The generation is the focal timescale in evolutionary-process studies because this is the scale of expression of successive descendent phenotypes, changing or unchanging, whether subject to natural selection or random drift. Evolution on longer timescales involves aggregation of repeated generation-to-generation steps of change. When random walks are considered models for evolutionary change, the step interval i_0, step displacement d_0, and step rate r_0 of the random walks must match i_0, d_0, and r_0 for the process being studied, and the step interval of the random walk is necessarily the biological generation time for the evolutionary sequence. Felsenstein (1985) considered time in his Brownian diffusion models to be time in generations, but this was only stated explicitly in an early appendix (Felsenstein 1973: 490), and the requirement is often overlooked or ignored.

Understating the step interval or overstating the step rate for a Brownian process systematically overestimates the associated variances, standard deviations, and disparity ranges. Similarly, overstating the step interval or understating the rate for a Brownian process systematically underestimates the associated variances, standard deviations, and disparity ranges. Since the variances will be proportional to interval length in every case, this proportionality by itself is no demonstration of appropriate scaling.

13.2 Carnivores and Ungulates: Worked Examples

Theodore Garland used a time-calibrated carnivore–ungulate phylogeny (Garland, 1992, figure 1; Garland and Janis, 1993) to calculate Felsenstein independent contrasts, and then used the contrasts to compare rates of evolution in the two groups. Garland (1992) calculated contrasts using differences in $\log_{10}$ values of body weight, and standardized these by dividing by the square root of the sum of relevant branch lengths (in years). Standardization using a square-root-of-branch-length denominator is a logical extension of finding, or assuming, that variances increase in proportion to time (Felsenstein, 1985). Phenotypic contrasts or differences will then increase with the square root of time. Garland's standardized contrasts are rates in units of $\log_{10}$ body weight per square root of time in years. He found that rates of body weight evolution in carnivores do not differ significantly from those in ungulates.

It is possible to make the same comparison in conventional and more broadly comparable units. The coefficient of variation (phenotypic standard deviation as a proportion of the mean) is typically about 0.15 for body weight in mammalian populations (Yablokov, 1974; Gingerich, 2000; McKellar and Hendry, 2009) — which is a close approximation to the phenotypic standard deviation of natural log (log base e or ln) transformed measurements (Lewontin, 1966). This means that carnivores and ungulates ranging in weight from 2.5 kg for *Mephitis mephitis* to 2,000 kg for *Ceratotherium simum* (Garland, 1992, figure 1) span a range of about

45 phenotypic standard deviation units (ln 2,000 − ln 2.5 = 7.60 − 0.91; 7.60 − 0.91 = 6.68 ln units; 6.68 / 0.15 = 44.6 phenotypic standard deviation units).

13.2.1 Conventional Differences and Rates

The carnivore species in the phylogeny shown in Garland's (1992) figure 1 have divergence times ranging from 0.5 to 58 million years before present, meaning that they have been evolving separately for 1 to 116 million years. The first two carnivores in Garland's phylogeny, the bears *Ursus horribilis* and *U. americanus*, have been separated for 5 + 5 = 10 million years. The generation time for a carnivore of median size included here is estimated to be about 4.4 years (Equation 3.2). An interval of 5 + 5 = 10 million years is then equivalent to about 2.27 million generations of separation. The two bears have body weights of 251.3 and 93.4 kg, respectively. Natural logs, 5.53 and 4.54, differ by 0.99 / 0.15 = 6.60 standard deviation units. This is a minimal cumulative difference of one from the other because the difference between them may have been greater at some time in the past. Similar estimates of separation time and weight can be made for all 120 pair-wise combinations of Garland's 16 carnivore species.

The ungulate species in the phylogeny of Garland's (1992) figure 1 have divergence times ranging from 0.5 to 66 million years, meaning that they have been evolving separately for 1 to 132 million years. The first two ungulates in Garland's phylogeny, the rhinoceroses *Diceros bicornis* and *Ceratotherium simum*, have been separated for 13 + 13 = 26 million years. The generation time for an ungulate of median size included here is estimated to be about 6.5 years (Equation 3.2). An interval of 13 + 13 = 26 million years is then equivalent to about 4.00 million generations of separation. The two rhinoceroses have body weights of 1,200 and 2,000 kg, respectively. Natural logs, 7.09 and 7.60, differ by 0.51 / 0.15 = 3.41 standard deviation units. This is again a minimal cumulative difference of one from the other because the difference between them may have been greater in the past. Similar estimates of separation time and weight can be made for all 351 pair-wise combinations of Garland's 27 ungulate species.

Figure 13.1a is a combined LDI–LRI plot for the distributions of conventional body weight differences and rates for Garland's (1992) carnivores. The median interval length separating species is $10^{7.397} = 2.49 \times 10^7$ generations, the median body weight difference between species is $10^{0.998} = 9.95$ standard deviation units, and the median haldane rate for body weight is $h_{7.397} = 10^{-6.227} = 5.93 \times 10^{-7}$ standard deviations per generation. The projected step rate for the differences and rates is $h_0 = 10^{-0.562} = 0.274$ standard deviations per generation on a timescale of one generation. Confidence intervals for the step rate range from about $h_0 = 10^{-2.50}$

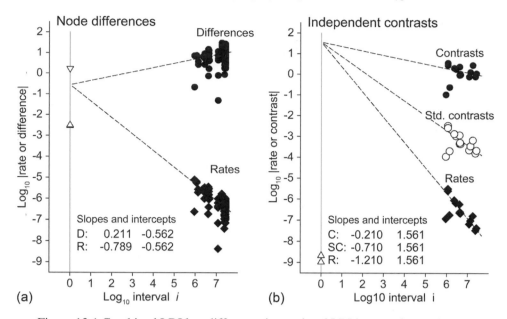

Figure 13.1 Combined LDI log-difference-interval and LRI log-rate-interval plots for body weight in 16 species of mammalian carnivores studied by Garland (1992). (a) Terminal node differences. Differences between species pairs are weight differences in standard deviation units. Rates are weight differences divided by the corresponding two-way interval in generations. Extrapolations of contrasts and rates point to a common step rate of $h_0 = 10^{-0.562} = 0.274$ standard deviations per generation. Median intervals, differences, and rates are $10^{7.397}$, $10^{0.998}$, and $10^{-6.227}$, respectively. (b) Phylogenetic independent contrasts. Contrasts are weight differences, in standard deviations, calculated using the *pic* function of Paradis (2012) and the phylogenetic tree of Garland (1992). Standardized contrasts are contrasts divided by the square root of the corresponding interval in generations. Rates are contrasts divided by the entire interval. Extrapolations of contrasts, standardized contrasts, and rates point to a common (and unrealistic) step rate of $h_0 = 10^{1.561} = 36.392$ standard deviations per generation. Median intervals, contrasts, standardized contrasts, and rates are $10^{6.612}$, $10^{0.054}$, $10^{-3.329}$, and $10^{-6.663}$, respectively. Paired triangles enclose a 95% confidence interval for each intercept.

to $h_0 = 10^{0.23}$. A step rate of $h_0 = 0.274$ is about double the median step rate of $h_0 = 0.153$ that we found for field studies (Table 9.1).

   Figure 13.2a is a combined LDI–LRI plot for the distributions of conventional body weight differences and rates for Garland's (1992) ungulates. The median interval length separating species is $10^{6.789} = 6.15 \times 10^6$ generations, the median body weight difference between species is $10^{0.948} = 8.87$ standard deviation units, and the median haldane rate for body weight is $h_{6.789} = 10^{-5.982} = 1.04 \times 10^{-6}$ standard deviations per generation. The projected step rate for the differences and rates is $h_0 = 10^{-2.070} = 0.009$ standard deviations per generation on a timescale of

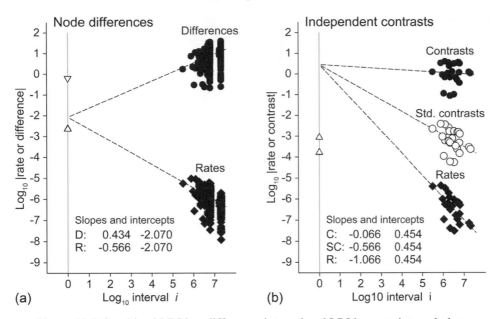

Figure 13.2 Combined LDI log-difference-interval and LRI log-rate-interval plots for body weight in 27 species of mammalian ungulates studied by Garland (1992). (a) Terminal node differences. Differences between species pairs are weight differences in standard deviation units. Rates are weight differences divided by the corresponding two-way interval in generations. Extrapolations of differences and rates point to a common step rate of $h_0 = 10^{-2.070} = 0.009$ standard deviations per generation. Median intervals, differences, and rates are $10^{6.789}$, $10^{0.948}$, and $10^{-5.982}$, respectively. (b) Phylogenetic independent contrasts. Contrasts are weight differences, in standard deviations, calculated using the *pic* function of Paradis (2012) and the phylogenetic tree of Garland (1992). Standardized contrasts are contrasts divided by the square root of the corresponding interval in generations. Rates are contrasts divided by the entire interval. Extrapolations of contrasts, standardized contrasts, and rates point to a common step rate of $h_0 = 10^{0.454} = 2.844$ standard deviations per generation. Median intervals, contrasts, standardized contrasts, and rates are $10^{6.333}$, $10^{0.024}$, $10^{-3.189}$, and $10^{-6.318}$, respectively. Paired triangles enclose a 95% confidence interval for each intercept.

one generation. Confidence intervals for the step rate range from about $h_0 = 10^{-2.68}$ to $h_0 = 10^{-0.23}$. A step rate of $h_0 = 0.009$ is less than one tenth of the median step rate of $h_0 = 0.153$ found for field studies (Table 9.1).

13.2.2 Contrasts, Standardized Contrasts, and Contrast Rates

Figure 13.1b is a combined LDI–LRI plot for the distributions of phylogenetic independent contrasts, standardized independent contrasts, and contrast rates for the body weights of carnivores, interpreted in the context of Garland's (1992) phylogenetic tree for carnivores. The median interval length separating species is $10^{6.612} = 4.09 \times 10^6$ generations, the median body weight contrast is $10^{0.054} = 1.13$ standard deviation units, the median standardized contrast is $10^{-3.329} = 4.69 \times 10^{-4}$, and the median haldane rate is $h_{6.612} = 10^{-6.663} = 2.17 \times 10^{-7}$ standard deviations per generation. The extrapolated step rate for all is $h_0 = 10^{1.561} = 36.392$ standard deviations per generation on a timescale of one generation. Confidence intervals for the step rate range from about $h_0 = 10^{-9.10}$ to $h_0 = 10^{4.25}$. The extrapolated step rate of $h_0 = 36.392$ is much greater than the maximum step rate of $h_0 = 5.41$ that we found for field studies (Table 9.1). It is implausible if not impossible as a step rate.

Figure 13.2b is a combined LDI–LRI plot for the distributions of phylogenetic independent contrasts, standardized independent contrasts, and contrast rates for the body weights of ungulates, interpreted in the context of Garland's (1992) phylogenetic tree for ungulates. The median interval length separating species is $10^{6.333} = 2.15 \times 10^6$ generations, the median body weight contrast is $10^{0.024} = 1.06$ standard deviation units, the median standardized contrast is $10^{-3.189} = 6.47 \times 10^{-4}$, and the median haldane rate is $h_{6.333} = 10^{-6.318} = 4.81 \times 10^{-7}$ standard deviations per generation. The extrapolated step rate for all is $h_0 = 10^{0.454} = 2.844$ standard deviations per generation on a timescale of one generation. Confidence intervals for the step rate range from about $h_0 = 10^{-3.10}$ to $h_0 = 10^{4.16}$. The extrapolated step rate of $h_0 = 2.884$ is nearly 20 times greater than the median step rate of $h_0 = 0.153$ that we found for field studies (Table 9.1), and again implausible if not impossible as a step rate.

Based on our experience in previous chapters, we expect both the differences between species and their corresponding rates of change to be dependent on the time intervals involved. This is clear for the carnivore and ungulate differences and rates plotted in Figures 13.1a and 13.2a. It will come as a surprise to some that independent contrasts, standardized contrasts, and contrast rates are similarly dependent on the time intervals involved. This is clear when we examine the contrasts, standardized contrasts, and contrast rates plotted for carnivores in Figure 13.1b and for ungulates in Figure 13.2b. Plotting conventional node differences and rates side by side with independent contrasts, standardized contrasts, and contrast rates shows the distortion and loss of constraining statistical power resulting from the latter.

13.2.3 Forward-Modeling Simulations

Figure 13.3a is the time-calibrated carnivore and ungulate phylogeny of Garland (1992, figure 1) drawn with the independent variable, geological time, on the *y*-axis, and the dependent variable, body weight, on the *x*-axis. Geological time is in

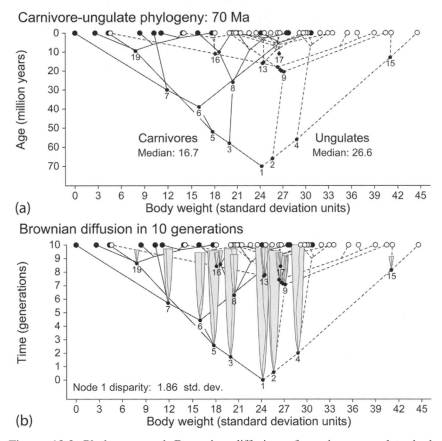

Figure 13.3 Phylogeny and Brownian diffusion of carnivore–ungulate body weights. (a) 70-million-year phylogeny of Garland (1992, figure 1) linking the body weights of mammalian carnivores and ungulates. Median weight for carnivores is 16.7 standard deviations greater than the weight of the smallest species. Median weight for ungulates is 26.6 standard deviations greater. Note that modern carnivores (closed circles linked by solid lines) and ungulates (open circles linked by dashed lines) overlap extensively in body weight in spite of their 70-million-year phylogenetic separation. (b) Carnivore–ungulate phylogeny scaled as if it evolved in 10 generations, showing random walks and Brownian diffusion envelopes for the 19 deepest nodes. Step rate is 0.15 standard deviations per generation; diffusion envelopes enclose 95% of expected variation. Node 1 disparity after 10 generations is much less than the observed disparity of body weight in carnivores and ungulates.

millions of years. Body weight is calibrated in phenotypic standard deviations starting with the smallest species, *Mephitis mephitis*, set to zero, and ending with the largest at about 45 standard deviation units. Carnivorous mammals, with a median at 16.7 standard deviation units, tend to be smaller, and herbivorous ungulates, with a median at 26.6 standard deviation units, tend to be larger. However, there is considerable overlap in body weights for the two groups.

There are 16 terminal taxa of carnivores, and an additional 15 nodes within the carnivore clade. There are 27 terminal taxa of ungulates, and an additional 26 nodes within the ungulate clade. The 19 deepest nodes, including the one linking carnivores and ungulates, serve as landmarks and are numbered for reference consecutively starting with the oldest. It is reasonable to assume, following Garland (1992), that the phylogeny in Figure 13.3a is representative of carnivores and ungulates in terms of morphological disparity and in terms of the order and relative timing of branching.

We can now begin a series of forward-modeling simulations starting at successive nodes and moving forward through time to see what Brownian diffusion looks like when modeled on a timescale of generations, and calibrated in terms of the median step rate, $h_0 = 0.15$, that we found in Chapter 8 field studies. In the simulations here h_0 is regarded as a constant, but a random variable with a mean $h_0 = 0.15$ and representative dispersion would yield the same result.

Brownian diffusion depends on the step rate and the time span. Dependence on time span is easily seen by holding the step rate constant and increasing the time span in successive simulations. Successive simulations here have time spans differing by an order of magnitude. The goal is to see how dependence on phylogeny behaves in relation to time. An arbitrary but conservative bench mark is the time it takes for Brownian diffusion to develop the morphological disparity, 45 standard-deviations, representative of the entire carnivore–ungulate radiation. When this happens, there is no longer any necessary dependence on phylogeny.

The x-axis in Figure 13.3b is the same as that in Figure 13.3a, but now the phylogeny has been fit to a new timescale and the y-axis represents just 10 generations of evolutionary time. In conjunction with the new timescale, a series of random walk simulations is superimposed on the phylogeny. A 10-step random walk that starts at node 1 on the x-axis is expected to end at the same value in terms of standard deviation units, but this becomes increasingly unlikely as the number of steps increases (Figure 4.2c). The variance of a random walk increases with the number of steps, as $h_0^2 \cdot t$, and the standard deviation of a random walk increases as the square root of the number of steps, as $h_0 \cdot \sqrt{t}$.

The phenotypic variance at any step will equal the step rate squared, multiplied by the number of generations. For $h_0 = 0.15$ and $t = 10$: $\sigma^2 = 0.15^2 \cdot 10 = 0.225$. The corresponding phenotypic standard deviation is equal to the square root of this or $\sigma = \sqrt{0.225} = 0.474$. Alternatively, and equivalently, the standard deviation is equal to the step rate multiplied by the square root of the number of generations: $\sigma = 0.15 \cdot \sqrt{10} = 0.474$.

A 95% confidence interval for the node 1 random walk after 10 generations is given by the node-1 initial value $\pm 1.96 \cdot 0.15 \cdot \sqrt{10}$. The corresponding disparity is then $2 \cdot 1.96 \cdot 0.15 \cdot \sqrt{10} = 1.86$ standard deviation units. The random walks and envelopes of Brownian diffusion in Figure 13.3b do not come close to the 45 standard deviations of body weight disparity in Garland's (1992) carnivore and ungulate phylogeny, nor do they link the deeper nodes of the phylogeny.

Figure 13.4a shows the same phylogeny as that in Figures 13.3a and 13.3b, but now the *y*-axis is scaled to 100 generations of evolutionary time. Random walk simulations have again been added for the deeper nodes. The resulting random walks and envelopes of Brownian diffusion do not approach the 45 units of body weight disparity displayed by the carnivores and ungulates, nor do they link the

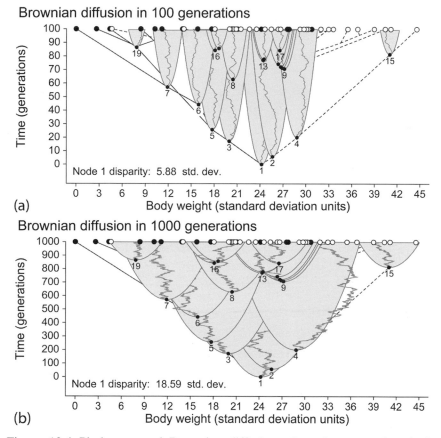

Figure 13.4 Phylogeny and Brownian diffusion of carnivore–ungulate body weights. (a) Carnivore–ungulate phylogeny scaled as if it evolved in 100 generations, showing random walks and Brownian diffusion envelopes for the 19 deepest nodes. (b) Carnivore–ungulate phylogeny scaled as if it evolved in 1,000 generations, showing random walks and Brownian diffusion envelopes for the 19 deepest nodes. Step rates are 0.15 standard deviations per generation; diffusion envelopes enclose 95% of expected variation. Note that node 1 disparities, 5.88 and 18.59, are less than the observed disparity of body weight in carnivores and ungulates.

deeper nodes of the phylogeny. Brownian diffusion scaled to 1,000 generations of evolutionary time (Figure 13.4b) explains more of the carnivore–ungulate disparity and the diffusion begins to link deeper nodes in the phylogeny.

A different picture emerges in Figure 13.5, where simulations for 10,000 and 100,000 generations of Brownian diffusion are superimposed on the carnivore–ungulate phylogeny. All of the deeper nodes are now linked by diffusion. The node 1 disparity for 10,000 generations of diffusion is 58.80 standard deviation units,

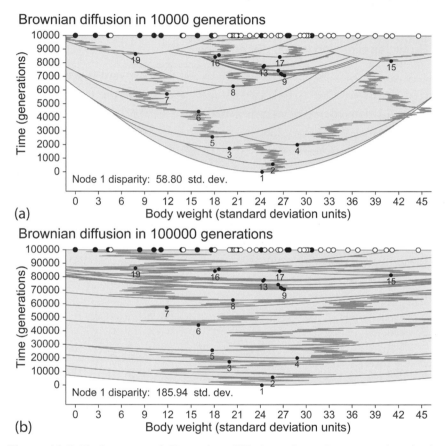

(a)

(b)

Figure 13.5 Phylogeny and Brownian diffusion of carnivore–ungulate body weights. (a) Carnivore–ungulate phylogeny scaled as if it evolved in 10,000 generations, showing random walks and Brownian diffusion envelopes for the 19 deepest nodes. (b) Carnivore–ungulate phylogeny scaled as if it evolved in 100,000 generations, showing random walks and Brownian diffusion envelopes for the 19 deepest nodes. Step rates are 0.15 standard deviations per generation; diffusion envelopes enclose 95% of expected variation. Note that node 1 disparities, 58.80 and 185.94, are greater than the observed disparity of body weight in carnivores and ungulates. Simulation shows that the influence of phylogeny on body weight was lost within about 10,000 generations — which is less than one tenth of 1% of carnivore-ungulate evolutionary history.

and the node 1 disparity for 100,000 generations is 185.94 units. Both are substantially greater than the 45 units of disparity displayed by the carnivores and ungulates combined. At 10,000 generations the phylogeny is no longer visible through the diffusion envelopes, and it can have little remaining influence on body weight. The median generation time for Garland's (1992) carnivores and ungulates is 5.7 years, so the limit for phylogeny's influence on carnivore–ungulate body weight is about 57,000 years — less than one tenth of 1 percent (less than 0.001) of the group's combined 70-million-year history.

13.3 Discussion

The simulations of Brownian diffusion illustrated by forward modeling in Figures 13.3–13.5 show how statistical independence of phenotypic traits emerges from their initial dependence on the hierarchical structure of a phylogeny. Brownian diffusion ensures that statistical independence of phenotypic traits is achieved predictably with the passage of time. The time required for this independence is linked to the variability of species, the size of the phylogenetic contrasts of interest, and the step rate of the Brownian diffusion process. Most species of carnivores and ungulates that are similar in body weight remain so after 10,000 generations because they have responded similarly to the environments they live in, not because of their phylogenetic relationships. This is another way of phrasing Joseph Felsenstein's disclaimer that *"statistical independence is not a problem* for species sampled from the same phylogeny *when species respond rapidly to natural selection"* (Felsenstein, 1985, p. 6; italics added) — laboratory selection experiments (Chapter 7) and field studies (Chapter 8) show that species do respond rapidly to artificial and natural selection.

The only rates we know recorded on a timescale of one generation to characterize natural selection and drift in nature are the step rates in Chapter 8 with a median of $h_0 = 0.15$ standard deviations per generation. Step rates from the three sets of "control" lineages in the selection experiments summarized in Table 7.2 have medians of $h_0 = 0.152$, $h_0 = 0.178$, and $h_0 = 0.250$, respectively, so there is no reason to expect drift rates to be lower than the median of $h_0 = 0.15$ for field rates.

In hindsight, the broad overlap of body weight in carnivores and ungulates, and the crisscrossing of lineages leading to different body weights within carnivores and ungulates (Figure 13.3a) were both indications that phylogeny's influence on body weight is limited. Modern cetaceans provide a similar example (Gingerich 2015). Body weight and change in body weight are unusual in being easy to measure and unusual in being well studied experimentally. The empirical step rate of $h_0 = 0.15$ standard deviations per generation is high, and speciation with divergent selection could, in theory, generate the full 45 standard deviation disparity of body sizes of carnivores and ungulates from an intermediate node in as few as $22.5/0.15 = 150$ generations. Brownian diffusion requires some 10,000 generations, but still this is a small fraction, less than one tenth of 1%, of the combined group's evolutionary history.

Body weight is a phenotypic characteristic that is easily quantified and varies greatly among carnivores and ungulates, but there may be traits that behave differently. There may be traits that retain statistical dependence on phylogeny through the whole history of a group. If so, the traits, the groups, and the associated rates need to be identified and quantified. Statistical dependence on phylogeny does not extend automatically to any characteristic, and dependence on phylogeny may never last very long for the morphological and life history traits that we most often study in comparative biology.

13.4 Summary

1. A realistic evolutionary model must match the generation-to-generation time-scale being modeled and employ rates that are representative on this timescale.
2. Time in a Brownian diffusion evolutionary model is time in generations, a requirement often overlooked or ignored.
3. Forward-modeling of Brownian diffusion shows that the time span of phylogeny's influence on body weight in mammalian carnivores and ungulates lasts no more than about 10,000 generations, or some 57,000 years, less than one tenth of 1% of the combined group's evolutionary history.
4. Statistical dependence on phylogeny does not extend automatically to any or all characteristics of interest, and dependence on phylogeny may never last long for the morphological and life history traits that we most often study in comparative biology.

14

Rate Perspective on Early Bursts of Evolution

> The basic differentiation of each [mammalian] order took a much shorter
> time than its later adjustment, spread, and diversification ... It follows
> that the basic differentiation must have proceeded, on the average, more
> rapidly than the later recorded evolution, almost surely twice as fast and
> probably more, quite possibly ten or fifteen times as rapidly in
> some cases.
>
> *Simpson, 1944,* Tempo and Mode in Evolution*, p. 121*

An "early burst" model of evolutionary diversification and morphological disparity
is well established in paleontology (Simpson, 1944; Foote, 1997; Gingerich, 2001;
Payne et al., 2009; Smith et al., 2010; Hughes et al., 2013) and in ecological theory
(Hutchinson, 1959; Levinton, 1979; Schluter, 2000; Gavrilets and Vose, 2005;
Ingram et al., 2012). When a lineage first enters a new adaptive zone, the net rate of
morphological evolution is rapid. Then, as ecological space becomes filled and
opportunity diminishes, the net rate of change necessarily slows down. Rate in this
context refers to the rate of increase of species richness or, alternatively, to the rate
of expansion of morphological disparity in relation to time. Species richness and
morphological disparity in a clade are often closely related and thus are, to some
degree, interchangeable.

Contradicting this established view, a recent review of comparative information
for living animals claimed to find that early bursts of body size and shape evolution
are rare (Harmon et al., 2010). This is worth examining from a rate perspective.

14.1 Comparative Approach to Early Bursts

Harmon et al. (2010) used phylogenetic methods to analyze the distribution of
morphological variation in different taxonomic groups in relation to the expectations of three models of change through time: (1) a baseline Brownian motion or,
here, Brownian diffusion ("BD") model, (2) an Ornstein–Uhlenbeck or single

stationary peak ("OU") model, and (3) an early burst ("EB") model. The BD model is expected to have less variance in subclades than in the whole clade because net variance σ^2 increases in proportion to time and subclades are younger. We saw this in the last chapter (e.g., Figure 13.4). The OU model is expected to have about the same net variance σ^2 in subclades as in the whole clade because variance accumulates in proportion to time but the variance is constrained in subclades and whole clades by the intensity $\alpha \geq 0$ described in Chapter 4. Rapid diversification is logical when new adaptive zones are colonized, and subsequent stability is logical when they are filled.

The EB model of Harmon et al. (2010, p. 2389) is a model "in which the net rate of evolution slows exponentially through time as the radiation proceeds." This describes, in general terms, the BD model. The distinction Harmon et al. make is that their restricted EB model, hereafter EB*, has net variance σ^2 increasing through time by an amount less than expected for the BD model, with the loss dependent on a parameter $r < 0$. For EB* to be recognizably different from the BD model, parameter r must be distinguishable from zero. According to Harmon et al. (2010), when $r = 0$, then EB* is exactly BD.

Harmon et al. (2010) fit these models to phylogenies, body sizes, and body shapes in a survey of clades of extant taxonomic groups. For body sizes in clades of at least 10 species, 25 clades were reported to favor the BD model, 12 clades were reported to favor the OU model, and only one clade was found to favor the EB* model. For body shapes in clades of at least 10 species, 14 clades were reported to favor the BD model, 14 clades were reported to favor the OU model, and again only one clade was found to favor the EB* model. This led Harmon et al. (2010) to conclude that early bursts of diversification are rare in evolution, contrary to the descriptions of paleontologists studying patterns of morphological disparity through time in the fossil record.

A broad and conflicting generalization like this deserves scrutiny. Here we focus on body size evolution in one clade studied by Harmon et al.: Darwin's finches or Geospizini, a clade in which Harmon et al. found the BD model to be strongly favored over both OU and EB*.

14.2 Rates of Evolution in Darwin's Finches

Darwin's finches, isolated geographically in the Galápagos Islands of Ecuador (14 species) and Cocos Island of Costa Rica (one species), represent a classic evolutionary radiation. They have been intensively studied by Lack (1947), Grant (1981; 1986), Schluter (1984), Petren et al. (1999), Grant and Grant (2014), and others. Darwin's finches have no fossil record, but molecular clocks suggest that the group originated 2.84 million years before present (Grant, 1994) or 2.3 million years before present (Sato et al., 2001). We calculated rates for two species of Galápagos finches in Chapter 8.

Harmon et al. (2010, table 1) chose tarsus length to represent body size in 14 species of Darwin's finches, they chose 2.5 million years before present to represent the time of initial divergence, and they chose one year to represent generation time. The birds, Geospizini, generally have a longer generation time, but this need not concern us here. Figure 14.1 shows tarsus length in relation to the phylogenetic tree of Petren et al. (1999), with the initial divergence set at 2.5 million years or 2.5 million generations. The tarsus lengths in the right-hand column are calibrated in phenotypic standard deviation units and expressed relative to the tarsus length of the smallest species *Geospiza fulginosa*.

Tarsus lengths can be compared in terms of conventional node differences and rates (Figure 14.2a). Differences can be calculated for all 91 pair-wise combinations

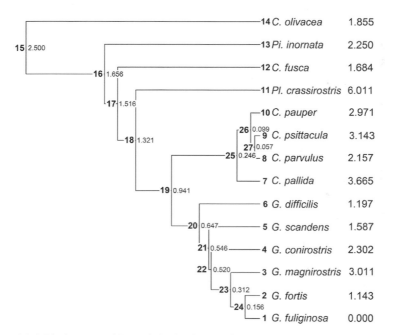

Figure 14.1 Phylogeny of Darwin's finches or Geospizini based on microsatellite DNA length variation (Petren et al., 1999). Fourteen species are represented, in the genera *Camarhynchus*, *Geospiza*, *Pinaroloxias*, and *Platyspiza*. All are from the Galápagos Islands except *Pi. inornata* from Cocos Island. Right-hand column of numbers is tarsus length in standard deviation units relative to the smallest species (*G. fulginosa*; trait measurements, within-species standard deviation, and generation time are those of Harmon et al., 2010). Node ages (inset 2.500, etc.) are in millions of years or millions of generations, interpolated from UPGMA branch lengths of Petren et al. (1999); phylogeny scaled from Harmon et al. (2010).

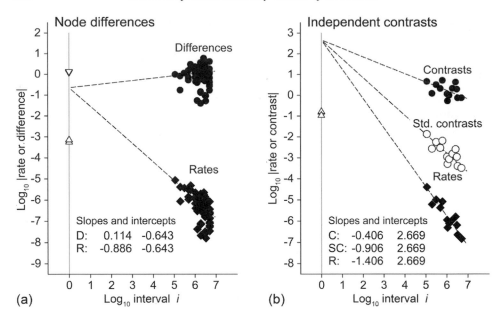

Figure 14.2 Combined LDI log-difference-interval and LRI log-rate-interval plots for tarsus length in 14 species of Darwin's finches (Figure 14.1). (a) Terminal node differences. Differences between species pairs are tarsus-length differences in standard deviation units. Rates are tarsus length differences divided by the corresponding two-way interval in generations. Extrapolations of differences and rates point to a common step rate of $h_0 = 10^{-0.643} = 0.228$ standard deviations per generation. Median intervals, differences, and rates are $10^{6.422}$, $10^{0.064}$, and $10^{-6.184}$, respectively. (b) Phylogenetic independent contrasts. Contrasts are tarsus-length differences, in standard deviations, calculated using the *pic* function of Paradis (2012) and the phylogenetic tree of Figure 14.1. Standardized contrasts are contrasts divided by the square root of the corresponding interval in generations. Rates are contrasts divided by the entire interval. Extrapolations of contrasts, standardized contrasts, and rates point to a common (unrealistic) step rate of $h_0 = 10^{2.669} = 466.7$ standard deviations per generation. Median intervals, contrasts, standardized contrasts, and rates are $10^{6.039}$, $10^{0.290}$, $10^{-2.804}$, and $10^{-5.942}$, respectively. Paired triangles enclose 95% confidence intervals for each intercept. Note that the vertical scale in (b) has been moved down by one unit compared to that in (a).

of the 14 species. Each difference is a minimal cumulative difference of one tarsal length from the other because differences may have been greater at some time in the past. Each of the 91 pairs of species has an associated two-way interval length separating them in terms of generations. A conventional rate, in standard deviations per generation, is then the ratio of the observed difference in tarsus length in standard deviation units to the associated interval length in generations.

Temporal-scaling extrapolations of differences and rates point to a common intercept and step rate of $h_0 = 10^{-0.643} = 0.228$ standard deviations per generation

(Figure 14.2a). This extrapolated step rate is a little higher than the median step rate of $h_0 = 0.153$ for field studies found in Chapter 8 (see Table 9.1).

Confidence intervals for the intercepts in Figure 14.2a range from -3.254 to $+0.139$ (open triangles), where $10^{-3.254} = 0.0006$ and $10^{0.139} = 1.376$.

Tarsus lengths can also be compared in terms of phylogenetic independent contrasts, standardized contrasts, and rates (Figure 14.2b), just as we compared carnivore and ungulate body weights in terms of contrasts and rates in Chapter 13. Contrasts here are tarsus-length differences, in standard deviation units, calculated for each of the 13 sister groups involving terminal taxa in Figure 14.1. Contrasts were calculated using the *pic* function of Paradis (2012), and the phylogenetic tree in Figure 14.1. Standardized contrasts are contrasts divided by the square root of the corresponding interval in generations. Contrast rates are contrasts divided by the entire corresponding interval, again expressed in standard deviations per generation.

Temporal-scaling extrapolations of contrasts, standardized contrasts, and rates point to a common intercept and step rate of $h_0 = 10^{2.669} = 466.7$ standard deviations per generation (Figure 14.2a). This extrapolation is virtually impossible as a step rate. Confidence intervals for the intercepts range from -0.966 to $+3.966$ (open triangles, some off the chart), where $10^{-0.966} = 0.108$ and $10^{3.966} = 9906.2$.

Harmon et al. (2010) reported a maximum likelihood estimate of $\sigma^2 = 1.74$ variance units per million years (million generations) for body size evolution in Geospizini. This is not a step rate, but a rate on a timescale of ca. $10^{6.039} = 1.09 \times 10^6$ years or generations (Figure 14.2b). In Chapter 13 we saw that the variance of a random walk increases with the number of steps, as $h_0^2 \cdot t$, and the standard deviation of a random walk increases as the square root of the number of steps, as $h_0 \cdot \sqrt{t}$. If we set $1.74 = h_{6.039}^2 \cdot 1{,}000{,}000$ and solve for $h_{6.039}$, we find $h_{6.039} = 0.0013$ and $\log_{10} h_{6.039} = -2.880$. The latter, $\log_{10} h_{6.039} = -2.880$, is very close to the median of the logged standardized contrasts (Figure 14.2b). This corroborates Garland's (1992) use of standardized contrasts as representative rates. However, these are not step rates but rates on a much longer timescale.

The reason for scaling the differences, contrasts, and rates in Figure 14.2 is to estimate the step rate for tarsus length evolution in Geospizini. The step rate can then be used in a forward-modeling exercise to illustrate Brownian diffusion and Ornstein–Uhlenbeck diffusion. Temporal-scaling extrapolation of conventional differences and rates point to a common intercept and step rate of $h_0 = 10^{-0.643} = 0.228$ standard deviations per generation (Figure 14.2a) that is close to the median step rate of $h_0 = 0.153$ for field studies found in Chapter 8 (Table 9.1). It would be preferable to use the latter for realistic modeling to avoid any extrapolation, but the lower rate of 0.0013 standard deviations per generation, derived from Harmon et al.'s (2010) fit to Geospizini, is better for illustrative purposes — even though it is a rate on much too long a timescale to represent a step rate.

14.3 Forward Modeling of Brownian Diffusion

Brownian diffusion is a standard model in evolutionary studies (Chapter 4). Figure 14.3a shows a 2.5 million-year history for Darwin's finches, Geospizini, inferred from allele length variation at 16 DNA microsatellite loci (Petren et al., 1999). Tarsus-length values expressed in standard deviation units are those listed in Figure 14.1. Ancestral node values were estimated using the *ace* function of

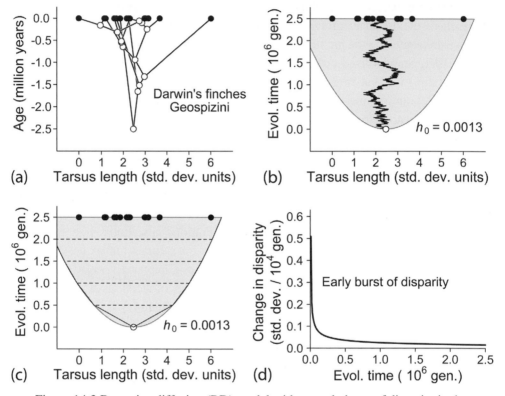

Figure 14.3 Brownian diffusion (BD) model with an early burst of disparity in the 2.5-million-year history of Darwin's finches, Geospizini. (a) Tarsus-length values in standard deviation units, measured (closed circles) or inferred (open circles), showing change through time on the phylogenetic tree of Figure 14.1. (b) 95% confidence envelope for BD (shaded) of tarsus length using $h_{6.039} = 0.0013$ as a step rate (see text). Diffusion would be more rapid for a more representative step rate. One 2.5-million-step random walk is superimposed on the diffusion envelope. After 2.5×10^6 generations of diffusion, the confidence envelope spans 8.06 standard deviation units of disparity. (c) Disparity increases through time at declining rates. Dashed lines represent disparity at intervals of 0.5×10^6 generations. Rates of change are proportional to the connecting line segments, which decrease in their deviation from vertical through time. (d) Summary showing declining rates of change in disparity, all on time scales of 10^4 generations. Rates are highest early in evolutionary time, reflecting the early burst of disparity inherent in a BD model.

Paradis (2012). One 2.5-million-step random walk with a constant step rate $h_0 = 0.0013$ is shown in Figure 14.3b. This starts at the base of the phylogenetic tree and proceeds forward in time to the present. A large number of random walks like this define the 95% confidence envelope for Brownian diffusion shown by the shaded parabolic distribution. Note that at the end, after 2.5×10^6 generations of diffusion, the confidence envelope spans just over eight standard deviations of disparity and includes all of the finch tarsal lengths.

Disparity increases through time. This is shown in simple form in Figure 14.3c, where dashed lines represent the disparity at successive 0.5 million-generation intervals. The most rapid increase in disparity shown in Figure 14.3c is that from 0 to 0.5 million generations. The slowest increase in disparity in Figure 14.3c is that from 2.0 to 2.5 million generations. The increase in disparity is most rapid early and slower later in time — which is the common understanding of an early burst of morphological evolution.

This is an early burst model because, following Simpson (1944) in this chapter's epigraph, early limits in a Brownian diffusion model increase more rapidly than any increase expected later, with the former easily 10 or 15 times more rapid than the latter. Figure 14.3d shows, in detail, rates of change in disparity for the Brownian diffusion model applied to Darwin's finches. All of the rates compared in Figure 14.3d are rates on a common timescale of 10^4 generations.

14.4 Forward Modeling of Ornstein–Uhlenbeck Diffusion

In Chapter 4 (Equation 4.6) we saw that Brownian diffusion is a special case of Ornstein–Uhlenbeck diffusion where the intensity parameter $\alpha = 0$. Figure 14.4a shows a 2.5-million-step Ornstein–Uhlenbeck random walk that starts at the base of the phylogenetic tree and proceeds forward in time to the present. The step rate in Figure 14.4a is a constant $h_0 = 0.0013$, the rate in Figure 14.3b. Here, however, there is an additional negative feedback term with intensity $\alpha = 0.0001$ regulating the random walk. This brings the random walk back toward the initial value, θ, at each step.

The intensity $\alpha = 0.0001$ in Figure 14.4a was chosen to make the random walk visible, while still constraining it to track θ. A larger α value relative to h_0 would make the random walk narrower and more difficult to see, and a smaller α value relative to h_0 would make the random walk more random and more like the random walk in Figure 14.3b. As in Figure 14.3c, a large number of random walks define the 95% confidence envelope for Ornstein–Uhlenbeck diffusion shown by the shaded distribution that is barely visible behind the random walk in Figure 14.4a. The confidence interval was constructed by simulation, and the "large number" of random walks that define it was limited to 50 because of their

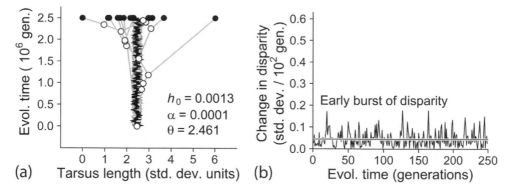

Figure 14.4 Ornstein–Uhlenbeck (OU) model with an initial burst of disparity in the 2.5 million-year history of Darwin's finches, Geospizini. (a) 2.5-million-step OU time series for tarsus length using the long-term net rate of $h_{6.039} = 0.0013$ as a step rate (Harmon et al. 2010; see text). This is superimposed on a 95% 0.04–0.06 standard deviation wide confidence envelope (shaded and barely visible in background). Diffusion would be more rapid for a more representative initial step rate. Confidence envelope is based on 50 2.5-million-step OU time series. (b) Summary showing initial burst of disparity within a few generations followed by more or less constant disparity (solid line), all on timescales of one generation. Disparity rises at the beginning, reflecting the early burst of disparity inherent in an OU model.

individual sizes. Here at the end, after 2.5×10^6 generations of diffusion, the Ornstein–Uhlenbeck confidence envelope spans only about 0.04–0.06 standard deviations of disparity — which does not begin to include all of the finch tarsal lengths. The confidence envelope could be widened by increasing h_0 to a representative rate, or by decreasing the value of α to make it even closer to zero.

Figure 14.4b is a plot of the change in disparity over time for the Ornstein–Uhlenbeck model of Figure 14.4a. The first 250 generations of disparity are shown so they can be seen in detail. Note that the only real change is the initial change from zero to 0.04–0.06 standard deviations of disparity, and Ornstein–Uhlenbeck disparity is more or less constant after the initial change. This is an early burst model because the change in disparity happens in a burst, however small, at the beginning of the time series. The size of the burst in absolute terms and the size of the burst relative to subsequent variation both depend on the relative sizes of α and h_0.

14.5 Forward Modeling of Diffusion with a Declining Step Rate

Brownian diffusion and Ornstein–Uhlenbeck diffusion are end members of a common model. The first expands rapidly, limited only by randomly fluctuating positive and negative signs. The second expands rapidly with fluctuating signs,

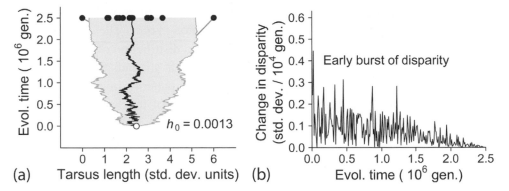

Figure 14.5 Early burst (EB*) model with an early burst of disparity in the 2.5-million-year history of Darwin's finches. (a) 2.5-million-step EB* time series is superimposed on a 95% confidence envelope for EB* diffusion (shaded in background) for tarsus length. Step rate is the long-term net rate of $h_{6.039}$ = 0.0013 (Harmon et al., 2010; see text). Confidence envelope, spanning about 5.0 standard deviation units, is based on 50 2.5-million-step EB* time series. Diffusion would be more rapid for a representative initial step rate. Note attenuation and stabilization of both the time series and the confidence envelope as h_0 decreases through time. (b) Summary showing initial burst of disparity within a few generations followed by declining disparity, all on timescales of 10^4 generations. Rates are highest early in evolutionary time, reflecting the early burst of disparity inherent in an EB* model.

further limited by early and often intense negative feedback. The early burst model of Harmon et al. (2010; here EB*) is a variation on the common model where attenuation is built into the rate parameter. The EB* model in Figure 14.5a has an initial step rate of h_0 = 0.0013, which declines systematically by 0.0013/(2.5×10^6) = 5.2×10^{-10} standard deviations in each generation to reach a step rate of zero standard deviations per generation at the end of the 2.5-million-generation time series.

One 2.5-million-step random walk is shown in Figure 14.5a. This starts at the base of the phylogenetic tree and proceeds forward in time to the present. A large number of random walks like this define the 95% confidence envelope shown by the shaded distribution behind the random walk. The confidence interval was constructed by simulation, and the "large number" of random walks that define it was again limited to 50 because of their individual sizes. Here at the end, after 2.5×10^6 generations of diffusion, the EB* confidence envelope spans about 5.0 standard deviations of disparity, which includes all but two of the finch tarsal lengths.

The random walk in Figure 14.5a is interesting in itself in the way it differs from those in Figures 14.3b and 14.4a. As mentioned above, the initial step rate is h_0 = 0.0013, which declines by 5.2×10^{-10} standard deviations in each generation to reach a step rate of zero at the end of the 2.5-million-generation time series.

The declining step rate is evident in attenuation of the random walk's roughness and in its increasing smoothness moving forward through time. The declining step rate also leads to a declining net rate for the expected distribution of disparity, and increasing disparity slows exponentially through time as the radiation proceeds.

The change in disparity in Figure 14.5b has a pattern similar to that for Brownian diffusion in Figure 14.3d, with the addition of much variation due to disparity estimation from only 50 random walk simulations. There is an early burst of disparity and then the change in disparity slows exponentially through time, illustrating the early burst of disparity inherent in an EB* model.

14.6 Discussion

Timescale matters in analyzing adaptive radiations and in modeling the evolution of morphological disparity. The appropriate timescale is one of generations (Felsenstein, 1973), a requirement seemingly forgotten or ignored in many subsequent studies. Rates matter too and the appropriate rates are haldane step rates, h_0, calculated in standard deviations per generation on the generation-to-generation timescale of the process being modeled. Rates calculated on some longer multi-generation or million-year timescale have to be scaled to represent the process being modeled.

Rates per generation on a one-generation timescale, h_0, can be estimated from rates on longer intervals by temporal scaling, not by averaging. When this is done for tarsus length in Darwin's finches, rates on a timescale of one generation average $h_0 = 10^{-0.643} = 0.228$ standard deviations per generation (Figure 14.2a). This is more than five orders of magnitude higher than conventional rates estimated from node differences, where $h_{6.422} = 10^{-6.184}$. Use of representative step rates such as $h_0 = 0.228$ or the slightly lower $h_0 = 0.153$ standard deviations per generation would make the patterns shown in Figures 14.3b, 14.4a, and 14.5a much more volatile and make the increase in disparity in each much more rapid. It would not, however, change the geometry of the patterns involved.

Brownian diffusion (BD), Ornstein–Uhlenbeck diffusion (OU), and the narrow early burst model (EB*) of Harmon et al. (2010) all produce early bursts of disparity. These corroborate the descriptions of evolutionary diversification and morphological disparity by Simpson (1944) and other paleontologists, and the characterizations by Hutchinson (1959) and other ecologists. McPeek (2008), Phillimore and Price (2008), and Rabosky and Lovette (2008) all found early bursts of speciation in comparative data. Comparative biologists reconstruct patterns retrospectively back from the present, and paleontologists generally trace patterns in the fossil record forward through time. Here both approaches yield similar results: Early bursts of speciation and disparity are common, not rare, in adaptive radiations.

14.7 Summary

1. Two models of diversification are commonly recognized in evolutionary radiations: Brownian diffusion (BD) and Ornstein–Uhlenbeck diffusion (OU). Harmon et al. (2010) added a third narrowly defined "early burst" model (here ER*), which is a particular form of Brownian diffusion with a step rate decreasing through time, leading to exponential slowing of net rates of change.
2. Darwin's finches, Geospizini, are analyzed as a case study for comparison of BD, OU, and ER*. The step rate of evolutionary change in tarsus length for Geospizini can be estimated from temporal scaling of node differences analyzed in the context of a phylogeny. This yields an estimated step rate of $h_0 = 10^{-0.643}$ = 0.228 standard deviations per generation that is close to the median step rate of $h_0 = 0.153$ for field studies found in Chapter 8.
3. Forward modeling of Brownian diffusion, Ornstein–Uhlenbeck diffusion, and the narrow early burst model through 2.5 million generations of geospizine diversification indicates that BD, OU, and EB* are all early burst models, with net disparity increasing most rapidly at the beginning of each radiation.
4. Early bursts of evolution are common, as paleontologists and ecologists have recognized for many years.

15

Summary and Conclusions

At the beginning of the book, in Chapter 1, we saw the challenge before us to be reconciliation of the Lamarck–Darwin thesis of slow gradual change with what we called the Lyell–Linnaeus antithesis involving mysterious creation of species followed by stasis. We found that quantification of rates, relative rates, allows an explicit comparison of Darwinian gradualism to Eldredge–Gould punctuated equilibria, and we identified a difference in rate distributions that enables the two to be distinguished. Rates of change in Darwinian evolution form a single distribution around a central mode of stasis, whereas rates of change in punctuated equilibria are trimodal, with speciation rates, positive and negative, well separated from the central stasis mode. *Rates tell us how change takes place, and rates are the key to reconciliation of the Lamarck–Darwin thesis of slow gradual change with the Lyell–Linnaeus antithesis involving "punctuation" and stasis.*

15.1 Variation, Time, and Random Walks

Study of rates requires that we quantify phenotypic variation, quantify evolutionary time, and learn to recognize random walks as null models. *Quantification of phenotypic variation shows it to be normal on a geometric or proportional scale (normal on a logarithmic scale).* The normality of biological variation is itself an indication that traits are polygenic with additive genetic variance underlying trait values. Natural logarithms (ln) are advantageous for transformation of trait measurements because, on a natural-log scale, standard deviations after transformation are interchangeable with the coefficients of variation of measurements before transformation. *After natural-log transformation, distributions of trait values for populations or samples are appropriately scaled in standard deviation units and compared in terms of their means and standard deviations.*

"Evolutionary time" is often used vaguely and ambiguously. What we are interested in is the timescale of the process underlying the evolution we see. This may be very different from the timescale of our observations. A generation is the average time between production of parental and parentally derived phenotypes, and a generation is the timescale for expression of new genotypes. *A generation is consequently the fundamental unit of time and the fundamental scale of time in the process of evolution by natural selection (and of evolution in the absence of selection). It follows that the random-walk models we construct for comparison to evolutionary time series must be random walks on a timescale of generations.*

Random walks are null models that tell us what to expect in an evolutionary time series in the absence of directional or stabilizing natural selection. Random walks have a *step* rate that is the change, positive or negative, expected in each step or generation. The sign of the change is what makes a random walk random. Random walks have *base* rates that may be step rates or, more often, correspond to the differences we observe in successive samples on longer non-overlapping intervals of time. *Net* rates include all of the rates corresponding to all of the differences we observe in our samples, whether the intervals overlap or not. *All of these rates are haldane rates, subscripted by log_{10} of their associated intervals (denominators) in generations, and expressed in standard deviation units per generation.* We are interested, for different reasons, in step rates, base rates we observe that are independent of each other, and net rates whether independent or not.

15.2 Temporal Scaling

Temporal scaling is analogous to allometric scaling as a way to compare one proportion to another on log–log axes. Temporal scaling is time-interval scaling — scaling differences or rates to the intervals of time over which they are calculated. Log-difference-interval or LDI plots have: (1) log_{10} values for intervals of time marked on the independent abscissa or *x*-axis; (2) log_{10} values for calculated differences in standard deviation units marked on the dependent ordinate or *y*-axis; and (3) all observed differences and their corresponding intervals plotted as points on these axes. Log-rate-interval or LRI plots have rates in standard deviations per generation substituted for differences. *Statistical analysis of LDI and LRI scatters enables quantification of proportional relationships between differences and intervals and between rates and intervals for an evolutionary time series.* Statistics of interest are temporal scaling slopes, temporal scaling intercepts, and confidence intervals for the slopes and intercepts.

Differences in a time series accumulate in relation to time, and rates reflect this accumulation. In a purely directional time series differences accumulate at a constant rate: The slope of an LDI plot is 1 and the slope of an LRI plot is 0. In a purely stationary time series differences average zero and do not accumulate: The slope of an LDI plot is 0 and the slope of an LRI plot is −1. One or the other, differences or rates, can be independent of corresponding interval lengths, but both cannot be independent. *Differences and rates are equally dependent on corresponding interval lengths in the special case of a random walk where the expected slope of an LDI plot is 0.5 and the expected slope of an LRI plot is −0.5. Temporal scaling provides an efficient way to identify randomness in a time series.* Confidence intervals enable randomness, directional change, and stasis to be identified when these can be distinguished.

The intercept of an LDI plot is always the same as the intercept of an LRI plot. *An LDI or LRI intercept provides an estimate of the underlying step rate for a random walk and the underlying step rate for an evolutionary time series.* Confidence intervals indicate how tightly such estimates are constrained.

15.3 Selection and Drift

Truncation and gradient forms of directional and stabilizing selection are effective in moving or stabilizing whole populations. Random drift, on the other hand, is acutely sensitive to sample size and inherently self-balancing. *Consequently, random drift has much less power to constrain a mean or to move it than any form of mass selection.*

15.4 Evolution in the Laboratory, Field, and Fossil Record

Chapters 7–9, on experimental selection in a laboratory setting, on natural selection in the field, and on change observed in the fossil record, are the empirical heart of this book:

- Analysis of 14 selection experiments in eight published studies yielded a total of 672 independent step rates quantifying high-line and low-line selection from one generation to the next. The median rate for this sample was found to be $h_0 = 0.33$ standard deviations per generation on a timescale of one generation.
- Analysis of 57 field studies yielded a total of 814 independent step rates quantifying change from one generation to the next. The median step rate for this sample was found to be $h_0 = 0.15$ standard deviations per generation on a timescale of one generation.
- Analysis of 34 paleontological studies in the fossil record yielded no step rates quantifying change from one generation to the next.

The median step rate in nature is $h_0 = 0.15$ standard deviations per generation on a timescale of one generation, and this is the best rate to use in genetic models of evolutionary change through time.

The log-rate-interval LRI slope for base rates in fossil studies lies between expectation for a stationary time series (-1.000) and expectation for a random time series (-0.500). It is significantly different from both, but closer to the slope expected for stasis. *Stasis predominates in the fossil record but change is significant too.*

15.5 Quantitative Synthesis

The long-standing Lamarck–Darwin thesis of slow and gradual evolutionary change through time can be reconciled with the Lyell–Linnaeus antithesis of no-change "equilibrium" supported by paleontologists for more than a century. Both the thesis and antithesis are deficient to some degree. First, evolution on the generational timescale of natural selection is gradual but not slow. And second, paleontologists working on timescales of thousands of generations cannot see change that is fast.

When field studies and fossil studies are combined, the resulting distribution of rates is continuous. The unshaded background rates in Figure 10.1 that unite generation-scale step rates from field studies with long-term rates from fossil studies are no less real than the shaded rates, and they represent a spectrum of change that is real as well. High step rates on a generational timescale promote rapid diversification and integration of biotas leading to relative stability, but such change, while rapid, is neither instantaneous nor mysterious. The change is gradual, *peu-à-peu*, step-by-step, as Lamarck and Darwin imagined. From a geological or paleontological perspective, where short-term change is censored, change averaged over thousands of generations resembles the no-change stasis of Lyell and Linnaeus.

Differences leading to competing Lamarck–Darwin and Lyell–Linnaeus perceptions of change in the past are not related to the evolutionary process itself but to what we see and do not see in field studies compared to fossil studies. High step rates are the key because they facilitate rapid change in individual lineages on short time scales and rapid stabilization of biotas on longer scales of time.

15.6 Consequences

The high step rates of change calculated in the field studies reviewed in Chapter 8 have consequences for a wide range of genetic models employed in evolutionary analyses. Four consequences are reviewed in Chapters 11–14. These include classic punctuated–equilibria cases in the fossil record, genetic models of evolution by drift and selection, the temporal duration of phylogeny's influence on phenotypes, and early bursts of change in adaptive radiations.

- The histories of Pleistocene *Poecilozonites* developed by Gould, Devonian *Phacops* developed by Eldredge, and Neogene *Metrarabdotos* developed by Cheetham are slightly complicated and include intervals of relative stasis, but all can be explained by gradual evolution at step rates documented in field studies.
- Genetic models are based on generation-to-generation change and mislead us when long-term net rates of evolution are substituted for the generation-scale step rates required. Random drift does not begin to equal its potential in evolving lineages — which is evidence for the near ubiquity of stabilizing natural selection.
- Forward-modeling of Brownian diffusion shows that the time span of phylogeny's influence on body weight in mammalian carnivores and ungulates lasts less than one tenth of 1 percent of their combined evolutionary history. Statistical dependence on phylogeny does not extend automatically to any or all characteristics of interest, and dependence on phylogeny may never last long for the morphological and life history traits we most often study in comparative biology.
- An early burst of evolution is commonly understood to mean that morphological disparity and taxonomic diversity increase most rapidly at the beginning of an adaptive radiation. The rates most rapid at the beginning are net rates, not step rates, and early bursts of evolution are common, as paleontologists and ecologists have recognized for many years.

15.7 Epilogue

"I do believe that natural selection will always act very slowly, often only at long intervals of time, and generally on only a very few of the inhabitants of the same region at the same time" (Charles Darwin, 1859, *Origin of Species*, quoted in Chapter 1). *Darwin was conservative and wrong to believe that evolution by natural selection is slow, intermittent, and weak. Finding evolution to be fast, as we have shown, means the natural selection Darwin advocated is persistent, ubiquitous, and more important than we knew.*

Appendix

Table A.1 *Generation times in bacteria, plants, and animals*

Classification	Species	Length (mm)	Weight (g)	Gen. time (yr)	Source
Bacteria	*Bacillus cereus*	0.0023		3.61E−05	Bonner (1965)
Bacteria	*Bacillus megaterium*			4.76E−05	Todor (2016)
Bacteria	*Escherichia coli*	0.0035	3.00E−13	3.23E−05	Yarwood (1956); Bonner (1965); Todor (2016)
Bacteria	*Lactobacillus acidophilus*			1.46E−04	Todor (2016)
Bacteria	*Mycobacterium tuberculosis*			1.64E−03	Todor (2016)
Bacteria	*Pseudomonas fluorescens*	0.003		5.71E−05	Bonner (1965)
Bacteria	*Rhizobium japonicum*			7.66E−04	Todor (2016)
Bacteria	*Spirochaeta sp.*	0.02		1.00E−03	Bonner (1965)
Bacteria	*Staphylococcus aureus*	0.001		5.14E−05	Bonner (1965); Todor (2016)
Bacteria	*Streptococcus lactis*			4.95E−05	Todor (2016)
Bacteria	*Streptococcus lactis*			9.13E−05	Todor (2016)
Bacteria	*Treponema pallidum*			3.77E−03	Todor (2016)
Alga	*Chlorella pyrenoidosa*		6.00E−11	3.81E−04	Yarwood (1956)
Alga	*Nereocystis luetkeana*	20,000		2.00E+00	Bonner (1965)
Fungus	*Erysiphe betae*		1.00E−06	9.51E−04	Yarwood (1956)
Fungus	*Monilinia fructicola*		8.00E−07	1.90E−03	Yarwood (1956)
Fungus	*Pichia anomala*		1.00E−10	1.90E−04	Yarwood (1956)

Table A.1 (*cont.*)

Classification	Species	Length (mm)	Weight (g)	Gen. time (yr)	Source
Fungus	*Saccharomyces cerevisiae*		2.00E−10	1.90E−04	Yarwood (1956)
Gymnospermae	*Abies balsamea*	16,000		15	Bonner (1965)
Gymnospermae	*Abies concolor*	45,000		35	Bonner (1965)
Gymnospermae	*Sequoia gigantea*	80,000	2.00E+09	60	Yarwood (1956); Bonner (1965)
Angiospermae	*Betula alleghaniensis*	22,000		40	Bonner (1965)
Angiospermae	*Cornus florida*	9,000		5	Bonner (1965)
Protista	*Colpidium colpoda*		5.00E−08	0.00114	Yarwood (1956)
Protista	*Didinium nasutum*	0.1		0.00075	Bonner (1965)
Protista	*Enchelys farcimen*		6.00E−08	0.00038	Yarwood (1956)
Protista	*Euglena gracilis*	0.04		0.00080	Bonner (1965)
Protista	*Paramecium caudatum*	0.2	5.00E−07	0.00114	Yarwood (1956); Bonner (1965)
Protista	*Stentor coeruleus*	1		0.00388	Bonner (1965)
Protista	*Tetrahymena geleii*	0.05		0.00029	Bonner (1965)
Rotifera	*Hydatina senta*		1.00E−05	0.00095	Yarwood (1956)
Bryozoa	*Bugulina flabellata*	1		0.33000	Grave (1930)
Crustacea	*Callinectes sapidus*	150		1.08	Bonner (1965)
Crustacea	*Daphnia longispina*	1.4		0.01	Bonner (1965)
Crustacea	*Limulus polyphemus*	200		10.00	Bonner (1965)
Insecta	*Apis mellifera*	15		0.05	Bonner (1965)
Insecta	*Brevicoryne brassicae*		1.00E−03	0.01	Yarwood (1956)
Insecta	*Drosophila melanogaster*			0.03	Carson (1987)
Insecta	*Lasiocampa quercus*			1.50	Bulmer (1977)
Insecta	*Magicicada spp.*			15.00	Bulmer (1977)
Insecta	*Melolontha hippocastani*			4.00	Bulmer (1977)
Insecta	*Melolontha melolontha*			3.50	Bulmer (1977)
Insecta	*Musca domestica*	7	2.00E−02	0.05	Yarwood (1956); Bonner (1965)
Insecta	*Tabanus atratus*	14		0.30	Bonner (1965)

Table A.1 (*cont.*)

Classification	Species	Length (mm)	Weight (g)	Gen. time (yr)	Source
Insecta	*Tribolium confusum*		1.00E−03	0.04	Yarwood (1956)
Mollusca	*Aequipecten irradians*	78		1.00	Bonner (1965)
Mollusca	*Crassostrea virginica*	30		1.00	Bonner (1965)
Mollusca	*Lymnaea stagnalis*	55		0.75	Bonner (1965)
Mollusca	*Physa gyrina*		1.00E+00	0.04	Yarwood (1956)
Mollusca	*Venus mercenaria*	6		1.50	Bonner (1965)
Echinoderma	*Asterias forbesi*	100		1.50	Bonner (1965)
Osteichthyes	*Menidia menidia*		3	1.00	Conover and Munch (2002)
Osteichthyes	*Engraulis encrasicolus*		70	3.10	Goodwin et al. (2006)
Osteichthyes	*Gadus morhua*		39,049	6.66	Goodwin et al. (2006)
Osteichthyes	*Micromesistius poutassou*		330	6.20	Goodwin et al. (2006)
Osteichthyes	*Melanogrammus aeglefinus*		2,762	6.30	Goodwin et al. (2006)
Osteichthyes	*Merluccius merluccius*		8,841	6.30	Goodwin et al. (2006)
Osteichthyes	*Clupea harengus*		282	6.49	Goodwin et al. (2006)
Osteichthyes	*Trachurus trachurus*		388	7.28	Goodwin et al. (2006)
Osteichthyes	*Scomber scombrus*		650	6.70	Goodwin et al. (2006)
Osteichthyes	*Pleuronectes platessa*		2,918	7.19	Goodwin et al. (2006)
Osteichthyes	*Pollachius virens*		21,090	7.90	Goodwin et al. (2006)
Osteichthyes	*Ammodytes sp.*		30	3.20	Goodwin et al. (2006)
Osteichthyes	*Sardina pilchardus*		80	3.30	Goodwin et al. (2006)
Osteichthyes	*Solea solea*		1,346	7.53	Goodwin et al. (2006)
Osteichthyes	*Merlangius merlangus*		1,476	4.65	Goodwin et al. (2006)
Amphibia	*Ambystoma tigrinum*	205		1.00	Bonner (1965)
Amphibia	*Diemictylus viridescens*	80		2.00	Bonner (1965)
Amphibia	*Rana pipiens*	60		1.50	Bonner (1965)

Table A.1 (*cont.*)

Classification	Species	Length (mm)	Weight (g)	Gen. time (yr)	Source
Reptilia	*Anolis carolinensis*	47		1.50	Bonner (1965)
Reptilia	*Masticophis taeniatus*	870		3.00	Bonner (1965)
Reptilia	*Terrapene carolina*	100		4.00	Bonner (1965)
Aves	*Anser caerulescens*		2,352	5.01	Niel and Lebreton (2005)
Aves	*Branta leucopsis*		1,642	7.55	Niel and Lebreton (2005)
Aves	*Ciconia ciconia*		3,521	4.98	Niel and Lebreton (2005); Gaillard et al. (1989)
Aves	*Fratercula artica*		373	11.81	Niel and Lebreton (2005); Gaillard et al. (1989)
Aves	*Fulmarus glacialis*		706	22.92	Niel and Lebreton (2005); Gaillard et al. (1989)
Aves	*Gyps fulvus*		7,400	12.33	Niel and Lebreton (2005)
Aves	*Larus argentatus*		934	9.80	Niel and Lebreton (2005); Gaillard et al. (1989)
Aves	*Larus ridibundus*		270	5.97	Niel and Lebreton (2005); Gaillard et al. (1989)
Aves	*Parus major*		20.19	1.80	Niel and Lebreton (2005); Gaillard et al. (1989)
Aves	*Petronia petronia*		19	1.56	Niel and Lebreton (2005)
Aves	*Phalacrocorax carbo*		2,819	6.28	Niel and Lebreton (2005)
Aves	*Rissa tridactyla*		393	9.18	Niel and Lebreton (2005); Gaillard et al. (1989)
Aves	*Sterna caspia*		644	4.22	Niel and Lebreton (2005)
Artiodactyla	*Aepyceros melampus*		44,000	4.36	Millar and Zammuto (1983)
Artiodactyla	*Cervus canadensis*	2,300	200,000	2.70	Bonner (1965); Eisenberg (1981)

Table A.1 (*cont.*)

Classification	Species	Length (mm)	Weight (g)	Gen. time (yr)	Source
Artiodactyla	*Cervus elaphus*		175,000	5.70	Millar and Zammuto (1983)
Artiodactyla	*Connochaetes taurinus*		165,000	6.29	Millar and Zammuto (1983)
Artiodactyla	*Hemitragus jemlahicus*		100,000	5.43	Millar and Zammuto (1983)
Artiodactyla	*Hippopotamus amphibius*		2,390,000	19.82	Millar and Zammuto (1983)
Artiodactyla	*Kobus defassa*		200,000	5.08	Millar and Zammuto (1983)
Artiodactyla	*Odocoileus virginianus*	1,100	47,730	2.00	Bonner (1965); Eisenberg (1981)
Artiodactyla	*Odocoileus hemionus*	1,345	55,767	4.00	Eisenberg (1981); Heppell et al. (2000)
Artiodactyla	*Ovis canadensis*	1,400	55,000	6.52	Eisenberg (1981); Millar and Zammuto (1983)
Artiodactyla	*Ovis dalli*		57,667	5.71	Heppell et al. (2000); Ernest (2003)
Artiodactyla	*Phacochoerus aethiopicus*		87,000	4.28	Millar and Zammuto (1983)
Artiodactyla	*Rangifer tarandus*	1,713	105,000	4.21	Eisenberg (1981); Heppell et al. (2000)
Artiodactyla	*Sus scrofa*		85,000	3.15	Millar and Zammuto (1983)
Artiodactyla	*Syncerus caffer*	2,500	490,000	6.98	Eisenberg (1981); Millar and Zammuto (1983)
Carnivora	*Lutra canadensis*		7,200	5.07	Millar and Zammuto (1983)
Carnivora	*Lynx rufus*		7,500	2.87	Millar and Zammuto (1983)
Carnivora	*Mephitis mephitis*		2,250	1.78	Millar and Zammuto (1983)
Carnivora	*Mustela nigripes*		809	2.42	Heppell et al. (2000); Ernest (2003)
Carnivora	*Panthera leo*	2,616	151,000	7.21	Eisenberg (1981); Heppell et al. (2000)

Appendix

Table A.1 (*cont.*)

Classification	Species	Length (mm)	Weight (g)	Gen. time (yr)	Source
Carnivora	*Taxidea taxus*	445	7,150	1.24	Eisenberg (1981); Millar and Zammuto (1983)
Carnivora	*Ursus americanus*	1,643	77,270	10.27	Eisenberg (1981); Heppell et al. (2000)
Carnivora	*Vulpes vulpes*	627	9,000	2.04	Eisenberg (1981); Heppell et al. (2000)
Lagomorpha	*Lepus americanus*	399	1,489	1.25	Eisenberg (1981); Heppell et al. (2000)
Lagomorpha	*Lepus europaeus*	640	4,472	2.98	Eisenberg (1981); Heppell et al. (2000)
Lagomorpha	*Ochotona princeps*		130	2.07	Millar and Zammuto (1983)
Lagomorpha	*Silvilagus floridanus*		1,250	1.29	Millar and Zammuto (1983)
Perissodactyla	*Diceros bicornis*	3,000	1,081,000	5.00	Bonner (1965); Eisenberg (1981)
Perissodactyla	*Equus burchelli*		270,000	8.74	Millar and Zammuto (1983)
Primates	*Cercopithecus aethiops*		3,733	7.84	Heppell et al. (2000); Ernest (2003)
Primates	*Gorilla beringei*		116,413	19.28	Langergraber et al. (2012)
Primates	*Homo sapiens*	1,700	58,169	29.10	Langergraber et al. (2012)
Primates	*Pan troglodytes*		44,764	24.63	Langergraber et al. (2012)
Primates	*Pongo pygmaeus*		35,700	26.70	Langergraber et al. (2012)
Primates	*Propithecus verreauxi*		2,800	17.50	Richard et al. (2000); Lawler (2011)
Proboscidea	*Loxodonta africana*	3,500	4,00,0000	25.80	Bonner (1965); Millar and Zammuto (1983)
Rodentia	*Castor canadensis*	680	18,000	4.87	Bonner (1965); Millar and Zammuto (1983)

Table A.1 (*cont.*)

Classification	Species	Length (mm)	Weight (g)	Gen. time (yr)	Source
Rodentia	*Clethrionomys glareolus*	108	25	0.33	Eisenberg (1981); Millar and Zammuto (1983)
Rodentia	*Marmota flaviventris*		2,930	3.92	Heppell et al. (2000); Ernest (2003)
Rodentia	*Myocastor coypus*		7,150	2.92	Heppell et al. (2000); Ernest (2003)
Rodentia	*Neotoma floridana*	250		0.58	Bonner (1965)
Rodentia	*Peromyscus leucopus*	100	20	0.27	Eisenberg (1981); Millar and Zammuto (1983)
Rodentia	*Peromyscus maniculatus*		20	0.35	Millar and Zammuto (1983)
Rodentia	*Sciurus carolinensis*	240	425	2.68	Charlesworth (1980); Eisenberg (1981)
Rodentia	*Sciurus carolinensis*		600	2.07	Millar and Zammuto (1983)
Rodentia	*Spermophilus armatus*		350	1.78	Millar and Zammuto (1983)
Rodentia	*Spermophilus beldingi*		250	1.56	Millar and Zammuto (1983)
Rodentia	*Spermophilus columbianus*		466	2.88	Heppell et al. (2000); Ernest (2003)
Rodentia	*Spermophilus dauricus*		200	2.58	Heppell et al. (2000); Ernest (2003)
Rodentia	*Spermophilus lateralis*		155	2.45	Millar and Zammuto (1983)
Rodentia	*Spermophilus parryii*		700	1.59	Millar and Zammuto (1983)
Rodentia	*Tamias striatus*	160	100	1.59	Eisenberg (1981); Millar and Zammuto (1983)
Rodentia	*Tamiasciurus hudsonicus*	221	189	1.95	Eisenberg (1981); Millar and Zammuto (1983)
Rodentia	*Zapus princeps*		29	1.96	Heppell et al. (2000); Ernest (2003)
Mysticeti	*Balaenoptera acutorostrata*	7,586	4,581,419	22.1	Taylor et al. (2007); Gingerich (2015)

Table A.1 (*cont.*)

Classification	Species	Length (mm)	Weight (g)	Gen. time (yr)	Source
Mysticeti	*Balaenoptera bonaerensis*	7,096	2,137,962	22.1	Taylor et al. (2007); Gingerich (2015)
Mysticeti	*Balaenoptera borealis*	13,002	15,275,661	23.3	Taylor et al. (2007); Gingerich (2015)
Mysticeti	*Balaenoptera edeni*	10,000	4,159,106	18.4	Taylor et al. (2007); Gingerich (2015)
Mysticeti	*Balaenoptera musculus*	23,714	79,615,935	30.8	Taylor et al. (2007); Gingerich (2015)
Mysticeti	*Balaenoptera physalus*	19,187	42,072,663	25.9	Taylor et al. (2007); Gingerich (2015)
Mysticeti	*Eschrichtius robustus*	11,376	15,922,087	22.9	Taylor et al. (2007); Gingerich (2015)
Mysticeti	*Eubalaena glacialis*	15,382	54,325,033	35.7	Taylor et al. (2007); Gingerich (2015)
Odontoceti	*Cephalorhynchus commersonii*	1,629	69,343	12.7	Taylor et al. (2007); Gingerich (2015)
Odontoceti	*Cephalorhynchus heavisidii*	1,690	70,958	14.4	Taylor et al. (2007); Gingerich (2015)
Odontoceti	*Delphinapterus leucas*	2,897	399,945	16.4	Taylor et al. (2007); Gingerich (2015)
Odontoceti	*Delphinus delphis*	1,945	82,414	14.8	Taylor et al. (2007); Gingerich (2015)
Odontoceti	*Globicephala macrorhynchus*	4,140	849,180	23.5	Taylor et al. (2007); Gingerich (2015)
Odontoceti	*Globicephala melas*	3,548	561,048	24	Taylor et al. (2007); Gingerich (2015)
Odontoceti	*Grampus griseus*	2,723	271,019	19.6	Taylor et al. (2007); Gingerich (2015)
Odontoceti	*Inia geoffrensis*	2,089	73,451	10.2	Taylor et al. (2007); Gingerich (2015)
Odontoceti	*Kogia breviceps*	2,858	319,154	12.1	Taylor et al. (2007); Gingerich (2015)
Odontoceti	*Kogia sima*	2,138	142,233	11.7	Taylor et al. (2007); Gingerich (2015)
Odontoceti	*Lagenodelphis hosei*	2,460	210,378	11.1	Taylor et al. (2007); Gingerich (2015)
Odontoceti	*Lagenorhynchus acutus*	2,234	14,7911	15.8	Taylor et al. (2007); Gingerich (2015)
Odontoceti	*Lagenorhynchus albirostris*	2,388	216,770	18.1	Taylor et al. (2007); Gingerich (2015)
Odontoceti	*Lagenorhynchus obliquidens*	1,954	103,514	21.2	Taylor et al. (2007); Gingerich (2015)
Odontoceti	*Lissodelphis borealis*	2,075	76,384	21.6	Taylor et al. (2007); Gingerich (2015)

Table A.1 (*cont.*)

Classification	Species	Length (mm)	Weight (g)	Gen. time (yr)	Source
Odontoceti	*Megaptera novaeangliae*	11,858	20,653,802	21.5	Taylor et al. (2007); Gingerich (2015)
Odontoceti	*Neophocaena phocaenoides*	1,622	33,037	16.5	Taylor et al. (2007); Gingerich (2015)
Odontoceti	*Orcinus orca*	4,477	1,364,583	25.7	Taylor et al. (2007); Gingerich (2015)
Odontoceti	*Phocoena dioptrica*	1,439	46,774	14.4	Taylor et al. (2007); Gingerich (2015)
Odontoceti	*Phocoena phocoena*	1,371	44,771	11.9	Taylor et al. (2007); Gingerich (2015)
Odontoceti	*Phocoenoides dalli*	1,832	97,275	15.1	Taylor et al. (2007); Gingerich (2015)
Odontoceti	*Physeter macrocephalus*	14,256	28,773,984	31.9	Taylor et al. (2007); Gingerich (2015)
Odontoceti	*Pontoporia blainvillei*	1,396	30,130	11.9	Taylor et al. (2007); Gingerich (2015)
Odontoceti	*Sotalia fluviatilis*	1,500	29,992	15.6	Taylor et al. (2007); Gingerich (2015)
Odontoceti	*Stenella attenuata*	1,849	59,429	23.1	Taylor et al. (2007); Gingerich (2015)
Odontoceti	*Stenella clymene*	1,795	62,951	14.7	Taylor et al. (2007); Gingerich (2015)
Odontoceti	*Stenella coeruleoalba*	1,986	86,298	22.5	Taylor et al. (2007); Gingerich (2015)
Odontoceti	*Stenella frontalis*	2,056	96,828	18.3	Taylor et al. (2007); Gingerich (2015)
Odontoceti	*Stenella longirostris*	1,901	61,376	13.7	Taylor et al. (2007); Gingerich (2015)
Odontoceti	*Tursiops aduncus*	2,506	179,887	21.1	Taylor et al. (2007); Gingerich (2015)
Odontoceti	*Tursiops truncatus*	2,600	200,909	21.1	Taylor et al. (2007); Gingerich (2015)
Pinnipedia	*Callorhinus ursinus*	1,340	55,000	10.38	Eisenberg (1981); Heppell et al. (2000)
Pinnipedia	*Eumetopias jubatus*	2,450	350,000	9.41	Eisenberg (1981); Heppell et al. (2000)
Pinnipedia	*Mirounga leonina*		579,400	8.33	Heppell et al. (2000); Ernest (2003)
Pinnipedia	*Phoca vitulina*	1,500	65,000	12.18	Eisenberg (1981); Heppell et al. (2000)

Table A.1 (*cont.*)

Classification	Species	Length (mm)	Weight (g)	Gen. time (yr)	Source
Sirenia	*Trichechus manatus*		387,500	13.23	Heppell et al. (2000); Ernest (2003)
Chiroptera	*Carollia perspicillata*	65	18	3.14	Eisenberg (1981); Heppell et al. (2000)
Chiroptera	*Myotis lucifugus*	47	9	4.76	Heppell et al. (2000); Eisenberg (1981)

References

Abel, O. (1918). Das Entwicklungstempo der Wirbeltierstämme. *Vereines zur Verbreitung Naturwissenschaftlicher Kenntnisse, Schriften, Wien*, 58, 91–120.

Ager, D. V. (1976). The nature of the fossil record. *Proceedings of the Geologists' Association, London*, 87, 131–159.

Aguirre, W. E. and Bell, M. A. (2012). Twenty years of body shape evolution in a threespine stickleback population adapting to a lake environment. *Biological Journal of the Linnean Society*, 105, 817–831.

Aguirre, W. E., Doherty, P. K., and Bell, M. A. (2004). Genetics of lateral plate and gillraker phenotypes in a rapidly evolving population of threespine stickleback. *Behaviour*, 141, 1465–1483.

Ashton, E. H. and Zuckerman, S. (1950). The influence of geographic isolation on the skull of the green monkey (*Cercopithecus aethiops sabacus*). I. A comparison between the teeth of the St. Kitts and the African Green monkey. *Proceedings of the Royal Society of London, Series B*, 137, 212–238.

Ayala, F. J., Serra, L. and Prevosti, A. (1989). A grand experiment in evolution: the *Drosophila subobscura* colonization of the Americas. *Genome*, 31, 246–255.

Babin-Fenske, J., Anand, M., and Alarie, Y. (2008). Rapid morphological change in stream beetle museum specimens correlates with climate change. *Ecological Entomology*, 33, 646–651.

Bader, R. S. (1955). Variability and evolutionary rate in the oreodonts. *Evolution*, 9, 119–140.

Baer, K. E. v. (1876). *Reden Gehalten in Wissenschaftlichen Versammlungen Und Kleinere Aufsätze Vermischten Inhalts. Zweiter Theil: Studien Aus Dem Gebiete Der Naturwissenschaften*. St. Petersburg, Schmitzdorff.

Bailey, D. W. (1959). Rates of subline divergence in highly inbred strains of mice. *Journal of Heredity*, 50, 26–30.

Baker, A. J. and Marshall, H. D. (1999). Population divergence in chaffinches *Fringilla coelebs* assessed with control-region sequences. In *Proceedings of the 22nd International Ornithological Congress, Durban*, eds. N. J. Adams and R. H. Slotow, Johannesburg, BirdLife South Africa, pp. 1899–1913.

Baker, A. J., Peck, M. K., and Goldsmith, M. A. (1990). Genetic and morphometric differentiation in introduced populations of common chaffinches (*Fringilla coelebs*) in New Zealand. *Condor*, 92, 76–88.

Baker, J. A., Heins, D. C., Foster, S. A., and King, R. W. (2008). An overview of life-history variation in female threespine stickleback. *Behaviour*, 145, 579–602.

Banerjee, M. K. and Basu, A. (1991). Uttar Pradesh: basic anthropometric data. *All India Anthropometric Survey, North Zone (Anthropological Survey of India, Calcutta)*, 10, 1–626.

Barrett, P. H., Gautrey, P. J., Herbert, S., Kohn, D., and Smith, S. (1987). *Charles Darwin's Notebooks, 1836–1844: Geology, Transmutation of Species, Metaphysical Enquiries.* Cambridge, British Museum (Natural History) and Cambridge University Press.

Basu, A., Ganguly, P., Ghosh, G. C., and Basu, S. K. (1989). Maharashtra: basic anthropometric data. *All India Anthropometric Survey, North Zone (Anthropological Survey of India, Calcutta)*, 4, 1–550.

Bather, F. A. (1920). Fossils and life. *Reports of the British Association for the Advancement of Science, London*, 1920, 61–86.

Bell, M. A. and Aguirre, W. E. (2013). Contemporary evolution, allelic recycling, and adaptive radiation of the threespine stickleback. *Evolutionary Ecology Research*, 15, 377–411.

Bell, M. A., Aguirre, W. E., and Buck, N. J. (2004). Twelve years of contemporary armor evolution in a threespine stickleback population. *Evolution*, 58, 814–824.

Bell, M. A., Baumgartner, J. V., and Olson, E. C. (1985). Patterns of temporal change in single morphological characters of a Miocene stickleback fish. *Paleobiology*, 11, 258–271.

Bell, M. A. and Haglund, T. R. (1982). Fine-scale temporal variation of the Miocene stickleback *Gasterosteus doryssus*. *Paleobiology*, 8, 282–292.

Bell, M. A., Travis, M. P. and Blouw, D. M. (2006). Inferring natural selection in a fossil threespine stickleback. *Paleobiology*, 32, 562–577.

Benson, R. B. J., Campione, N. E., Carrano, M. T., Mannion, P. D., Sullivan, C., Upchurch, P., and Evans, D. C. (2014). Rates of dinosaur body mass evolution indicate 170 Million years of sustained ecological innovation on the avian stem lineage. *PLoS Biology*, 12, e1001853.

Berg, H. C. (1983). *Random Walks in Biology*. Princeton, Princeton University Press.

Berg, L. S. (1926). *Nomogenesis, or Evolution Determined by Law*. London, Constable and Company.

Berger, J. (1986). *Wild Horses of the Great Basin: Social Competition and Population Size*. Chicago, University of Chicago Press.

Berry, R. J. (1964). The evolution of an island population of the house mouse. *Evolution*, 18, 468–483.

Beurlen, K. (1932). Funktion und Form in der organischen Entwicklung. *Naturwissenschaften, Berlin*, 20, 73–80.

Blanckenhorn, W. U. (2015a). Investigating yellow dung fly body size evolution in the field: Response to climate change? *Evolution*, 69, 2227–2234.

Blanckenhorn, W. U. (2015b). Data from: Investigating yellow dung fly body size evolution in the field: response to climate change? *Dryad Digital Repository*, https://doi.org/10.5061/dryad.3v4hn.

Boag, P. T. and Grant, P. R. (1981). Intense natural selection in a population of Darwin's finches (Geospizinae) in the Galapagos. *Science*, 214, 82–85.

Bonner, J. T. (1965). *Size and Cycle: An Essay on the Structure of Biology*. Princeton, Princeton University Press.

Bookstein, F. L. (1987). Random walk and the existence of evolutionary rates. *Paleobiology*, 13, 446–464.

Bookstein, F. L., Gingerich, P. D., and Kluge, A. G. (1978). Hierarchical linear modeling of the tempo and mode of evolution. *Paleobiology*, 4, 120–134.

Bowles, G. T. (1932). *New Types of Old Americans at Harvard and at Eastern Women's Colleges*. Cambridge, Harvard University Press.

Brett, C. E. and Baird, G. C. (1995). Coordinated stasis and evolutionary ecology of Silurian to Middle Devonian faunas in the Appalachian Basin. In *New Approaches to Speciation in the Fossil Record,* eds. D. H. Erwin and R. L. Anstey, New York, Columbia University Press, pp. 285–315.

Brinkmann, R. (1929). Statistisch-biostratigraphische Untersuchungen an mitteljurassischen Ammoniten über Artbegriff und Stammesentwicklung. *Abhandlungen der Gesellschaft der Wissenschaften zu Göttingen, Mathematisch-Physikalische Klasse, Neue Folge*, 13 (3), 1–249.

Brown, C. R. and Brown, M. B. (1998a). Intense natural selection on body size and wing and tail asymmetry in cliff swallows during severe weather. *Evolution*, 52, 1461–1475.

Brown, C. R. and Brown, M. B. (1998b). Fitness components associated with alternative reproductive tactics in cliff swallows. *Behavioral Ecology*, 9, 158–171.

Brown, M. B. and Brown, C. R. (2011). Intense natural selection on morphology of cliff swallows (*Petrochelidon pyrrhonota*) a decade later: did the population move between adaptive peaks? *Auk*, 128, 69–77.

Bruce, R. C. (1988). An ecological life table for the salamander *Eurycea wilderae. Copeia*, 1988, 15–26.

Buckman, S. S. (1909). *Yorkshire Type Ammonites, Volume 1, Part 1*. London, Wheldon and Wesley.

Bulmer, M. G. (1977). Periodical insects. *American Naturalist*, 111, 1099–1117.

Bumpus, H. C. (1899). The elimination of the unfit as illustrated by the introduced sparrow, *Passer domesticus. Biological Lectures Delivered at the Marine Biology Laboratory, Woods Hole*, 1898, 209–226.

Butler, M. A. and King, A. A. (2004). Phylogenetic comparative analysis: a modeling approach for adaptive evolution. *American Naturalist*, 164, 683–695.

Calder, W. A. (1984). *Size, Function, and Life History*. Cambridge, Massachusetts, Harvard University Press.

Calhoun, J. B. (1947). The role of temperature and natural selection in relation to the variations in the size of the English sparrow in the United States. *American Naturalist*, 81, 203–228.

Carroll, S. P. and Boyd, C. (1992). Host race radiation in the soapberry bug: natural history with the history. *Evolution*, 46, 1052–1069.

Carroll, S. P., Dingle, H., and Klassen, S. (1997). Genetic differentiation of fitness-associated traits among rapidly evolving populations of the soapberry bug. *Evolution*, 51, 1182–1188.

Carroll, S. P., Hendry, A. P., Reznick, D. N., and Fox, C. W. (2007). Evolution on ecological time-scales. *Functional Ecology*, 21, 387–393.

Carson, H. L. (1975). The genetics of speciation at the diploid level. *American Naturalist*, 109, 83–92.

Carson, H. L. (1987). Population genetics, evolutionary rates and Neo-Darwinism. In *Rates of Evolution*, eds. K. S. W. Campbell and M. F. Day, London, Allen and Unwin, pp. 209–217.

Caruso, N. M., Sears, M. W., Adams, D. C., and Lips, K. R. (2014). Widespread rapid reductions in body size of adult salamanders in response to climate change. *Global Change Biology*, 20, 1751–1759.

Cavalli-Sforza, L. L. and Bodmer, W. F. (1971). *The Genetics of Human Populations*. San Francisco, W. H. Freeman and Company.

Chaline, J. (1984). Le concept d'évolution polyphasée et ses implications. *Geobios, Lyon*, 17, 783–795.

Charlesworth, B. (1980). *Evolution in Age-Structured Populations*. Cambridge, Cambridge University Press.

Charlesworth, B. (1984). Some quantitative methods for studying evolutionary patterns in single characters. *Paleobiology*, 10, 308–318.

Charmantier, A., Kruuk, L. E. B., Blondel, J., and Lambrechts, M. M. (2004). Testing for microevolution in body size in three blue tit populations. *Journal of Evolutionary Biology*, 17, 732–743.

Cheetham, A. H. (1968). Morphology and systematics of the bryozoan genus *Metrarabdotos*. *Smithsonian Miscellaneous Collections*, 153, 1–121.

Cheetham, A. H. (1986). Tempo of evolution in a Neogene bryozoan: rates of morphologic change within and across species boundaries. *Paleobiology*, 12, 190–202.

Cheetham, A. H. and Jackson, J. B. C. (1995). Process from pattern: tests for selection versus random change in punctuated bryozoan speciation. In *New Approaches to Speciation in the Fossil Record*, eds. D. H. Erwin and R. L. Anstey, New York, Columbia University Press, pp. 184–207.

Cheetham, A. H., Sanner, J. and Jackson, J. B. C. (2007). *Metrarabdotos* and related genera (Bryozoa: Cheilostomata) in the late Paleogene and Neogene of tropical America. Journal of Paleontology 81(1), Supplement, *Paleontological Society Memoir*, 67, 1–96.

Chiba, S. (1996). A 40,000-year record of discontinuous evolution of island snails. *Paleobiology*, 22, 177–188.

Chiba, S. (2007). Morphological and ecological shifts in a land snail caused by the impact of an introduced predator. *Ecological Research*, 22, 884–891.

Clark, D. B. (1980). Population ecology of *Rattus rattus* across a desert-montane forest gradient in the Galápagos Islands. *Ecology*, 61, 1422–1433.

Clutton-Brock, T. H., Guinness, F. E., and Albon, S. D. (1982). *Red Deer: Behaviour and Ecology of Two Sexes*. Chicago, University of Chicago Press.

Clutton-Brock, T. H. and Pemberton, J. M. eds. (2004). *Soay Sheep: Dynamics and Selection in an Island Population*, Cambridge, Cambridge University Press.

Clyde, W. C. and Gingerich, P. D. (1994). Rates of evolution in the dentition of early Eocene *Cantius*: comparison of size and shape. *Paleobiology*, 20, 506–522.

Clyde, W. C., Hamzi, W., Finarelli, J. A., Wing, S. L., Schankler, D. M., and Chew, A. (2007). Basin-wide magnetostratigraphic framework for the Bighorn Basin, Wyoming. *Geological Society of America Bulletin*, 119, 848–859.

Colbert, E. H. (1948). Evolution of the horned dinosaurs. *Evolution*, 2, 145–163.

Conant, S. (1988). Geographic variation in the Laysan finch (*Telespyza cantans*). *Evolutionary Ecology*, 2, 270–282.

Conover, D. O. and Munch, S. B. (2002). Sustaining fisheries yields over evolutionary time scales. *Science*, 297, 94–96.

Cooch, E. G., Lank, D. B., Rockwell, R. F., and Cooke, F. (1991). Long-term decline in body size in a snow goose population: evidence of environmental degradation? *Journal of Animal Ecology*, 60, 483–496.

Cooke, F., Rockwell, R. F., and Lank, D. B. (1995). *The Snow Geese of La Pérouse Bay: Natural Selection in the Wild*. Oxford, Oxford University Press.

Cooper, G. A. and Williams, A. (1952). Significance of the stratigraphic distribution of brachiopods. *Journal of Paleontology*, 26, 326–337.

Cope, E. D. (1887). *The Origin of the Fittest, Essays on Evolution*. New York, D. Appleton and Co. 467.

Cope, E. D. (1896). *The Primary Factors of Organic Evolution*. Chicago, Open Court.

COSEWIC. (2012). COSEWIC assessment and status report on the grizzly bear *Ursus arctos. Committee on the Status of Endangered Wildlife in Canada, Ottawa*, 1–84.

COSEWIC. (2013a). COSEWIC assessment and status report on the Plains Bison *Bison bison bison* and the Wood Bison *Bison bison athabascae* in Canada. *Committee on the Status of Endangered Wildlife in Canada, Ottawa*, 1–109.

COSEWIC. (2013b). COSEWIC assessment and status report on the giant threespine stickleback *Gasterosteus aculeatus* and the unarmoured threespine stickleback *Gasterosteus aculeatus* in Canada. *Committee on the Status of Endangered Wildlife in Canada, Ottawa*, 1–62.

Coulson, T. and Crawley, M. J. (2004). Appendix 3: How average life tables can mislead. In *Soay Sheep: Dynamics and Selection in an Island Population*, eds. T. H. Clutton-Brock and J. M. Pemberton, Cambridge, Cambridge University Press, pp. 328–331.

Crow, J. F. and Kimura, M. (1979). Efficiency of truncation selection. *Proceedings of the National Academy of Sciences USA*, 76, 396–399.

Curry, G. B. (1982). Ecology and population structure of the recent brachiopod *Terebratulina* from Scotland. *Palaeontology*, 25, 227–246.

Cuvier, G. (1812). *Recherches sur les ossemens fossiles de quadrupèdes. Tomes I-IV.* Paris,

Cuvier, G. (1817). *Le règne animal, distribué d'aprés son organisation, pour servir De base à l'histoire naturelle des animaux et d'introduction à l'anatomie comparée, Tome I.* Paris, Deterville.

Cuvier, G. (1825). *Discours sur les révolutions de la surface du globe, et sur les changemens qu'elles ont produits dans le règne animal.* Dufour et d'Ocagne, Paris (reprinted 1969 by Culture et Civilisation, 115 Avenue Gabriel Lebon, Brussels) 400.

D'Agostino, R. B. (1986). Tests for the normal distribution. In *Goodness-of-Fit Techniques*, eds. R. B. D'Agostino and M. A. Stephens, New York, Marcel Dekker, pp. 367–419.

Dall, W. H. (1877). On a provisional hypothesis of saltatory evolution. *American Naturalist*, 11, 135–137.

Darwin, C. R. (1839). *Journal of Researches into the Geology and Natural History of the Various Countries Visited by the H. M. S. Beagle.* London, Henry Colburn.

Darwin, C. R. (1859). *On the Origin of Species by Means of Natural Selection.* London, John Murray.

Darwin, C. R. (1875). *The Variation of Animals and Plants under Domestication, Second Edition, Volumes 1 and 2.* London, John Murray.

Darwin, E. (1794–1796). *Zoonomia; or the Laws of Organic Life, Volumes I and II.* London, J. Johnson.

Dawkins, R. (1976). *The Selfish Gene.* Oxford, Oxford University Press.

de Beer, G. R. (1960). Darwin's notebooks on transmutation of species. Part I. First notebook (July 1837–February 1838). *Bulletin of the British Museum (Natural History), Historical Series*, 2, 25–73.

Depéret, C. (1907). *Les transformations du monde animal.* Paris, Ernest Flammarion

de Vries, H. (1901). *Die Mutationstheorie. Versuch Und Beobachtungen Über Die Entstehung Der Arten Im Pflanzenreich. Erster Band. Die Entstehung Der Arten Durch Mutation.* Leipzig, Verlag von Veit.

de Vries, H. (1905). The evidence of evolution. *Annual Report of the Smithsonian Institution*, 1904, 389–396.

de Vries, H. (1909). *The Mutation Theory: Experiments and Observations on the Origin of Species in the Vegetable Kingdom.* Chicago, Open Court Publishing.

Diamond, J. M., Pimm, S. L., Gilpin, M. E., and LeCroy, M. (1989). Rapid evolution of character displacement in myzomelid honeyeaters. *American Naturalist*, 134, 675–708.

Dobzhansky, T. (1937). *Genetics and the Origin of Species*. New York, Columbia University Press.

Dobzhansky, T. and Spassky, B. (1967). Effects of selection and migration on geotactic and phototactic behaviour of Drosophila. I. *Proceedings of the Royal Society of London, Series B*, 168, 27–47.

Dudley, J. W. and Lambert, R. J. (2004). 100 generations of selection for oil and protein in corn. *Plant Breeding Reviews*, 24, 79–110.

Eastman, L. M., Morelli, T. L., Rowe, K. C., Conroy, C. J. and Moritz, C. (2012). Size increase in high elevation ground squirrels over the last century. *Global Change Biology*, 18, 1499–1508.

Edwards, A. W. F. (1972). *Likelihood*. Cambridge, Cambridge University Press.

Edwards, A. W. F. (1992). *Likelihood, Expanded Edition*. Baltimore, Maryland, Johns Hopkins University Press.

Efron, B. and Tibshirani, R. J. (1986). Bootstrap methods for standard errors, confidence intervals, and other measures of statistical accuracy. *Statistical Science*, 1, 54–77.

Efron, B. and Tibshirani, R. J. (1993). *An Introduction to the Bootstrap. New York, Chapman and Hall* 436.

Eisenberg, J. F. (1981). *The Mammalian Radiations: An Analysis of Trends in Evolution, Adaptation, and Behavior*. Chicago, University of Chicago Press.

Eldredge, N. (1971). The allopatric model and phylogeny in Paleozoic invertebrates. *Evolution*, 25, 156–167.

Eldredge, N. (1972). Systematics and evolution of *Phacops rana* (Green, 1832) and *Phacops iowensis* Delo, 1935 (Trilobita) from the middle Devonian of North America. *Bulletin of the American Museum of Natural History*, 147, 45–114.

Eldredge, N. and Gould, S. J. (1972). Punctuated equilibria: an alternative to phyletic gradualism. In *Models in Paleobiology*, ed. T. J. M. Schopf, San Francisco, Freeman, Cooper and Company, pp. 82–115.

Elliott, E. B. (1863). *On the Military Statistics of the United States of America*. Berlin, International Statistical Congress: R. v. Decker.

Elliott, E. B. (1865). On the military statistics of the United States of America. In *Fünfte Internationalen Statistischen Congresses*, ed. E. Engel, Berlin, Königlichen Geheimen Ober-Hofbuchdruckerei, pp. 715–758.

Eloy de Amorim, M., Schoener, T. W., Santoro, G. R. C. C., Lins, A. C. R., Piovia-Scott, J., and Brand+úo, R. A. (2017). Lizards on newly created islands independently and rapidly adapt in morphology and diet. *Proceedings of the National Academy of Sciences USA*, 114, 8812–8816.

Endler, J. A. (1986). *Natural Selection in the Wild*. Princeton, Princeton University Press 336.

Erickson, G. M., Currie, P. J., Inouye, B. D., and Winn, A. A. (2006). Tyrannosaur life tables: an example of nonavian dinosaur population biology. *Science*, 313, 213–217.

Erickson, G. M., Makovicky P. J., Inouye B. D., Zhou C.–F., and Gao, K. −Q. (2009). A life table for *Psittacosaurus lujiatunensis*: initial insights into ornithischian dinosaur population biology. *Anatomical Record*, 292, 1514–1521.

Erickson, W. A. and Halvorson, W. L. (1990). Ecology and control of the roof rat (*Rattus rattus*) in Channel Islands National Park. *Cooperative National Park Resources Studies Unit, Institute of Ecology, University of California, Davis*, 38, 1–90.

Ernest, S. K. M. (2003). Life history characteristics of placental nonvolant mammals. *Ecology*, 84, 3402.

Estes, S. and Arnold, S. J. (2007). Resolving the paradox of stasis: models with stabilizing selection explain evolutionary divergence on all timescales. *American Naturalist*, 169, 227–244.

Falconer, D. S. (1973). Replicated selection for body weight in mice. *Genetical Research, Cambridge*, 22, 291–321.

Falconer, D. S. (1977). Why are mice the size they are? In *International Conference on Quantitative Genetics*, eds. E. Pollack, O. Kempthorne, and T. B. Bailey, Ames, Iowa State University Press, pp. 19–21.

Falconer, D. S. (1981). *Introduction to Quantitative Genetics, Second Edition*. London, Longman Group Limited.

Falconer, H. (1863). On the American fossil elephant of the regions bordering the Gulf of Mexico (*E. columbi* Falc.); with general observations on the living and extinct species. *Natural History Review, Dublin*, 3, 43–114.

Fechner, G. T. (1860). *Elemente Der Psychophysik*. Leipzig, Breitkopf and Härtel.

Feller, W. (1968). *An Introduction to Probability Theory and Its Applications*, Third Edition, Volume I. New York, John Wiley and Sons 509.

Felsenstein, J. (1973). Maximum-likelihood estimation of evolutionary trees from continuous characters. *American Journal of Human Genetics*, 25, 471–492.

Felsenstein, J. (1985). Phylogenies and the comparative method. *American Naturalist*, 125, 1–15.

Festing, M. F. W. (1973). A multivariate analysis of subline divergence in the shape of the mandible in C57BL/Gr mice. *Genetical Research, Cambridge*, 21, 121–132.

Fisher, R. A. and Ford, E. B. (1950). The "Sewall Wright effect." *Heredity*, 4, 117–119.

Foote, M. (1997). The evolution of morphological diversity. *Annual Review of Ecology and Systematics*, 28, 129–152.

Forstén, A.-M. (1990). Dental size trends in an equid sample from the Sandalja II cave of northwestern Yugoslavia. *Paläontologische Zeitschrift*, 64, 153–160.

Franks, S. J., Sim, S., and Weis, A. E. (2007). Rapid evolution of flowering time by an annual plant in response to a climate fluctuation. *Proceedings of the National Academy of Sciences USA*, 104, 1278.

Freudenthal, M. and Martín-Suárez, E. (2013). Estimating body mass of fossil rodents. *Scripta Geologica*, 145, 1–130.

Froese, D., Stiller, M., Heintzman, P. D., Reyes, A. V., Zazula, G. D., Soares, A. E. R., Meyer, M., Hall, E., Jensen, B. J. L., Arnold, L. J., MacPhee, R. D. E., and Shapiro, B. (2017). Fossil and genomic evidence constrains the timing of bison arrival in North America. *Proceedings of the National Academy of Sciences USA*, 114, 3457.

Gaillard, J.-M., Pontier, D., Allaine, D., Lebreton, J.-D., Trouvilliez, J. and Clobert, J. (1989). An analysis of demographic tactics in birds and mammals. *Oikos*, 56, 59–76.

Galton, F. (1869). *Hereditary Genius: An Inquiry into Its Laws and Consequences*. London, Macmillan.

Galton, F. (1879). The geometric mean in vital and social statistics. *Proceedings of the Royal Society of London*, 29, 365–367.

Galton, F. (1889). *Natural Inheritance*. London, MacMillan.

Galton, F. (1894). Discontinuity in evolution. *Mind, a Quarterly Review of Psychology and Philosophy, London*, 3, 362–372.

Garland, T. (1992). Rate tests for phenotypic evolution using phylogenetically independent contrasts. *American Naturalist*, 140, 509–519.

Garland, T. and Janis, C. M. (1993). Does metatarsal/femur ratio predict maximal running speed in cursorial mammals? *Journal of Zoology, London*, 229, 133–151.

Gates, R. R. (1911). The mutation theory [book review]. *American Naturalist*, 45, 254–256.

Gaudry, A. (1896). *Essai De Paléontologie Philosophique*. Paris, Masson.

Gauss, C. F. (1809). *Theoria Motus Corporum Celestium*. Hamburg, Perthes et Besser.

Gavrilets, S. and Vose, A. (2005). Dynamic patterns of adaptive radiation. *Proceedings of the National Academy of Sciences USA*, 102, 18040–18045.

Geary, R. C. (1935). The ratio of the mean deviation to the standard deviation as a test of normality. *Biometrika*, 27, 310–332.

Geiger, M., Sánchez-Villagra, M., and Lindholm, A. K. (2018). A longitudinal study of phenotypic changes in early domestication of house mice. *Royal Society Open Science*, 5, 172099.

Geoffroy Saint-Hilaire, É. (1828). Mémoire où l'on se propose de rechercher dans quels rapports de structure organique et de parenté sont entre eux les animalux des ages historiques et vivant actuellement et les espèces antédiluviennes et perdues. *Mémoires du Muséum National d'Histoire Naturelle, Paris*, 17, 209–229.

Geoffroy Saint-Hilaire, I. (1836). *Histoire générale et particulière des anomalies de l'organisation chez l'homme et les animaux, ou traité de tératologie,* tome troisième. Paris, J.-B. Baillière.

George, T. N. (1958). Rates of change in evolution. *Science Progress, a Quarterly Review of Current Scientific Investigations, London*, 46, 409–428.

Gilchrist, G. W., Huey, R. B., Balanyà, J., Pascual, M., and Serra, L. (2004). A time series of evolution in action: a latitudinal cline in wing size in South American *Drosophila subobscura*. *Evolution*, 58, 768–780.

Gilchrist, G. W., Huey, R. B., and Serra, L. (2001). Rapid evolution of wing size clines in *Drosophila subobscura*. *Genetica*, 112, 273–286.

Gillespie, J. H. (1984). The molecular clock may be an episodic clock. *Proceedings of the National Academy of Sciences USA*, 81, 8009–8013.

Gingerich, P. D. (1974). Stratigraphic record of early Eocene *Hyopsodus* and the geometry of mammalian phylogeny. *Nature*, 248, 107–109.

Gingerich, P. D. (1976). Paleontology and phylogeny: patterns of evolution at the species level in early Tertiary mammals. *American Journal of Science*, 276, 1–28.

Gingerich, P. D. (1983). Rates of evolution: effects of time and temporal scaling. *Science*, 222, 159–161.

Gingerich, P. D. (1984). Punctuated equilibria – where is the evidence? *Systematic Zoology*, 33, 335–338.

Gingerich, P. D. (1989). New earliest Wasatchian mammalian fauna from the Eocene of northwestern Wyoming: composition and diversity in a rarely sampled high-floodplain assemblage. *University of Michigan Papers on Paleontology*, 28, 1–97.

Gingerich, P. D. (1993). Quantification and comparison of evolutionary rates. *American Journal of Science*, 293, 453–478.

Gingerich, P. D. (1994). New species of *Apheliscus*, *Haplomylus*, and *Hyopsodus* (Mammalia, Condylarthra) from the late Paleocene of southern Montana and early Eocene of northwestern Wyoming. *Contributions from the Museum of Paleontology, University of Michigan*, 29, 119–134.

Gingerich, P. D. (1995). Statistical power of EDF tests of normality and the sample size required to distinguish geometric-normal (lognormal) from arithmetic-normal distributions of low variability. *Journal of Theoretical Biology*, 173, 125–136.

Gingerich, P. D. (2000). Arithmetic or geometric normality of biological variation: an empirical test of theory. *Journal of Theoretical Biology*, 204, 201–221.

Gingerich, P. D. (2001). Rates of evolution on the time scale of the evolutionary process. *Genetica*, 112/113, 127–144.

Gingerich, P. D. (2003). Mammalian responses to climate change at the Paleocene-Eocene boundary: Polecat Bench record in the northern Bighorn Basin, Wyoming. In *Causes and consequences of globally warm climates in the early Paleogene*, eds. S. L. Wing, P. D. Gingerich, B. Schmitz, and E. Thomas, Geological Society of America, Special Papers, 369: 463–478.

Gingerich, P. D. (2010). Mammalian faunal succession through the Paleocene-Eocene thermal maximum (PETM) in western North America. *Vertebrata PalAsiatica, Beijing*, 48, 308–327.

Gingerich, P. D. (2014). Species in the primate fossil record. *Evolutionary Anthropology*, 23, 33–35.

Gingerich, P. D. (2015). Body weight and relative brain size (encephalization) in Eocene Archaeoceti (Cetacea). *Journal of Mammalian Evolution*, 23, 17–31.

Gingerich, P. D. (2019). Data from: Rates of Evolution: A Quantitative Synthesis (Cambridge University Press). *Dryad Digital Repository*, https://doi.org/10.5061/dryad.1tn7123.

Gingerich, P. D., Smith, B. H. and Rosenberg, K. R. (1982). Allometric scaling in the dentition of primates and prediction of body weight from tooth size in fossils. *American Journal of Physical Anthropology*, 58, 81–100.

Gingerich, P. D. and Smith, T. (2006). Paleocene-Eocene land mammals from three new latest Clarkforkian and earliest Wasatchian wash sites at Polecat Bench in the northern Bighorn Basin, Wyoming. *Contributions from the Museum of Paleontology, University of Michigan*, 31, 245–303.

Goldschmidt, R. (1933). Some aspects of evolution. *Science*, 78, 539–547.

Goldschmidt, R. (1940). *The Material Basis of Evolution*. New Haven, Yale University Press 436.

Goodwin, N. B., Grant, A., Perry, A. L., Dulvy, N. K., and Reynolds, J. D. (2006). Life history correlates of density-dependent recruitment in marine fishes. *Canadian Journal of Fisheries and Aquatic Sciences*, 63, 494–509.

Gould, S. J. (1969). An evolutionary microcosm: Pleistocene and Recent history of the land snail *P. (Poecilozonites)* in Bermuda. Bulletin of the Museum of Comparative Zoology, *Harvard University*, 138, 407–532.

Gould, S. J. (1984). Smooth curve of evolutionary rate: a psychological and mathematical artifact. *Science*, 226, 994–995.

Gould, S. J. (1985). The paradox of the first tier: an agenda for paleobiology. *Paleobiology*, 11, 2–12.

Gould, S. J. (1988). Trends as changes in variance: a new slant on progress and directionality in evolution. *Journal of Paleontology*, 62, 319–329.

Gould, S. J. and Eldredge, N. (1977). Punctuated equilibria: the tempo and mode of evolution reconsidered. *Paleobiology*, 3, 115–151.

Gowe, R. S. and Fairfull, R. W. (1985). The direct response to long-term selection for multiple traits in egg stocks and changes in genetic parameters with selection. In *Poultry Genetics and Breeding*, eds. W. G. Hill, J. M. Manson, and D. Hewitt, Harlow, British Poultry Science Ltd., pp. 125–146.

Gowe, R. S., Johnson, A. S., Downs, J. H., Gibson, R., Mountain, W. F., Strain, J. H., and Tinney, B. F. (1959). Environment and poultry breeding problems: 4. The value of a random-bred control strain in a selection; study. *Poultry Science*, 38, 443–462.

Gradstein, F. M., Ogg, J. G., Schmitz, M. D., and Ogg, G. M. eds. (2012). *The Geological Time Scale 2012*, Amsterdam, Elsevier.

Grant, B. R. and Grant, P. R. (2010). Songs of Darwin's finches diverge when a new species enters the community. *Proceedings of the National Academy of Sciences*, 107, 20156–20163.

Grant, P. R. (1981). Speciation and the adaptive radiation of Darwin's finches. *American Scientist*, 69, 653–663.

Grant, P. R. (1986). *Ecology and Evolution of Darwin's Finches*. Princeton, Princeton University Press.

Grant, P. R. (1994). Population variation and hybridization: comparison of finches from two archipelagos. *Evolutionary Ecology*, 8, 598–617.

Grant, P. R. and Grant, B. R. (1993). Evolution of Darwin's finches caused by a rare climatic event. *Proceedings of the Royal Society B: Biological Sciences*, 251, 111–117.

Grant, P. R. and Grant, B. R. (2002). Unpredictable evolution in a 30-year study of Darwin's finches. *Science*, 296, 707–711.

Grant, P. R. and Grant, B. R. (2014). *40 Years of Evolution: Darwin's Fionches on Daphne Major Island*. Princeton, Princeton University Press.

Grant, V. (1963). *The Origin of Adaptations*. New York, Columbia University Press.

Grave, B. H. (1930). The natural history of *Bugula flabellata* at Woods Hole, Massachusetts, including the behavior and attachment of the larva. *Journal of Morphology*, 49, 355–383.

Haeckel, E. H. P. A. (1866). *Generelle Morphologie Der Organismen. Zweiter Band. Allgemeine Entwicklungsgeschichte Der Organismen*. Berlin, Georg Reimer.

Hairston, N. G. (1987). *Community Ecology and Salamander Guilds*. Cambridge, Cambridge University Press.

Haldane, J. B. S. (1949). Suggestions as to quantitative measurement of rates of evolution. *Evolution*, 3, 51–56.

Haldane, J. B. S. (1958). The theory of evolution, before and after Bateson. *Journal of Genetics*, 56, 11–27.

Haller, B. C. and Hendry, A. P. (2014). Solving the paradox of stasis: squashed stabilizing selection and the limits of detection. *Evolution*, 68, 483–500.

Hansen, T. F. (1997). Stabilizing selection and the comparative analysis of adaptation. *Evolution*, 51, 1341–1351.

Hansen, T. F. and Martins, E. P. (1996). Translating between microevolutionary process and macroevolutionary patterns: the correlation structure of interspecific data. *Evolution*, 50, 1404–1417.

Hargenvilliers, A.-A. (1817). *Recherches et considérations sur la formation et le recrutement de l'armée en France*. Paris, F. Didot.

Harmon, L. J., Losos, J. B., Jonathan Davies, T., Gillespie, R. G., Gittleman, J. L., Bryan Jennings, W., Kozak, K. H., McPeek, M. A., Moreno-Roark, F., Near, T. J., Purvis, A., Ricklefs, R. E., Schluter, D., Schulte II, J. A., Seehausen, O., Sidlauskas, B. L., Torres-Carvajal, O., Weir, J. T., and Mooers, A. Ø. (2010). Early bursts of size and shape evolution are rare in comparative data. *Evolution*, 64, 2385–2396.

Haugen, T. O. and Vøllestad, L. A. (2000). Population differences in early life-history traits in grayling. *Journal of Evolutionary Biology*, 13, 897–905.

Haugen, T. O. and Vøllestad, L. A. (2001). A century of life-history evolution in grayling. *Genetica*, 112, 475–491.

Hayami, I. (1984). Natural history and evolution of *Cryptopecten* (a Cenozoic-Recent pectinid genus). *Bulletin of the University Museum, University of Tokyo*, 24, 1–149.

Hendry, A. P. and Kinnison, M. T. (1999). The pace of modern life: measuring rates of contemporary microevolution. *Evolution*, 53, 1637–1653.

Hendry, A. P. and Quinn, T. P. (1997). Variation in adult life history and morphology among Lake Washington sockeye salmon (*Oncorhynchus nerka*) populations in relation to habitat features and ancestral affinities. *Canadian Journal of Fisheries and Aquatic Sciences*, 54, 75–84.

Heppell, S. S., Caswell, H., and Crowder, L. B. (2000). Life histories and elasticity patterns: perturbation analysis for species with minimal demographic data. *Ecology*, 81, 654–665.

Herschel, J. (1850). Quetelet on probabilities. *Edinburgh Review*, 92, 1–57.

Hilgen, F. J. Lourens, L. J. Van Dam, J. A. Beu, A. G. Boyes, A. F. Cooper, R. A. Krijgsman, W. Ogg, J. G. Piller, W. E., and Wilson, D. S. (2012). The Neogene period. In *The Geological Time Scale 2012*, eds. F. M. Gradstein, J. G. Ogg, M. D. Schmitz, and G. M. Ogg, Amsterdam, Elsevier, pp. 923–978.

Hoekstra, H. E., Hoekstra, J. M., Vignieri, S. N., Hoang, A., Beerli, P., and Kingsolver, J. G. (2002). Strength and tempo of directional selection in the wild. *Proceedings of the National Academy of Sciences USA*, 98, 9157–9160.

Holmes, M. W., Boykins, G. K. R., Bowie, R. C. K., and Lacey, E. A. (2016). Cranial morphological variation in *Peromyscus maniculatus* over nearly a century of environmental change in three areas of California. *Journal of Morphology*, 277, 96–106.

Hooks, A. P. (2013). *Prey plasticity responses to a native and nonnative predator*. M.Sc. thesis, Stony Brook, Stony Brook University, 45 pp.

Hopkins, C. G. (1899). Improvement in the chemical composition of the corn kernel. *Illinois Agriculture Experimental Station Bulletin*, 55, 205–240.

House, M. R. (1963). Bursts in evolution. *Report of the British Association for the Advancement of Science, London*, 19, 499–507.

Huber, P. J. (1981). *Robust Statistics*. New York, John Wiley and Sons.

Hudson, J. D. and Martill, D. M. (1994). The Peterborough Member (Callovian, Middle Jurassic) of the Oxford Clay Formation at Peterborough, UK. *Journal of the Geological Society*, 151, 113.

Hughes, M., Gerber, S., and Wills, M. A. (2013). Clades reach highest morphological disparity early in their evolution. *Proceedings of the National Academy of Sciences USA*, 110, 13875–13879.

Hunt, G. (2004). Phenotypic variation in fossil samples: modeling the consequences of time-averaging. *Paleobiology*, 30, 426–443.

Hunt, G. (2006). Fitting and comparing models of phyletic evolution: random walks and beyond. *Paleobiology*, 32, 578–601.

Hunt, G. (2007). Evolutionary divergence in directions of high phenotypic variance in the ostracode genus *Poseidonamicus*. *Evolution*, 61, 1560–1576.

Hunt, G. and Roy, K. (2006). Climate change, body size evolution, and Cope's Rule in deep-sea ostracodes. *Proceedings of the National Academy of Sciences USA*, 103, 1347–1352.

Hurst, H. E. (1951). Long-term storage capacity of reservoirs. *Transactions of the American Society of Civil Engineers*, 116, 770–808.

Hutchinson, G. E. (1959). Homage to Santa Rosalia or why are there so many kinds of animals? *American Naturalist*, 93, 145–159.

Huxley, J. S. (1924). Constant differential growth-ratios and their significance. *Nature*, 114, 895.

Huxley, J. S. (1932). *Problems of Relative Growth*. London, Methuen.

Huxley, J. S. (1942). *Evolution, the Modern Synthesis*. New York, Harper and Brothers.

Huxley, J. S. and Teissier, G. (1936). Terminalogy of relative growth. *Nature*, 137, 780–781.

Huxley, L. (1900). *Life and Letters of Thomas Henry Huxley, Volume 1*. London, Macmillan.

Huxley, T. H. (1859). Letter to Charles Lyell dated 25 June 1859. In *Life and Letters of Thomas Henry Huxley, Volume 1 (1900)*, ed. L. Huxley, London, Macmillan, pp. 185–187.

Ingram, T., Harmon, L. J., and Shurin, J. B. (2012). When should we expect early bursts of trait evolution in comparative data? Predictions from an evolutionary food web model. *Journal of Evolutionary Biology*, 25, 1902–1910.

Jablonski, D. (1997). Body-size evolution in Cretaceous molluscs and the status of Cope's rule. *Nature*, 385, 250–252.

Jaeckel, O. (1902). *Über Verschiedene Wege Phylogenetischer Entwicklung*. Jena, Gustav Fischer.

Jancović, I., Ahern, J. C. M., Karavanić, I., Stockton, T. ,and Smith, F. H. (2012). Epigravettian human remains and artifacts from Sandalja II, Istria, Croatia. *PaleoAnthropology*, 2012, 87–122.

Janis, C. M. (1990). Correlation of cranial and dental variables with body size in ungulates and macropodids. In *Body Size in Mammalian Paleobiology: Estimation and Biological Implications*, eds. J. D. Damuth and B. J. MacFadden, Cambridge, Cambridge University Press, pp. 255–299.

Jensen, H., Steinsland, I., Ringsby, T. H., and Saether, B.-E. (2008). Evolutionary dynamics of a sexual ornament in the house sparrow (*Passer domesticus*): the role of indirect selection within and between sexes. *Evolution*, 62, 1275–1293.

Jepsen, G. L., Mayr, E., and Simpson, G. G. eds. (1949). *Genetics, Paleontology, and Evolution*, Princeton, Princeton University Press.

Johannsen, W. L. (1909). *Elemente Der Exakten Erblichkeitslehre*. Jena, Gustav Fischer.

Johnston, R. F. and Selander, R. K. (1964). House sparrows: rapid evolution of races in North America. *Science*, 144, 548–550.

Johnston, R. F. and Selander, R. K. (1971). Evolution in the house sparrow. II. Adaptive differentiation in North American populations. *Evolution*, 25, 1–28.

Kaufmann, R. (1933). Variationsstatistische Untersuchungen über die "Artabwandlung" und "Artumbildung" an der oberkambrischen Trilobitengattung *Olenus* Dalm. *Abhandlungen aus dem Geologisch-Palaeontologisches Institut der Universität Greifswald*, 10, 1–55.

Kellogg, D. E. (1975). The role of phyletic change in the evolution of *Pseudocubus vema* (Radiolaria). *Paleobiology*, 1, 359–370.

Kennett, J. P. (1966). The *Globorotalia crassiformis* bioseries in north Westland and Marlborough, New Zealand. *Micropaleontology*, 12, 235–245.

Kerk, M. v. d., de Kroon, H., Conde, D. A., and Jongejans, E. (2013). Carnivora population dynamics are as slow and as fast as those of other mammals: implications for their conservation. *PLoS One*, 8, e70354.

Kimura, M. (1964). Diffusion models in population genetics. *Journal of Applied Probability*, 1, 177–232.

Kimura, Y., Flynn, L. J. ,and Jacobs, L. L. (2015). A palaeontological case study for species delimitation in diverging fossil lineages. *Historical Biology*, 28, 189–198.

Klepaker, T. (1993). Morphological changes in a marine population of threespined stickleback, *Gasterosteus aculeatus*, recently isolated in fresh water. *Canadian Journal of Zoology*, 71, 1251–1258.

Koontz, T. L., Shepherd, U. L., and Marshall, D. (2001). The effect of climate change on Merriam's kangaroo rat, *Dipodomys merriami*. *Journal of Arid Environments*, 49, 581–591.

Kristjánsson, B. K. (2005). Rapid morphological changes in threespine stickleback, *Gasterosteus aculeatus*, in freshwater. *Environmental Biology of Fishes*, 74, 357–363.

Kristjánsson, B. K., Skúlason, S., and Noakes, D. L. G. (2002). Rapid divergence in a recently isolated population of threespine stickleback (*Gasterosteus aculeatus* L.). *Evolutionary Ecology Research*, 4, 659–672.

Kruuk, L. E. B., Merilä, J., and Sheldon, B. C. (2001). Phenotypic selection on a heritable size trait revisited. *American Naturalist*, 158, 557–571.

Kruuk, L. E. B., Slate, J., Pemberton, J. M., Brotherstone, S., Guinness, F., and Clutton-Brock, T. H. (2002). Antler size in red deer: heritability and selection but no evolution. *Evolution*, 56, 1683–1695.

Kubitschek, H. E. (1974). Operation of selection pressure on microbial populations. *Symposia of the Society for General Microbiology*, 24, 105–130.

Kurtén, B. (1955). Sex dimorphism and size trends in the cave bear, *Ursus spelaeus* Rosenmüller and Heinroth. *Acta Zoologica Fennica*, 90, 1–48.

Kurtén, B. (1959). Rates of evolution in fossil mammals. *Cold Spring Harbor Symposia on Quantitative Biology*, 24, 205–215.

Kurtén, B. (1963). The rate of evolution. In *Science in Archaeology: A Comprehensive Survey of Progress and Research*, eds. D. R. Brothwell and E. Higgs, Bristol, Thames and Hudson, pp. 217–223.

Kurtén, B. (1965). Evolution in geological time. In *Ideas in Modern Biology,* ed. J. A. Moore, Garden City, New York, Natural History Press, pp. 329–354.

Lack, D. (1940). Variation in the introduced English sparrow. *Condor*, 42, 239–241.

Lack, D. (1947). *Darwin's Finches, an Essay on the General Biological Theory of Evolution*. Cambridge, Cambridge University Press.

Lamarck, J.-B. (1809). *Philosophie Zoologique, Tome Premier, Tome Second*. Paris, Librarie Dentu.

Lande, R. (1976). Natural selection and random genetic drift in phenotypic evolution. *Evolution*, 30, 314–334.

Lande, R. (1977). On comparing coefficients of variation. *Systematic Zoology*, 26, 214–217.

Langergraber, K. E., Prüfer, K., Rowney, C., Boesch, C., Crockford, C., Fawcett, K., Inoue, E., Inoue-Muruyama, M., Mitani, J. C., Muller, M. N., Robbins, M. M., Schubert, G., Stoinski, T. S., Viola, B., Watts, D., Wittig, R. M., Wrangham, R. W., Zuberbühler, K., Pääbo, S., and Vigilant, L. (2012). Generation times in wild chimpanzees and gorillas suggest earlier divergence times in great ape and human evolution. *Proceedings of the National Academy of Sciences*, 109, 15716–15721.

Larsson, K., Jeugd, H. P. v. d., and Veen, I. T. v. d. F. P. (1998). Body size declines despite positive directional selection on heritable size traits in a barnacle goose population. *Evolution*, 52, 1169–1184.

Lawler, R. R. (2011). Historical demography of a wild lemur population (*Propithecus verreauxi*) in southwest Madagascar. *Population Ecology*, 53, 229–240.

Lee, A. H. and Werning, S. (2008). Sexual maturity in growing dinosaurs does not fit reptilian growth models. *Proceedings of the National Academy of Sciences USA*, 105, 582–587.

Legendre, S. (1989). Les communautés de mammifères du Paléogène (Eocène supérieur et Oligocène) d'Europe occidentale: structures, milieux et évolution. *Münchner Geowissenschaftliche Abhandlungen, Reihe A, Geologie und Paläontologie*, 16, 1–110.

Lerner, I. M. (1954). *Genetic Homeostasis*. Edinburgh, Oliver and Boyd.

Lerner, I. M. and Dempster, E. R. (1951). Attenuation of genetic progress under continued selection in poultry. *Heredity*, 5, 75–94.

Levin, S. A. (1992). The problem of pattern and scale in ecology. *Ecology*, 73, 1943–1967.

Levinton, J. S. (1979). A theory of diversity equilibrium and morphological evolution. *Science*, 204, 335–336.

Lewontin, R. C. (1966). On the measurement of relative variability. *Systematic Zoology*, 15, 141–142.

Lexis, W. (1877). *Zur Theorie Der Massenerscheinungen in Der Menschlichen Gesellschaft*. Freiburg im Briesgau, Friedrich Wagner.

Lich, D. K. (1990). *Cosomys primus*: a case for stasis. *Paleobiology*, 16, 384–395.

Linnaeus, C. (1735). *Systema Naturae, Sive Regna Tria Naturae Systematice Proposita Per Classes, Ordines, Genera, Species*. Leiden, Lugduni Batavorum.

Lister, A. M. (1989). Rapid dwarfing of red deer on Jersey in the last interglacial. *Nature*, 342, 539–542.

Lovtrup (Løvtrup), S. (1974). *Epigenetics, a Treatise on Theoretical Biology*. New York, John Wiley and Sons.

Lyell, C. (1832). *Principles of Geology, Volume II*. London, John Murray.

Lynch, M. (1988). The rate of polygenic mutation. *Genetical Research, Cambridge*, 51, 137–148.

Lynch, M. (1990). The rate of morphological evolution in mammals from the standpoint of the neutral expectation. *American Naturalist*, 136, 727–741.

Mac Gillavry, H. J. (1968). Modes of evolution mainly among marine invertebrates. *Bijdragen tot de Dierkunde*, 38, 69–74.

MacFadden, B. J. (1985). Patterns of phylogeny and rates of evolution in fossil horses: hipparions from the Miocene and Pliocene of North America. *Paleobiology*, 11, 245–257.

MacFadden, B. J. (1986). Fossil horses from "Eohippus" (*Hyracotherium*) to *Equus*: scaling, Cope's law, and the evolution of body size. *Paleobiology*, 12, 355–369.

MacFadden, B. J. (1988). Fossil horses from "Eohippus" (Hyracotherium) to Equus, 2: rates of dental evolution revisited. *Biological Journal of the Linnean Society, London*, 35, 37–48.

MacFadden, B. J. (1992). *Fossil Horses: Systematics, Paleobiology, and Evolution of the Family Equidae*. Cambridge, Cambridge University Press.

MacFadden, B. J. and Hulbert, R. C. (1990). Body size estimates and size distribution of ungulate mammals form the Late Miocene Love Bone Bed of Florida. In *Body Size in Mammalian Paleobiology: Estimation an Biological Implications*, eds. J. D. Damuth and B. J. MacFadden, Cambridge, Cambridge University Press, pp. 337–363.

MacLeod, N. (1991). Punctuated anagenesis and the importance of stratigraphy to paleobiology. *Paleobiology*, 17, 167–188.

Maglio, V. J. (1973). Origin and evolution of the Elephantidae. *Transactions of the American Philosophical Society*, 63, 1–149.

Malmgren, B. A., Berggren, W. A. and Lohmann, G. P. (1983). Evidence for punctuated gradualism in the late Neogene *Globorotalia tumida* lineage of planktonic foraminifera. *Paleobiology*, 9, 377–389.

Malmgren, B. A. and Kennett, J. P. (1981). Phyletic gradualism in a late Cenozoic planktonic foraminiferal lineage; DSDP site 284, southwest Pacific. *Paleobiology*, 7, 230–240.

Mandelbrot, B. B. (1967). How long is the coast of Britain? Statistical self-similarity and fractional dimension. *Science*, 156, 636–638.

Mandelbrot, B. B. (1983). *Fractal Geometry of Nature*. San Francisco, W. H. Freeman.

Manser, A., Lindholm, A. K., König, B., and Bagheri, H. C. (2011). Polyandry and the decrease of a selfish genetic element in a wild house mouse population. *Evolution*, 65, 2435–2447.

Mayr, E. (1942). *Systematics and the Origin of Species*. New York, Columbia University Press.

Mayr, E. (1954). Change of genetic environment and evolution. In *Evolution As a Process*, ed. J. S. Huxley, A. C. Hardy and E. B. Ford, London, George Allen and Unwin, pp. 157–180.

Mayr, E. (1963). *Animal Species and Evolution*. Cambridge, Massachusetts, Harvard University Press.

McAlister, D. (1879). The law of the geometric mean. *Proceedings of the Royal Society of London*, 29, 367–376.

McCluskey, J., Olivier, T. J., Freedman, L., and Hunt, E. (1974). Evolutionary divergences between populations of Australian wild rabbits. *Nature*, 249, 278–279.

McDonald, J. N. (1981). *North American Bison: Their Classification and Evolution.* Berkeley, University of California Press.

McKellar, A. E. and Hendry, A. P. (2009). How humans differ from other animals in their levels of morphological variation. *PLoS One*, 4, e6876.

McPeek, M.-A. (2008). The ecological dynamics of clade diversification and community assembly. *American Naturalist*, 172, E270–E284.

McPhail, J. D. (1984). Ecology and evolution of sympatric sticklebacks (*Gasterosteus*): morphological and genetic evidence for a species pair in Enos Lake, British Columbia. *Canadian Journal of Zoology*, 62, 1402–1408.

Meehan, T. (1875). *Change by Gradual Modification Not the Universal Law.* Salem, Massachusetts, Salem Press.

Mendel, G. (1866). Versuche über Pflanzen-Hybriden. *Verhandlungen des naturforschenden Vereines in Brünn, Abhandlungen*, 4, 3–47.

Michaux, B. (1988). Organotaxism: an alternative "way of seeing" the fossil record. *Journal of Theoretical Biology*, 133, 397–408.

Millar, J. S. and Zammuto, R. M. (1983). Life histories of mammals: an analysis of life tables. *Ecology*, 64, 631–635.

Miller, G. S. (1912). *Catalogue of the Mammals of Western Europe.* London, British Museum (Natural History).

Millien, V., Ledevin, R., Boué, C., and Gonzalez, A. (2017). Rapid morphological divergence in two closely related and co-occurring species over the last 50 years. *Evolutionary Ecology*, 31, 847–864.

Milner, J. M., Albon, S. D., Illius, A. W., Pemberton, J. M., and Clutton-Brock, T. H. (1999). Repeated selection of morphometric traits in the Soay sheep on St Kilda. *Journal of Animal Ecology*, 68, 472–488.

Miracle, P. T. (1995). *Broad-Spectrum Adaptations Re-Examined: Hunter-Gatherer Responses to Late Glacial Environmental Changes in the Eastern Adriatic.* Ph.D., Ann Arbor, University of Michigan, 577 pp.

Mivart, St. G. J. (1871). *On the Genesis of Species. London, Macmillan* 342.

Møller, A. P. and Szép, T. (2005). Rapid evolutionary change in a secondary sexual character linked to climatic change. *Journal of Evolutionary Biology*, 18, 481–495.

Moorjani, P., Amorim, C. E. G., Arndt, P. F., and Przeworski, M. (2016). Variation in the molecular clock of primates. *Proceedings of the National Academy of Sciences USA*, 113, 10607–10612.

Moss, C. J. (2001). The demography of an African elephant (*Loxodonta africana*) population in Amboseli, Kenya. *Journal of Zoology*, 255, 145–156.

Müller, F. (1886). Ein Züchtungsvesuch an Mais. *Kosmos, Stuttgart*, 19, 22–26.

Nengovhela, A., Baxter, R. M., and Taylor, P. J. (2015). Temporal changes in cranial size in South African vlei rats (*Otomys*): evidence for the "third universal response to warming," *African Zoology*, 50, 233–239.

Neumayr, M. and Paul, C. M. (1875). Die Congerien- und Paludinenschichten Slavoniens und deren Faunen. Ein Beitrag zur Descendenz-Theorie. *Abhandlungen der Kaiserlich-Königlichen Geologischen Reichsanstalt, Wien*, 7(3), 1–113.

Newell, N. D. (1949). Phyletic size increase – an important trend illustrated by fossil invertebrates. *Evolution*, 3, 103–124.

Niel, C. and Lebreton, J.-D. (2005). Using demographic invariants to detect overharvested bird populations from incomplete data. *Conservation Biology*, 19, 826–835.

Ogg, J. G. Hinnov, L. A., and Huang, C. (2012). Jurassic. In *The Geological Time Scale 2012*, eds. F. M. Gradstein, J. G. Ogg, M. D. Schmitz, and G. M. Ogg, Amsterdam, Elsevier, pp. 731–791.

Orbigny, A. D. d. (1851). *Cours Élémentaire De Paléontologie Et De Géologie Strati-graphiques, Tome Second*. Paris, Victor Masson.

Osborn, H. F. (1905). Present problems in paleontology. *Popular Science Monthly*, 66, 226–242.

Osborn, H. F. (1925). The origin of species as revealed by vertebrate palaeontology. *Nature*, 115, 925–926, 961–963.

Ovcharenko, V. N. (1969). Transitional forms and species differentiation of brachiopods. *Paleontological Journal*, 3, 57–63.

Owen, R. (1868). *Anatomy of Vertebrates. Volume III. Mammals*. London, Longmans, Green, and Co. 915.

Palmer, A. R. (1965). Biomere – a new kind of biostratigraphic unit. *Journal of Paleontology*, 39, 149–153.

Paradis, E. (2012). *Analysis of Phylogenetics and Evolution With R,* Second Edition. New York, Springer.

Patton, J. L., Yang, S. Y. ,and Myers, P. (1975). Genetic and morphologic divergence among introduced rat populations (*Rattus rattus*) of the Galapagos archipelago, Ecuador. *Systematic Zoology*, 24, 296–310.

Payne, J. L., Boyer, A. G., Brown, J. H., Finnegan, S., Kowalewski, M., Krause, R. A., Lyons, S. K., McClain, C. R., McShea, D. W., Novack-Gottshall, P. M., Smith, F. A., Stempien, J. A., and Wang, S. C. (2009). Two-phase increase in the maximum size of life over 3.5 billion years reflects biological innovation and environmental opportunity. *Proceedings of the National Academy of Sciences*, 106, 24–27.

Pearson, E. S. and Hartley, H. O. (1966). *Biometrika Tables for Statisticians, Volume I,* Third Edition. Cambridge, Cambridge University Press.

Pearson, K. (1894). Contributions to the mathematical theory of evolution. I. On the dissection of asymmetrical frequency curves. Philosophical Transactions of the Royal Society of London*, Series A*, 185, 71–110.

Pearson, K. (1895). Contributions to the mathematical theory of evolution. II. Skew variation in homogeneous material. Philosophical Transactions of the Royal Society of London*, Series A*, 186, 343–414.

Pearson, K. (1905). The problem of the random walk. *Nature*, 72, 294, 342.

Pearson, K. (1924). *The Life, Letters and Labours of Francis Galton. Volume II–Researches of Middle Life*. London, Cambridge University Press.

Pegueroles, G., Papaceit, M., Quintana, A., Guillén, A., Prevosti, A., and Serra, L. (1995). An experimental study of evolution in progress: clines for quantitative traits in colonizing and Palearctic populations of *Drosophila*. *Evolutionary Ecology*, 9, 453–465.

Peirce, C. S. (1873). On the theory of errors of observations. *U. S. Coast Survey Report for 1870, Appendix 21*, 1870, 200–224.

Pergams, O. R. W. and Ashley, M. V. (1999). Rapid morphological change in island deer mice. *Evolution*, 53, 1573–1581.

Pergams, O. R. W., Byrn, D., Lee, K. L. Y., and Jackson, R. (2015). Rapid morphological change in black rats (*Rattus rattus*) after an island introduction. *Peer J*, 3, e812.

Peters, R. H. (1983). *Ecological Implications of Body Size*. Cambridge, Cambridge University Press.

Petren, K., Grant, B. R., and Grant, P. R. (1999). A phylogeny of Darwin's finches based on microsatellite DNA length variation. *Proceedings of the Royal Society of London, Series B*, 266, 321–329.

Pfeifer, S. P. (2017). The demographic and adaptive history of the African green monkey. *Molecular Biology and Evolution*, 34, 1055–1065.

Philiptschenko, J. (1927). *Variabilität Und Variation*. Berlin, Gebrüder Borntraeger.

Phillimore, A. B. and Price, T. D. (2008). Density-dependent cladogenesis in birds. *PLoS Biology*, 6, e71.

Playfair, J. (1802). *Illustrations of the Huttonian Theory of the Earth*. Edinburgh, William Creech.

Porter, W. T. (1894). The growth of St. Louis children. *Transactions of the Academy of Science of St. Louis*, 6, 263–380.

Press, W. H., Flannery, B. P., Teukolsky, S. A., and Vetterling, W. T. (1989). *Numerical Recipes: The Art of Scientific Computing (Fortran Version)*. Cambridge, Cambridge University Press.

Przybylo, R., Sheldon, B. C., and Merilä, J. (2000). Climatic effects on breeding and morphology: evidence for phenotypic plasticity. *Journal of Animal Ecology*, 69, 395–403.

Quetelet, A. (1835). *Sur l'homme et le développement de ses facultés, ou essai de physique sociale,* Tome Premier. Paris, Bachelier.

Quetelet, A. (1846). *Lettres sur la théorie des probabilities, appliquée aux sciences morales et politiques*. Bruxelles, M. Hayez.

Quetelet, A. (1849). *Letters Addressed to H.R.H. the Grand Duke of Saxe Coburg and Gotha, on the Theory of Probabilities As Applied to the Moral and Political Sciences Translated From the French by Olinthus Gregory Downes*. London, C. and E. Layton.

Quetelet, A. (1869). *Physique sociale, ou essay sur le développement des facultés,* Tome I. Brussels, C. Muquardt.

Quetelet, A. (1870). *Anthropométrie ou mesure des différentes facultés de l'homme*. Brussels, C. Muquardt.

R Core Team. (2017). *R: A Language and Environment for Statistical Computing*. Vienna, Austria, R Foundation for Statistical Computing.

Rabosky, D. L. and Lovette, I. J. (2008). Explosive evolutionary radiations: decreasing speciation or increasing extinction through time? *Evolution*, 62, 1866–1875.

Raup, D. M. and Crick, R. E. (1981). Evolution of single characters in the Jurassic ammonite *Kosmoceras*. *Paleobiology*, 7, 200–215.

Reiss, M. J. (1989). *The Allometry of Growth and Reproduction*. Cambridge, Cambridge University Press.

Rensch, B. (1943). Die paläontologischen Evolutionsregeln in zoologischer Betrachtung. *Biologia Generalis, Vienna*, 17, 1–55.

Rensch, B. (1947). *Neuere Probleme Der Abstammungslehre: Die Transspezifische Evolution*. Stuttgart, Ferdinand Enke Verlag.

Reznick, D. N., Shaw, F. H., Rodd, F. H., and Shaw, R. G. (1997). Evaluation of the rate of evolution in natural populations of guppies (*Poecilia reticulata*). *Science*, 275, 1934–1937.

Richard, A. F., Dewar, R. E., Schwartz, M., and Ratsirarson, J. (2000). Mass change, environmental variability and female fertility in wild Propithecus verreauxi. *Journal of Human Evolution*, 39, 381–391.

Richard-Hansen, C., Vié, J.-C., Vidal, N., and Kéravec, J. (1999). Body measurements on 40 species of mammals from French Guiana. *Journal of Zoology*, 247, 419–428.

Richardson, L. F. (1961). The problem of contiguity: an appendix to Statistics of Deadly Quarrels. *General Systems Yearbook, Society for the Advancement of General Systems Theory*, 6, 139–187.

Rickwood, A. E. (1977). Age, growth and shape of the intertidal brachiopod *Waltonia inconspicua* Sowerby, from New Zealand. *American Zoologist*, 17, 63–73.

Roden, M. K., Parrish, R. R., and Miller, D. S. (1990). The absolute age of the Eifelian Tioga Ash bed, Pennsylvania. *Journal of Geology*, 98, 282–285.

Sander, P. M., Christian, A., Clauss, M., Fechner, R., Gee, C. T., Griebeler, E.-M., Gunga, H.-C., Hummel, J., Mallison, H., Perry, S. F., Preuschoft, H., Rauhut, O. W. M., Remes, K., Tütken, T., Wings, O., and Witzel, U. (2011). Biology of the sauropod dinosaurs: the evolution of gigantism. *Biological Reviews*, 86, 117–155.

Sato, A., Tichy, H., O'hUigin, C., Grant, P. R., Grant, B. R., and Klein, J. (2001). On the origin of Darwin's finches. *Molecular Biology and Evolution*, 18, 299–311.

Saunders, W. B. (1984). *Nautilus* growth and longevity: evidence from marked and recaptured animals. *Science*, 224, 992–990.

Schindewolf, O. H. (1936). *Paläontologie, Entwicklungslehre Und Genetik: Kritik Und Synthese*. Berlin, Gebrüder Borntraeger.

Schindewolf, O. H. (1950). *Grundfragen Der Paläontologie: Geologische Zeitmessung, Organische Stammesentwinklung, Biologische Systematik*. Stuttgart, Erwin Nägele.

Schluter, D. (1984). Morphological and phylogenetic relations among the Darwin's finches. *Evolution*, 38, 921–930.

Schluter, D. (2000). *The Ecology of Adaptive Radiation*. Oxford, U.K., Oxford University Press.

Schmalhausen, I. I. (1935). Determination of basic concepts and methods of investigations of growth [in Russian]. In *The Growth of Animals*, ed. S. Kaplansky, Moscow, pp. 8–60.

Schmalhausen, I. I. (1943). [Rate of evolution and factors which determine it]. *Zhurnal Obshchei Biologii, Moskva*, 4, 253–285 [Russian with English summary].

Schmidt-Nielsen, K. (1984). *Scaling: Why Is Animal Size So Important?* Cambridge, Cambridge University Press.

Schneider, D. C. (1994). Scale-dependent patterns and species interactions in marine nekton. In *Aquatic Ecology: Scale, Pattern, and Process*, eds. P. Giller and D. H. A. Rafaelli, London, Blackwell, pp. 441–467.

Secord, R., Bloch, J. I., Chester, S. G. B., Boyer, D. M., Wood, A. R., Wing, S. L., Kraus, M. J., McInerney, F. A., and Krigbaum, J. (2012). Evolution of the earliest horses driven by climate change in the Paleocene-Eocene thermal maximum. *Science*, 335, 959–962.

Seeley, R. H. (1986). Intense natural selection caused by a rapid morphological transition in a living marine snail. *Proceedings of the National Academy of Sciences USA*, 83, 6897–6901.

Selander, R. K. and Johnston, R. F. (1967). Evolution in the house sparrow. *I. Intra-population variation in North America. Condor*, 69, 217–258.

Sewertzoff, A. N. (1931). *Morphologische Gesetzmässigkeiten Der Evolution*. Jena, Gustav Fischer Verlag.

Sheets, H. D. and Mitchell, C. E. (2001). Uncorrelated change produces the apparent dependence of evolutionary rate on interval. *Paleobiology*, 27, 429–445.

Simberloff, D. S., Dayan, T., Jones, C., and Ogura, G. (2000). Character displacement and release in the small Indian mongoose, *Herpestes javanicus. Ecology*, 81, 2086–2099.

Simpson, G. G. (1943). Criteria for genera, species, and subspecies in zoology and paleontology. *Annals of the New York Academy of Sciences*, 44, 145–178.

Simpson, G. G. (1944). *Tempo and Mode in Evolution*. New York, Columbia University Press.

Simpson, G. G. (1953). *The Major Features of Evolution*. New York, Columbia University Press 434.

Slobodkin, L. B. (1961). *Growth and Regulation of Animal Populations*. New York, Holt, Reinhart, and Winston.

Smith, F. A., Boyer, A. G., Brown, J. H., Costa, D. P., Dayan, T., Ernest, S. K. M., Evans, A. R., Fortelius, M., Gittleman, J. L., Hamilton, M. J., Harding, L. E., Lintulaakso, K.,

Lyons, S. K., McCain, C., Okie, J. G., Saarinen, J. J., Sibly, R. M., Stephens, P. R., Theodor, J., and Uhen, M. D. (2010). The evolution of maximum body size of terrestrial mammals. *Science*, 330, 1216–1219.

Smith, F. A., Browning, H., and Shepherd, U. L. (1998). The influence of climate change on the body mass of woodrats *Neotoma* in an arid region of New Mexico, USA. *Ecography*, 21, 140–148.

Smith, T. B., Freed, L. A., Lepson, J. K., and Carothers, J. H. (1995). Evolutionary consequences of extinctions in populations of a Hawaiian honeycreeper. *Conservation Biology*, 9, 107–113.

Sokal, R. R. and Rohlf, F. J. (1981). *Biometry,* Second Edition. San Francisco, W. H. Freeman.

Spanbauer, T. L., Fritz, S. C., and Baker, P. A. (2018). Punctuated changes in the morphology of an endemic diatom from Lake Titicaca. *Paleobiology*, 44, 89–100.

Spitzer, F. L. (1964). *Principles of Random Walk*. Princeton, Van Nostrand.

St. Louis, V. L. and Barlow, J. C. (1991). Morphometric analysis of introduced and ancestral populations of the Eurasian tree sparrow. *VVilson Bulletin*, 103, 1:12.

Stanley, S. M. (1973). An explanation for Cope's rule. *Evolution*, 27, 1–26.

Stanley, S. M. (1975). A theory of evolution above the species level. *Proceedings of the National Academy of Sciences USA*, 72, 646–650.

Stanley, S. M. (1989). The empirical case for the punctuational model of evolution. *Journal of Social and Biological Structures,*

Stanley, S. M. and Yang, X. (1987). Approximate evolutionary stasis for bivalve morphology over millions of years: a multivariate, multilineage study. *Paleobiology*, 13, 113–139.

Stearns, S. C. (1983). The genetic basis of differences in life history traits among six populations of mosquitofish (*Gambusia affinis*) that shared ancestors in 1905. *Evolution*, 37, 618–627.

Stigler, S. M. (1986). *The History of Statistics: Measurement of Uncertainty Before 1900*. Cambridge, MA, Harvard University Press.

Stuart, Y. E., Campbell, T. S., Hohenlohe, P. A., Reynolds, R. G., Revell, L. J., and Losos, J. B. (2014). Rapid evolution of a native species following invasion by a congener. *Science*, 346, 463–466.

Sylvester-Bradley, P. C. (1959). Iterative evolution in fossil oysters. In *Proceedings of the 15th International Congress of Zoology*, pp. 193–197.

Sylvester-Bradley, P. C. (1977). Biostratigraphical tests of evolutionary theory. In *Concepts and Methods of Biostratigraphy*, eds. E. G. Kauffman and J. E. Hazel, Stroudsburg, Pa., Dowden, Hutchinson and Ross, pp. 41–63.

Szuma, E. (2003). Microevolutionary trends in the dentition of the Red fox (*Vulpes vulpes*). *Journal of Zoological Systematics and Evolutionary Research*, 41, 47–56.

Taylor, B. L., Chivers, S. J., Larese, J., and Perrin, W. F. (2007). *Generation Length and Percent Mature Estimates for IUCN Assessments of Cetaceans*. Administrative Report LJ-07–01, La Jolla, California, Southwest Fisheries Science Center.

Taylor, J. M. (1974). Morphological skull variation in the Australian wild rabbit: a multivariate analysis. Bachelor of Medical Science thesis, Perth, University of Western Australia, 182 pp.

Taylor, J. M., Freedman, L., Olivier, T. J., and McCluskey, J. (1977). Morphometric distances between Australian wild rabbit populations. *Australian Journal of Zoology*, 25, 721–732.

Theriot, E. C., Fritz, S. C., Whitlock, C., and Conley, D. J. (2006). Late Quaternary rapid morphological evolution of an endemic diatom in Yellowstone Lake, Wyoming. *Paleobiology*, 32, 38–54.

Todor, K. (2016). The growth of bacterial populations. In *Online Textbook of Bacteriology*, Madison, University of Wisconsin, http://textbookofbacteriology.net/growth_3.html.

Tseng, M., Kaur, K. M., Soleimani Pari, S., Sarai, K., Chan, D., Yao, C. H., Porto, P., Toor, A., Toor, H. S., and Fograscher, K. (2018a). Decreases in beetle body size linked to climate change and warming temperatures. *Journal of Animal Ecology*, 87, http://dx.doi.org/10.1111/1365-2656.12789.

Tseng, M., Kaur, K. M., Soleimani Pari, S., Sarai, K., Chan, D., Yao, C. H., Porto, P., Toor, A., Toor, H. S., and Fograscher, K. (2018b). Decreases in beetle body size linked to climate change and warming temperatures. *Dryad Digital Repository*, https://doi.org/10.5061/dryad.5164v.

Turner, J. R. G. (1986). The genetics of adaptive radiation: a neo-Darwinian theory of punctuational evolution. In *Patterns and Processes in the History of Life*, eds. D. M. Raup and D. Jablonski, Berlin, Springer, pp. 183–207.

Uyeda, J. C., Hansen, T. F., Arnold, S. J., and Jason Pienaar. (2011). The million-year wait for macroevolutionary bursts. *Proceedings of the National Academy of Sciences USA*, 108, 15908–15913.

Van Valen, L. M. (1974). Two modes of evolution. *Nature*, 252, 298–300.

Van Valkenburgh, B. (1990). Skeletal and dental predictors of body mass in carnivores. In *Body Size in Mammalian Paleobiology: Estimation and Biological Implications*, eds. J. D. Damuth and B. J. MacFadden, Cambridge, Cambridge University Press, pp. 181–205.

Vandenberghe, N. Hilgen, F. J. Speijer, R. P. Ogg, J. G. Gradstein, F. M. Hammer, O. Hollis, C. J., and Hooker, J. J. (2012). The Paleogene period. In *The Geological Time Scale 2012*, eds. F. M. Gradstein, J. G. Ogg, M. D. Schmitz, and G. M. Ogg, Amsterdam, Elsevier, pp. 855–921.

Vermeij, G. J. (1977). The Mesozoic marine revolution: evidence from snails, predators, and grazers. *Paleobiology*, 3, 245–258.

Vermeij, G. J. (1982). Environmental change and the evolutionary history of the periwinkle *Littorina littorea* in North America. *Evolution*, 36, 561–580.

Villermé, L. R. (1829). Mémoire sur la taille de l'homme en France. *Annales d'Hygiene Publique et de Medecine Legale*, 1, 351–399.

Vrba, E. S. (1985). Environment and evolution: alternative causes of the temporal distribution of evolutionary events. *South African Journal of Science*, 81, 229–236.

Waagen, W. H. (1869). Die Formenreihe des *Ammonites subradiatus*: Versuch einer paläontologischen Monographie. Geognostisch-Paläontologische Beiträge, *München*, 2, 179–256.

Waddington, C. H. (1939). *An Introduction to Modern Genetics*. London, George Allen and Unwin.

Waddington, C. H. (1942). Canalization of development and the inheritance of acquired characters. *Nature*, 150, 563–565.

Wedekind, R. (1920). Über Virenzperioden (Blüteperioden). *Sitzungsberichte der Gesellschaft zur Beförderung der gesammten Naturwissenschaften zu Marburg*, 1920 (2), 18–31.

Weidenreich, F. (1943). The skull of *Sinanthropus pekinensis*: a comparative study on a primitive hominid skull. *Palaeontologica Sinica, New Series D*, 10, 1–485.

Weismann, A. (1872). *Ueber Den Einfluss Der Isolirung Auf Die Artbildung. Leipzig, Wilhelm Engelmann* 108.

Weismann, A. (1883). *Über Die Vererbung*. Jena, Gustav Fischer.

Weldon, W. F. R. (1893). On certain correlated variations in *Carcinus maenas*. *Proceedings of the Royal Society of London*, 54, 318–329.

Weldon, W. F. R. (1895). Attempt to measure the death-rate due to the selective destruction of *Carcinus moenas* with respect to a particular dimension. *Proceedings of the Royal Society of London*, 57, 360–379.

Wiedmann, J. (1973). Evolution or revolution of ammonoids at Mesozoic system boundaries? *Biological Reviews, Cambridge*, 48, 159–194.

Wiens, J. A. (1989). Spatial scaling in ecology. *Functional Ecology*, 3, 385–397.

Williams, H. S. (1910). Persistence of fluctuating variations as illustrated by the fossil genus *Rhipidomella*. *Geological Society of America Bulletin*, 21, 295–312.

Williamson, P. G. (1981). Palaeontological documentation of speciation in Cenozoic molluscs from Turkana Basin. *Nature*, 293, 437–443.

Williamson, S. H. (2017). *Daily Closing Value for the Dow Jones Average, 1885 to Present*. https://measuring worth.com/DJA/.

Wood, A. R., Zelditch, M. L., Rountrey, A. N., Eiting, T. P., Sheets, H. D., and Gingerich, P. D. (2007). Multivariate stasis in the dental morphology of the Paleocene-Eocene condylarth *Ectocion*. *Paleobiology*, 33, 248–260.

Wright, S. (1931). Evolution in Mendelian populations. *Genetics*, 16, 97–159.

Wright, S. (1968). *Evolution and the Genetics of Populations. Volume 1: Genetic and Biometric Foundations*. Chicago, University of Chicago Press.

Wright, S. (1969). *Evolution and the Genetics of Populations. Volume 2: The Theory of Gene Frequencies*. Chicago, University of Chicago Press.

Yablokov, A. V. (1974). *Variability of Mammals*. New Delhi, Amerind Publishing.

Yarwood, C. E. (1956). Generation time and the biological nature of viruses. *American Naturalist*, 40, 97–102.

Yule, G. U. (1924). A mathematical theory of evolution, based on the conclusions of Dr. J. C. Willis, F.R.S. *Philosophical Transactions of the Royal Society of London, Series B*, 213, 21–87.

Zakrzewski, R. J. (1969). The rodents from the Hagerman local fauna, upper Pliocene of Idaho. *Contributions from the Museum of Paleontology, University of Michigan*, 23, 1–36.

Zeleny, C. (1922). The effect of selection for eye facet number in the white bar-eye race of *Drosophila melanogaster*. *Genetics*, 7, 1–115.

Zeng, Z. and Brown, J. H. (1987). Population ecology of a desert rodent: *Dipodomys merriami* in the Chihuahuan Desert. *Ecology*, 68, 1328–1340.

Zeuner, F. E. (1946). *Dating the Past: An Introduction to Geochronology*. London, Methuen and Company.

Ziegler, A. M. (1966). The Silurian brachiopod *Eocoelia hemisphaerica* (J. de C. Sowerby) and related species. *Palaeontology*, 9, 523–543.

Zimmermann, W. (1953). *Evolution: Die Geschichte Ihrer Probleme Und Erkenntnisse*. Freiburg, Karl Alber.

Index

Index